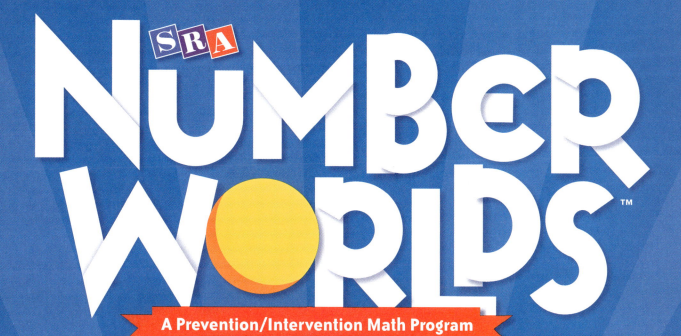

NUMBER WORLDS™

A Prevention/Intervention Math Program

Teacher Edition
Level F

Sharon Griffin

with **Building Blocks** **Douglas H. Clements**
Julie Sarama

SRA
Columbus, OH

Author
Sharon Griffin
Associate Professor of Education and
Adjunct Associate Professor of Psychology
Clark University
Worcester, Massachusetts

Building Blocks Authors

Douglas H. Clements
Professor of Early Childhood
and Mathematics Education
University at Buffalo
State University of New York, New York

Julie Sarama
Associate Professor of Mathematics Education
University at Buffalo
State University of New York, New York

Contributing Writers
Sherry Booth, *Math Curriculum Developer,* Raleigh, North Carolina
Elizabeth Jimenez, *English Language Learner Consultant,* Pomona, California

Program Reviewers

Jean Delwiche
Almaden Country School
San Jose, California
Cheryl Glorioso
Santa Ana Unified School District
Santa Ana, California
Sharon LaPoint
School District of Indian River County
Vero Beach, Florida

Leigh Lidrbauch
Pasadena Independent School District
Pasadena, Texas
Dave Maresh
Morongo Unified School District
Yucca Valley, California
Mary Mayberry
Mon Valley Education Consortium, AIU 3
Clairton, Pennsylvania

Lauren Parente
Mountain Lakes School District
Mountain Lakes, New Jersey
Juan Regalado
Houston Independent School District
Houston, Texas
M. Kate Thiry
Dublin City School District
Dublin, Ohio
Susan C. Vohrer
Baltimore County Public Schools
Baltimore, Maryland

SRAonline.com

SRA

Send all inquiries to:
SRA/McGraw-Hill
4400 Easton Commons
Columbus, OH 43219

ISBN 0-07-605340-7

10 11 12 13 14 RRMN 13 12 11 10 09

Acknowledgments
Development of the **Number Worlds** program was made possible by generous grants from the James S. McDonnell Foundation. The author gratefully acknowledges this support as well as the contributions of all the teachers and children who used the program in various stages of development and who helped shape its current form.

Building Blocks was supported in part by the National Science Foundation under Grant No. ESI-9730804, "Building Blocks-Foundations for Mathematical Thinking, Pre-Kindergarten to Grade 2: Research-based Materials Development" to Douglas H. Clements and Julie Sarama. The curriculum was also based partly upon work supported in part by the Institute of Educational Sciences (U.S. Dept. of Education, under the Interagency Educational Research Initiative, or IERI, a collaboration of the IES, NSF, and NICHHD) under Grant No. R305K05157, "Scaling Up TRIAD: Teaching Early Mathematics for Understanding with Trajectories and Technologies" and by the IERI through a National Science Foundation NSF Grant No. REC-0228440, "Scaling Up the Implementation of a Pre-Kindergarten Mathematics Curricula: Teaching for Understanding with Trajectories and Technologies." Any opinions, findings, and conclusions or recommendations expressed in this material are those of the authors and do not necessarily reflect the views of the funding agencies.

Photo Credit
C12 ©PhotoDisc/Getty Images, Inc.

The **McGraw·Hill** Companies

Build a solid foundation with

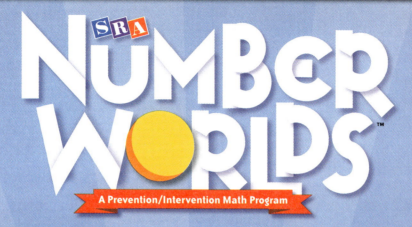

NUMBER WORLDS™

A Prevention/Intervention Math Program

Make a world of difference in your students' **math skills** with a versatile program that has **proven results** in the classroom.

Number Worlds is an intensive intervention program that focuses on students who are one or more grade levels behind in elementary mathematics. It provides all the tools teachers need to assess students' abilities, individualize instruction, build foundational skills and concepts, and make learning fun. And only *Number Worlds* includes a prevention program for Grades Pre-K–1. It's a unique course full of activities that build foundational math skills and prepare younger children to understand more complex concepts later.

Targeted instruction
Through intense identification and development of core concepts, *Number Worlds* creates competency that quickly puts students on-level with their peer groups. The program provides hands-on activities proven effective with even the lowest-level students, including:
- Computer activities
- Discussion activities
- Paper-and-pencil activities

Precise assessment
Number Worlds' easy-to-use assessment component pinpoints the exact unit in which students should begin the curriculum. Weekly and unit tests monitor progress with open response and/or multiple-choice questions to identify when students are ready to return to the main math curriculum.

Flexibility for teachers and students
Number Worlds' lessons are flexible for use in many settings:
- Resource room
- After school
- Summer school

Teacher's aides and parents can use *Number Worlds* after class or at home.

Comprehensive, fully-integrated program
Complete *Number Worlds* program kits are available at every level and include:
- Teacher Edition
- Student workbooks (Levels C–H)
- Student worksheet blackline masters
- Manipulatives (Levels A–F)
- Software for assessment, placement, professional development, and activities

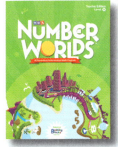

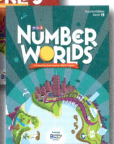
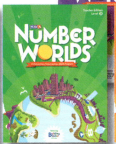
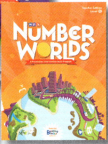
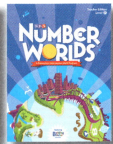

Number Worlds
Teacher Editions

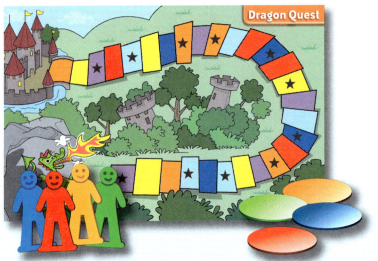

Number Worlds
Manipulatives and Games

Building Blocks
Screen Capture

Program Components
- Teacher Edition
- Student Workbooks
- Assessment
- Manipulatives
- Games
- Technology featuring **Building Blocks** software

Number Worlds Results

Number Worlds has been developed and refined since the mid-1980s and has been the only such program to show proven results through years of rigorous field testing. These tests show how students who began at a disadvantage surpassed the performance of students who began on-level with their peers, simply with the help of the **Number Worlds** program.

One of the tests was a longitudinal study conducted to measure the progress of three groups of children from the beginning of Kindergarten to the end of Grade 2. The treatment and control groups both tested one to two years behind normative measures in mathematical knowledge, while the normative group was on track. The treatment group received the **Number Worlds** program while the other two groups used a variety of other mathematical programs during the entire course of the study.

The chart below shows the progress of each group in mean developmental mathematics-level scores as measured by the **Number Worlds** test. The treatment group using the **Number Worlds** program met and exceeded normative mean developmental-level scores by the end of Grade 2. Meanwhile, the control group continued to fall behind their peers.

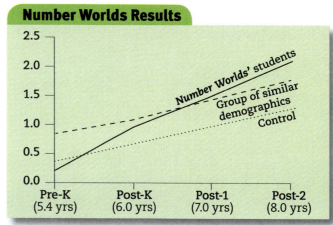

Longitudinal study showing mean developmental scores in mathematical knowledge during Grades K–2

Building Blocks Software Results

The *Building Blocks* software, incorporated into the **Number Worlds** program, is the result of National Science Foundation-funded research. *Building Blocks* includes research-based computer tools with activities and a management system that guides children through research-based learning trajectories.

The program is designed to
• Build upon young children's experiences with mathematics with activities that integrate ways to explore and represent mathematics
• Involve children in "doing mathematics"
• Establish a solid foundation
• Develop a strong conceptual framework
• Emphasize the development of children's mathematical thinking and reasoning abilities
• Develop learning in line with state and national standards

In research studies, *Building Blocks* software was shown to increase young children's knowledge of multiple essential mathematical concepts and skills. One study tested *Building Blocks* against a comparable preschool math program and a no-treatment control group. All classrooms were randomly assigned, the "gold standard" of scientific evaluation. *Building Blocks* children significantly outperformed both the comparison group and control group of children. Results indicate strong positive effects with achievement gains near or exceeding those recorded for individual tutoring.

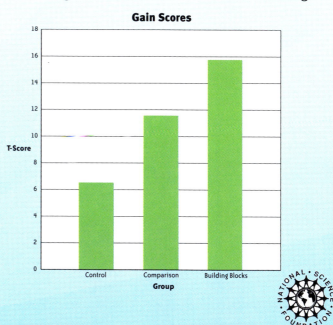

Get students back on track with confidence

Number Worlds is the only program that includes a prevention instruction section for students in **Grades Pre-K–1**. This unique 30-week course of daily instruction improves students' grasp of the world of math so they can move forward with the head start they need.

For students in **Grades 2–6** who are one or more grade levels behind in math, *Number Worlds* intervention program is an invaluable tool. It builds on students' current level of understanding with six 4-week intensive units per grade.

With each daily lesson, teachers have the opportunity to reach the program's knowledge objectives through problem-solving activities, small-group interaction, and discussion. By asking good questions in the classroom, teachers can encourage learning while defining areas that require extra work, ultimately helping to bring students to their appropriate grade level.

Prevention

Fundamental Concepts Levels A–C

Level A	Level B	Level C
Children acquire well-developed counting and quality schemas.	Children develop a well-consolidated central conceptual structure for single-digit numbers.	Children link their central conceptual structure of number to the formal symbol system.

Intervention

Core Content Topics Levels D–H

Level D	Level E	Level F	Level G	Level H
Math Intervention for Grades				
1–3	2–4	3–5	4–6	5–7
Develops Concepts Covered in Grades				
K–1	1–2	2–3	3–4	4–5
Number Sense	Number Sense	Number Sense	Number Sense	Number Sense
Number Patterns and Relationships (Algebra)	Number Patterns and Relationships (Algebra)	Number Patterns and Relationships (Algebra)	Number Patterns and Relationships (Algebra)	Number Patterns and Relationships (Algebra)
Addition	Addition	Addition & Subtraction	Multiplication	Fractions, Decimals & Percents
Subtraction	Subtraction	Multiplication and Beginning Division	Division	Multiplication & Division
Geometry & Measurement	Geometry & Measurement	Geometry & Measurement	Geometry & Measurement	Geometry & Measurement
Data Analysis & Applications	Data Analysis & Applications	Data Analysis & Applications	Data Analysis & Applications	Data Analysis & Applications

Engage your students with step-by-step lessons

Number Worlds Lesson Planner provides a wide array of helpful information before lessons even begin. Background overviews, activity ideas, and tips prepare teachers to the fullest extent.

Weekly Planners map out an entire week of lessons, complete with pacing options, goals, and the resources necessary to get the most out of every class period.

Manipulative-rich lessons are proven to help students turn **abstract concepts into concrete understanding**.

Math Background gives teachers math **context for the lesson**.

Get **insight into your students' capabilities** and how their minds work.

Building Blocks' **research-based software** gives teachers over 150 activity choices that go hand-in-hand with the lessons.

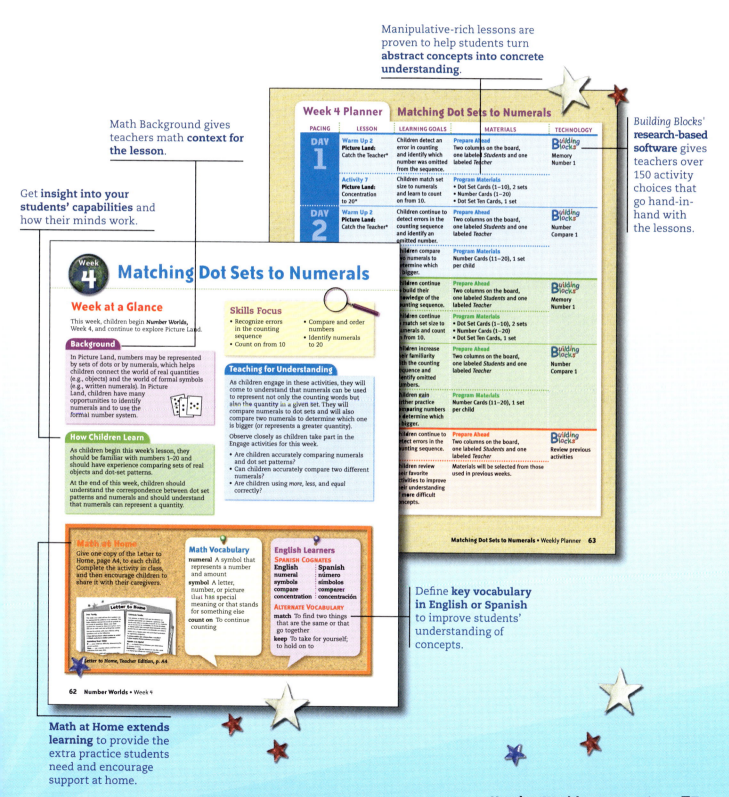

Week 4 Planner — Matching Dot Sets to Numerals

PACING	LESSON	LEARNING GOALS	MATERIALS	TECHNOLOGY
DAY 1	**Warm Up 2** **Picture Land:** Catch the Teacher*	Children detect an error in counting and identify which number was omitted from the sequence.	**Prepare Ahead** Two columns on the board, one labeled *Students* and one labeled *Teacher*	Building Blocks Memory Number 1
	Activity 7 **Picture Land:** Concentration to 20*	Children match set size to numerals and learn to count on from 10.	**Program Materials** • Dot Cards (1–10), 2 sets • Number Cards (1–20) • Dot Set Ten Cards, 1 set	
DAY 2	**Warm Up 2** **Picture Land:** Catch the Teacher*	Children continue to detect errors in the counting sequence and identify an omitted number.	**Prepare Ahead** Two columns on the board, one labeled *Students* and one labeled *Teacher*	Building Blocks Number Compare 1
		Children compare two numerals to determine which is bigger.	**Program Materials** Number Cards (11–20), 1 set per child	
		Children continue to build their knowledge of the counting sequence.	**Prepare Ahead** Two columns on the board, one labeled *Students* and one labeled *Teacher*	Building Blocks Memory Number 1
		Children continue to match set size to numerals and count on from 10.	**Program Materials** • Dot Set Cards (1–10), 2 sets • Number Cards (1–20) • Dot Set Ten Cards, 1 set	
		Children increase their familiarity with the counting sequence and identify omitted numbers.	**Prepare Ahead** Two columns on the board, one labeled *Students* and one labeled *Teacher*	Building Blocks Number Compare 1
		Children gain further practice comparing numbers to determine which is bigger.	**Program Materials** Number Cards (11–20), 1 set per child	
		Children continue to detect errors in the counting sequence.	**Prepare Ahead** Two columns on the board, one labeled *Students* and one labeled *Teacher*	Building Blocks Review previous activities
		Children review their favorite activities to improve their understanding of more difficult concepts.	Materials will be selected from those used in previous weeks.	

Matching Dot Sets to Numerals • Weekly Planner **63**

Week 4 — Matching Dot Sets to Numerals

Week at a Glance

This week, children begin *Number Worlds*, Week 4, and continue to explore Picture Land.

Background

In Picture Land, numbers may be represented by sets of dots or by numerals, which helps children connect the world of real quantities (e.g., objects) and the world of formal symbols (e.g., written numerals). In Picture Land, children have many opportunities to identify numerals and to use the formal number system.

How Children Learn

As children begin this week's lesson, they should be familiar with numbers 1–20 and should have experience comparing sets of real objects and dot-set patterns.

At the end of this week, children should understand the correspondence between dot set patterns and numerals and should understand that numerals can represent a quantity.

Skills Focus
• Recognize errors in the counting sequence
• Count on from 10
• Compare and order numbers
• Identify numerals to 20

Teaching for Understanding

As children engage in these activities, they will come to understand that numerals can be used to represent not only the counting words but also the quantity in a given set. They will compare numerals to dot sets and will also compare two numerals to determine which one is bigger (or represents a greater quantity).

Observe closely as children take part in the Engage activities for this week.

• Are children accurately comparing numerals and dot set patterns?
• Can children accurately compare two different numerals?
• Are children using *more, less,* and *equal* correctly?

Math at Home

Give one copy of the Letter to Home, page A4, to each child. Complete the activity in class, and then encourage children to share it with their caregivers.

Letter to Home, Teacher Edition, p. A4

Math Vocabulary

numeral A symbol that represents a number and amount

symbol A letter, number, or picture that has special meaning or that stands for something else

count on To continue counting

English Learners

SPANISH COGNATES

English	Spanish
numeral	número
symbols	símbolos
compare	comparer
concentration	concentración

ALTERNATE VOCABULARY

match To find two things that are the same or that go together

keep To take for yourself; to hold on to

62 Number Worlds • Week 4

Define **key vocabulary in English or Spanish** to improve students' understanding of concepts.

Math at Home extends learning to provide the extra practice students need and encourage support at home.

Thorough lesson plans guide you every step of the way

Every comprehensive **Number Worlds** lesson is divided into four distinct sections for simplified time management in the classroom. Whether it's time for concept building or skill building, in-depth discussion or assessment, **Number Worlds** always helps you keep learning objectives within reach.

FIN-tastic!

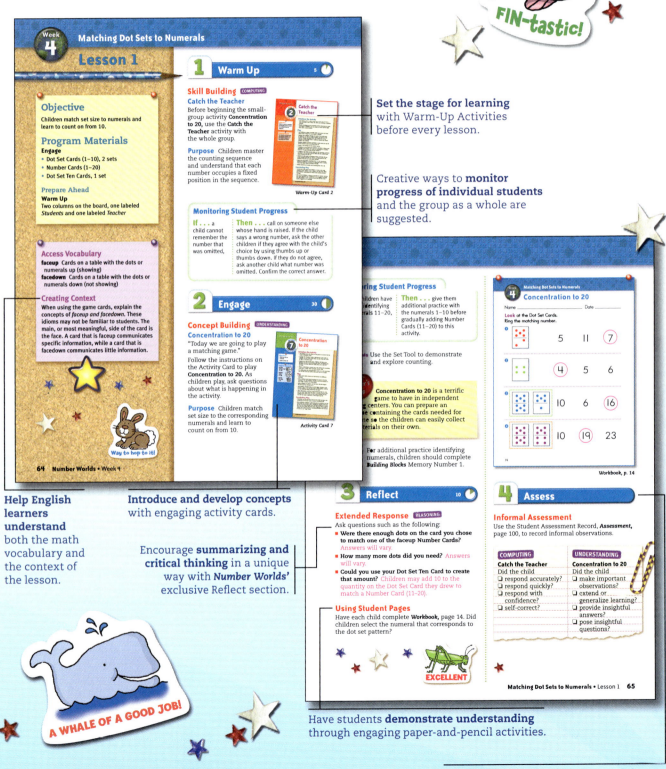

Set the stage for learning with Warm-Up Activities before every lesson.

Creative ways to **monitor progress of individual students** and the group as a whole are suggested.

Help English learners understand both the math vocabulary and the context of the lesson.

Introduce and develop concepts with engaging activity cards.

Encourage **summarizing and critical thinking** in a unique way with **Number Worlds'** exclusive Reflect section.

A WHALE OF A GOOD JOB!

Have students **demonstrate understanding** through engaging paper-and-pencil activities.

Assess student progress after each lesson.

Program Authors

Sharon Griffin, Ph.D., is a Professor of Education and Psychology at Clark University. She specializes in child development and mathematics education and has been studying how playing games that involve numbers helps children structure and understand the world. She conducted research on the development of math competence in the preschool and early school years and used this theoretical work as the basis to create the *Number Worlds* curriculum. Dr. Griffin has worked closely with teachers as they have introduced this curriculum into their classrooms. She has also worked collaboratively with schools around the country and in Canada as they have sought to systematically reform their mathematics programs.

Dr. Griffin's work has been widely published, and she is the author of the chapter "Fostering the Development of Whole-Number Sense: Teaching Mathematics in the Primary Grades" that appeared in a recent book published by the National Research Council. She is a member of the Mathematical Sciences Education Board in the National Academies of Science. She is also involved in an Organization of Economic Cooperation and Development project that brings together leading researchers in neuroscience and cognitive science from several countries to allow each to inform and advance the others' work.

Dr. Griffin holds a B.A. in Psychology from McGill University, a M.Ed. in Learning and Instruction from the University of New Hampshire, and a Ph.D. in Cognitive Science from the University of Toronto.

Douglas H. Clements, Professor of Early Childhood, Mathematics, and Computer Education at the University at Buffalo, State University of New York, has conducted research and published widely on the learning and teaching of geometry, computer applications in mathematics education, the early development of mathematical ideas, and the effects of social interactions on learning. Along with Julie Sarama, Dr. Clements has directed several research projects funded by the National Science Foundation and the U.S. Department of Education's Institute of Educational Sciences, one of which resulted in the mathematics software and activities included in *Building Blocks*.

Julie Sarama is an Associate Professor of Mathematics Education at the University at Buffalo, State University of New York. She conducts research on the implementation and effects of software and curricula in mathematics classrooms, young children's development of mathematical concepts and competencies, implementation and scale-up of educational reform, and professional development. Dr. Sarama has taught secondary mathematics and computer science, gifted mathematics at the middle school level, and preschool and Kindergarten mathematics methods and content courses for elementary to secondary teachers.

Sherry Booth is a mathematics curriculum specialist. Her past projects include the JASON web-based mathematics courses, the ATLAS Project, and the Math Partners project funded by the National Science Foundation. She has collaborated with researchers and designers to develop mathematics curricula that includes software, video, teacher guides, and student materials.

Contents

Contents

Addition and Subtraction

Multiplication and Beginning Division

Contents

Contents

Appendix

Getting Started

Preparing to Use *Number Worlds*

This section provides an overview of classroom management issues and explanations of the Number Worlds program elements and how to use them.

Program Goal

Number Worlds was designed to foster the development of good intuitions about number and number environments. It also was designed to give children a desire to explore such environments and a sense of confidence in moving around within them. Children who develop a solid base of Number Sense have developed the core foundation on which all higher-order understandings are built.

Levels A–C (PreKindergarten, Kindergarten, and First Grade) are intended to serve as **Prevention** for later problems in mathematics. The thoroughly tested, engaging activities bring all students up to level so that they are ready for elementary mathematics. The activities are specifically targeted to address foundational understandings, but are engaging for all students.

Levels D–H (Grades 2–6) provide **Intervention** for students who are 1–2 years behind their peers in mathematics. Targeted units focus on computing, understanding, reasoning, applying, and engaging—the specific concepts and skills that build mathematics proficiency.

Program Objectives

- Teach concepts and skills that build a strong foundation for later learning
- Expose children to the major ways numbers are represented in the world—as objects, symbols, horizontal and vertical lines, and on dials
- Ensure that children acquire the interconnected knowledge that underlies number sense
- Include hands-on and computer activities that provide concrete representations of concepts
- Encourage communication using the language and vocabulary of formal mathematics
- Develop concept understanding with activities that are engaging and appropriate for children from all social and cultural backgrounds

Building Blocks

In addition to hands-on and workbook activities, *Number Worlds* includes *Building Blocks* software to provide additional exposure to and practice with foundational math concepts. *Building Blocks* software is an essential element of the *Number Worlds* curriculum.

Building Blocks software has these advantages

- It combines visual displays, animated graphics, and speech.
- It links "concrete" (graphical) and symbolic (e.g., numerals or spoken words) representations, which build understanding.
- It provides feedback.
- It provides opportunities to explore.
- It focuses children's attention and increases their motivation.
- It individualizes—gives children tasks at children's own ability levels.
- It provides undivided attention, proceeding at the child's pace.
- It keeps a variety of records.
- It provides more manageable manipulatives (e.g., manipulatives "snap" into position).
- It offers more flexible and extensible manipulatives (e.g., manipulatives can be cut apart).
- It provides more manipulatives (you never run out!).
- It stores and retrieves children's work so they can work on it again and again, which facilitates reflection and long-term projects.
- It records and replays children's actions.
- It presents clearer mathematics (e.g., using tools such as a "turn tool," helps children become aware of mathematical processes).

Building Blocks software stores records of how children are doing on every activity. It assigns them to just the right difficulty level. You can also view records of how the whole group or any individual is doing at any time.

The **Number Worlds** program is designed for flexible use. Each lesson is designed to take from 45–60 minutes. It is highly recommended that students spend at least one hour daily using the **Number Worlds** program.

Intervention Models

The program can be effectively used in each of the following environments:

• During Class Time

The Prevention program is designed for whole-class implementation at PreK–1. It can also be used in the same way as the Intervention program with a teacher or teacher's aide working with small groups apart from the rest of the class.

• Math Resource Room

Number Worlds is effective used in a math resource room with a teacher or teacher's aide working with a small group of students.

• After School

Because the activities are engaging, after school programs have used **Number Worlds** effectively for both intervention and regular education students. Both benefit from the intensive math activities.

• Intervention Classrooms

Number Worlds is very effective at every grade level for classes that need intervention in mathematics.

• Summer School

Number Worlds is ideal for summer school programs with an intensive focus on mathematics. Depending on the length of the summer school session, students can work through more than one lesson a day and make substantial progress.

• Tutoring

Number Worlds is also very effective when used in a one-on-one tutoring situation in which a teacher or aide participates in activities with the child.

Many teachers who have used the **Number Worlds** program have found that a teaching assistant— a student teacher, a teacher aide, or a parent volunteer—can help students become extremely familiar with the participatory structures that are required to teach the program so students are able to assume more control over their own learning and become less reliant on the help of a guide or coach. In fact, many **Number Worlds** teachers find it useful to institute "mini-teachers," students who have experience with an activity and understand its concepts, to help their peers.

Although **Number Worlds** activities can be effective with on-level students, they are truly beneficial for students who are not making adequate progress in their core program.

Supplemental Intervention

Number Worlds is geared for students identified with math deficiencies and who have not responded to reteaching efforts. It provides scientifically-based math instruction emphasizing the five critical elements of mathematics proficiency: understanding, computing, applying, reasoning, and engaging. This can be accomplished in a variety of environments for 45–90 minutes per day with a teacher or teacher's aide.

Intensive Intervention

Number Worlds is also effective for students with low skills and a sustained lack of adequate progress in mathematics. The program provides intensive focus on developing mathematical understanding and skills, and includes explicit instruction designed to meet the individual needs of struggling students. This can be accomplished when **Number Worlds** is used as replacement of core lessons for 60–90 minutes a day. The program must be implemented by a teacher or specialized math teacher or teacher's aide.

Understanding Students

For successful math intervention, it is imperative that teachers understand what students know and don't know about math. Teachers also need to know where students' current knowledge fits within the expected developmental sequence, what knowledge students have available to build upon, and what knowledge—the next steps in the sequence—students have yet to master. Finally, teachers need to become familiar with the problem-solving strategies students are using and the range of strategies they need to acquire to become efficient mathematics thinkers and problem-solvers. Using the Number Worlds program and the assessment tools it provides will help teachers acquire a rich understanding of their students along all of these dimensions.

Grades PreK–1 (Levels A–C)

Children with impoverished math backgrounds may come to preschool already 1–2 years behind their peers. Adults frequently overestimate children's understanding of number. These children may not have had experience playing board games, singing counting songs, or participating in number experiences at home. These children may not realize that the same number on a line and on a clock face represent the same quantity.

In the Levels A–C **Number Worlds** program, children explore five different ways number and quantity is represented. Each of the five "lands" of **Number Worlds** that children encounter in the lower grades exposes children to a different representation of number and helps children learn the language used to talk about number in that context.

In **Object Land** students explore the world of counting numbers by counting and comparing sets of objects or pictures of objects. In Object Land you might ask:

- **How many or few do you have?**
- **Which is bigger or smaller?**

In **Picture Land** numbers are represented as sets of stylized, semi-abstract dot-set patterns such as in a die and also as tally marks and numerals. In Picture Land you might ask:

- **What did you roll/pick?**
- **Which has more or less?**

In **Line Land** number is represented as a position on a path or a line. The language used for numbers in Line Land refers to a particular place on a line and also to the moves along a line. These types of questions are asked in Line Land:

- **Where are you now? How far did you go?**
- **Who is farther or less far along the line?**
- **Do you go forward or backward?**

In **Sky Land** number is represented as a position on a vertical scale such as on a thermometer or a bar graph. Sky Land inspires these questions:

- **How high or low are you now?**
- **What number or amount is higher or lower?**

In **Circle Land** number is represented as a point on a dial, such as a clock face or a sundial. In Circle Land you might ask:

- **How many times did you go around the dial?**
- **Which number is farther or less far around?**

Grades 2–6 (Levels D–H)

At grades 2–6 (Levels D–H) students may have difficulty with one, two, or many different math concepts. The goal of the upper grades is to develop foundational understandings in each concept so that students develop mathematical proficiency. Every lesson involves activities that actively engage students in understanding, computing, applying, reasoning, and engagement with the concept. At the end of each week, an assessment will help teachers determine whether a student has reached proficiency. The units are carefully sequenced to develop concepts across grade levels and can be used flexibly to meet student needs.

A variety of program materials are designed to help teachers provide a quality mathematics curriculum.

Teacher Edition

The *Teacher Edition* is the heart of the *Number Worlds* curriculum. It provides background for teachers and complete lesson plans with thorough instructions on how to develop math concepts. It explains when and how to use the program resources.

Activity Cards

These cards are packaged so that each can be carried around and used to direct or teach an activity. The activities, particularly at grades PreK–1 are employed again and again. The Activity Cards provide detailed descriptions of each activity, with suggested questions to ask during the activity.

Activity Sheets

In the *Teacher Edition,* you will find numbered Activity Sheets in blackline master form that accompany activities developed for *Number Worlds.* These sheets can be copied in sufficient quantity for the number of children in your class who will be using that activity on a particular day.

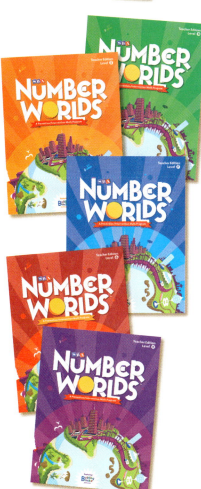

Student Workbook

The *Student Workbook* includes developmental activities to help students develop higher-order thinking skills and practice basic skills.

Assessment

Number World's flexible assessment component allows for prescriptive placement and reliable curriculum and criterion assessment.

Each level of the program includes the following:

- **Placement Tests** to identify where students should begin the *Number Worlds* curriculum.
- **Weekly Tests** to measure student comprehension of the week's five daily lessons.
- **Unit Tests** to evaluate concept acquisition for the entire unit. These tests are available in both open-response and multiple-choice formats.
- **Rubrics** to informally evaluate student understanding of each lesson.
- The **Number Knowledge Test,** designed to measure students' intuitive knowledge of number—the knowledge that helps children make sense of quantitative problems.

Manipulatives and Technology

Number Worlds provides a wealth of manipulative resources. Establishing procedures for use of materials will simplify management issues and allow students to spend more time developing mathematical understanding.

Intervention Packages

A *Number Worlds Intervention Package* is available at every level and includes everything small groups need to use the program.

- Teacher Guide
- Assessment
- Student Workbooks
- Activity Manipulatives and Props
- Software

The package provides storage space for the program so it can be stored and moved as convenient. Activity game mats, manipulatives, demonstration props, playing cards, and pieces are all included. Students will benefit from a demonstration on how you want students to remove materials and return materials to the kit.

Manipulative Modules

Manipulatives are also available for specific topics: Counting, Base-Ten, Fractions, Geometry, Measurement, Money, and Time. Every unit of *Number Worlds* uses manipulative material. If these manipulatives are not already an integral part of your math curriculum, they are available from SRA.

Manipulative Topic Modules Grades K–6

Base-Ten: Cubes, Flats, Rods, Units
ISBN 0-07-605418-7

Geometry I: Pattern Blocks, Attribute Blocks, Solids, Mirror Cards
ISBN 0-07-605414-4

Geometry II: Geoboards, Protractors, Safety Compass, Gummed Tape
ISBN 0-07-605415-2

Measurement I: Rulers, Tape Measures, Measuring Cups, Liter Pitcher, Double-Pan Balance, Thermometer
ISBN 0-07-605416-0

Measurement II: Platform Scale, Metric Weight Set, Customary Weight Set
ISBN 0-07-605417-9

Money: Pennies, Nickles, Dimes, Quarters, Half Dollars, Money Packet
ISBN 0-07-605419-5

Counting: Craft Sticks, Rubber Bands, Panda Bear Counters, Math-Link Cubes
ISBN 0-07-605412-8

Probability: Spinners, Counters
ISBN 0-07-605420-9

Time: Clock Faces, Stopwatches
ISBN 0-07-605421-7

Fractions: Fraction Tiles, Fraction Circles
ISBN 0-07-605413-6

Technology Resources

For Students

Building Blocks activities are engaging, research-based software activities designed to reinforce levels of mathematical development in different strands of mathematics.

eMathTools are electronic tools to help students solve problems and explore and demonstrate concepts.

For Teachers

eAssess An assessment tool to grade, track, and report student progress

For more information on *Number Worlds* manipulatives and software see Appendix C.

Levels A–C
(Grades PreK–1)

Includes 30 weeks of daily lessons.

Levels D–H (Grades 2–6)

Each level includes 6 4-week units that address specific concepts and skills for a total of 24 weeks of instruction. Units are carefully sequenced to develop concepts for students who are one to two grade levels behind their peers. Placement Tests in the **Assessment** Book help teachers place students at the appropriate level and concept based on their demonstrated understanding.

Weekly Overview

- **Teaching for Understanding** provides the big ideas of the chapter.
- **Background** provides a refresher of the mathematics principles relevant to the chapter.
- **How Children Learn** offers insight into how children learn and gives research-based teaching strategies.
- **Weekly Planner** includes objectives that explain how the key concepts are developed lesson by lesson and which resources can be used with each lesson.

- **Lessons** provide overview, ideas for differentiating instruction, complete lesson plans, teaching strategies, and assessments that inform instruction.
- **Cumulative Tests** are provided incrementally to allow you to evaluate whether students are retaining previously developed concepts and skills.

Teacher Edition, Level C

Routines

Lesson Plans

Every lesson throughout *Number Worlds* is structured the same way.

1 **Warm Up**
2 **Engage**
3 **Reflect**
4 **Assess**

Routines for each part of the lesson are explained in the following discussions:

1 Warm Up

Warm Up exercises **provide** cumulative review and computation practice for students and give you opportunities to assess students' skills quickly. Warm Up is an essential component of *Number Worlds* because it helps students review concepts they will need in the Engage activities. It also gives students daily opportunities to sharpen their counting and mental math skills.

Begin every day with the lesson Warm Up with the entire group. These short activities are used in a whole-group format at the start of each day's lesson.

2 Engage

Engage is the heart of the lesson instruction. Here are suggestions for how to introduce lesson concepts, ideas for Guided Discussion, Skill Building, Game Demonstrations, and Strategy Building activities to develop student understanding. Before beginning the Engage activity,

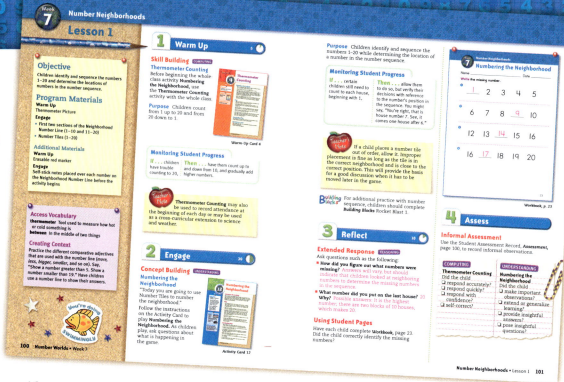

familiarize yourself with the lesson objective, and make sure materials are available for student use. Introduce the activity, and use math talk to help students explore the lesson concepts.

Math Talk is expected in every *Number Worlds* lesson. When students speak the language of mathematics, they communicate mathematically, explain their thinking, and demonstrate understanding. Routines or rules for Math Talk established at the beginning of the year can make discussions more productive and promote listening and speaking skills.

1. **Pay attention to others.** Give your full attention to the person who is speaking. This includes looking at the speaker and nodding to show that you understand.

2. **Wait for speakers to answer and complete their thoughts.** Sometimes teachers and other students get impatient and move on and ask someone else or give the answer before someone has

a chance to think and speak. Giving students time to answer is a vital part of teaching for understanding.

3. **Listen.** Let yourself finish listening before you begin to speak. You can't listen if you are busy thinking about what you want to say next.

4. **Respect speakers.** Take turns and make sure that everyone gets a chance to speak and that no one dominates the conversation.

5. **Build on others' ideas.** Make connections, draw analogies, or expand on the idea.

6. **Ask questions.** Asking questions of another speaker shows that you were listening. Ask if you are not sure you understand what the speaker has said, or ask for clarification or explanation. It is a good idea to repeat in your own words what the speaker said so you can be sure your understanding is correct.

The **Activity Cards** include suggestions for some of the many questions that can be asked during an activity. The questions were selected because they are the kinds of questions children will be able to learn "by heart" and ask themselves. Eventually, the children can and should assume the role of asking the questions during a game. This will help them to become independent learners, responsible for their own learning.

The kinds of questions you ask will depend upon the way in which number is represented, the stage during the activity that you are asking the questions, and the level of difficulty the children are ready to explore. Whatever the combination of these three elements, you should ask the children as often as applicable, "How do you know?" and "How did you figure it out?"

Familiarizing children with these different ways of talking about number and quantity is a major goal of the **Number Worlds** program. It will enable children to realize, for example, that words such as *bigger, more, farther, higher,* and *further around* can all refer to an increase in quantity of the same magnitude, even though this change is expressed in different words, and very likely is represented in a different form. This understanding lies at the heart of number sense.

Using Student Pages

In every week of levels C–H, you will assign student pages to be completed during class. Students will finish at different times and should know what they can do to use their time productively until the Reflect part of the lesson. Students should not feel penalized for finishing early but should do something that is mathematically rewarding.

Student Workbook exercises in Number Worlds are primarily non-mechanical. Students cannot do all the problems on a page in a mechanical, non-thinking manner. Student workbook pages help the students learn to think about the problems.

Because student workbook exercises are non-mechanical, they sometimes require your active participation.

1. Make sure students know what pages to work on and any special requirements of those pages.
2. Tell students whether they should work independently or in small groups as they complete the pages.
3. Tell students how long they have to work on the student pages before you plan to begin the Reflect part of the lesson.

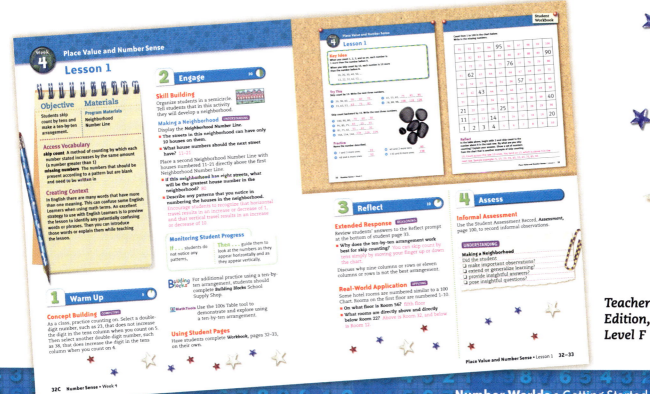

Teacher Edition, Level F

4. Tell students what their options are if they finish early. Technology resources are listed under the Assign Student Pages heading in each lesson. These activities further explore and develop lesson concepts. They include
   a. suggested *eMathTools* to use.
   b. *Building Blocks* activities to do.

5. As students work on the student pages, circulate around the room to monitor their progress. Use the Monitoring Student Progress suggestions for ideas on what to look for. Comment positively on student work, and stop to ask exploring, synthesizing, clarifying, or refocusing questions.

6. You may also use this time to work with English Learners or students who need intervention.

7. Because activities are an important and integral part of the program that provide necessary practice in traditional basic skills as well as higher-order thinking skills, when activities are included in a lesson, be sure to stop work on student pages early enough to leave enough time to play the game. Students may have to complete the student pages outside school in that case.

8. Complete the Informal Assessment Checklists on the last page of each lesson.

3 Reflect

Reflect is a vital part of the lesson that offers ways to help students summarize and reflect on their understanding of lesson concepts. Engaging children in reflection is as important as assessing—and a good way to assess. When children talk about their thinking, using their own words, they engage in mathematical generalizing and communicating. Allowing children to discuss what they did during an activity helps build mathematical reasoning but also develops social skills such as turn taking, listening, and speaking. At the designated time, have students stop working and direct their attention to reflecting on the lesson.

Use the suggested questions in Reflect or ask students to consider these ideas:

A. Summarize their ideas about the lesson concepts
B. Compare how the lesson concept or skill is like or different from other skills
C. Ask how students have seen or can apply the lesson in other curricular areas, other strands of mathematics, or in the world outside of school
D. Think about related matters that go beyond the scope of the lesson

Discuss student solutions to Extended Response questions.

Reflect Questions

A powerful reflection question is, "How do you know?" or "How did you figure that out?" Children may or may not answer you and often cannot provide reasons for their answers. They may shrug their shoulders and say, "I don't know," "Because," or, "Because I'm smart." As the year progresses, children become accustomed to explaining their ideas and their answers give more insight into their mathematical thinking. Young children who have such discussions with teachers and with each other begin to question and correct each other. Incorporate time for reflection into your classroom mathematics activities to develop deep understanding. The following are good questions and challenges.

- **How do you know?**
- **How did you figure that out?**
- **Why?**
- **Show me how you did that.**
- **Tell me about . . .**
- **How is that the same?**
- **How is that different?**

Assess

Assess helps you use informal and formal assessments to summarize and analyze evidence of student understanding and plan for differentiating instruction.

Goals of Assessment

1. **Improve instruction** by informing teachers about the effectiveness of their lessons
2. **Promote growth** of students by identifying where they need additional instruction and support
3. **Recognize accomplishments**

Phases of Assessment

Planning As you develop lesson plans, you can consider how you might assess the instruction, determining how you will tell if students have grasped the material.

Gather Evidence Throughout the instructional phase, you can informally and formally gather evidence of student understanding. The Informal Assessment Checklist and **Student Assessment Record** are provided to help you record data. The end of every lesson is designed to help in conducting meaningful assessments.

Summarize Findings Taking time to reflect on the assessments to summarize findings and make plans for follow-up is a critical part of any lesson.

Use Results Use the results of your findings to differentiate instruction or to adjust or confirm future lessons.

Number Worlds is rich in opportunities to monitor student progress to accomplish these goals.

Informal Daily Assessment

Informal Daily assessments evaluate students' math proficiencies in computational fluency, reasoning, understanding, applying, and engagement.

Warm-Up exercises, activities, and *Student Workbook* pages can be used for day-to-day observation and assessment of how well each student is learning skills and grasping concepts. Because of their special nature, these activities are an effective and convenient means of monitoring students.

Activities, for example, allow you to watch students practice particular skills under conditions more natural to them than most classroom activities. Warm-Up exercises allow you to see individual responses, give immediate feedback, and involve the entire class.

Simple rubrics enable teachers to record and track their observations. These can later be recorded by hand on the Student Assessment Record or in *eAssess* to help provide a more complete view of student proficiency.

Formal Assessment

The *Student Workbook* and the *Assessment Book* provide formal assessments for each chapter. Included are Entry Tests, Weekly Tests, and Cumulative Reviews to evaluate students' understanding of chapter concepts.

Classroom Management

The majority of the *Number Worlds* activities involve small groups and materials that will be new and intriguing to your students. Children as young as five years can learn to function well in semi-autonomous small learning groups. A clear-cut set of classroom expectations must be in place to support and encourage independent learning behaviors. Creating this set of expectations and preparing children to work effectively in small groups will be a task you may wish to address right from the start, as early as the first week of school.

Hints for Starting Small-Group Activities

The following suggestions will help you handle several small groups at once. You might decide to use alternative suggestions with different activities.

- Arrange the class into groups, and have the materials ready and divided ahead of time. Lead the activity with all the groups following along at the same time.

- Have the whole group sit in a semicircle around you. Have volunteers demonstrate the activity, giving many children an opportunity to participate. Once the children are familiar with the activity (perhaps on another day), organize them into small groups and have each group do the activity.

- Have a class helper lead a whole-group activity while you take one group at a time.

- Concentrate on one group while the other groups are involved with another, independent activity.

Because managing several small learning groups at once can be a challenge, several aids and devices have been included in the *Number Worlds* program to make this easier.

Activity Cards are available for each activity in the program. They can be carried around by the teacher or teacher's helper and used, on the spot, to direct the activity. The **Activity Cards** have a built-in categorization system that will tell you how difficult an activity is, how many children can play, what the children will be doing, what they can learn, and your role in the activity.

Activity Card 48, Level C

Activity Sheets are available in this *Teacher Edition*. These activity sheets accompany activities developed for each *Number Worlds* land. These sheets are intended to be copied in sufficient quantities for the number of children in your class who will be using that activity on a particular day.

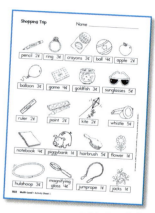

page B22, Level C

Manipulatives are included for all the activities in the five lands. Once the children are familiar with the materials, they can be responsible for gathering them for activities in small groups.

The materials themselves should be organized within the **Number Worlds Intervention Package** so the children can gather the supplies themselves. It is important that you establish a procedure for the management of the materials so that your students will be able to collect and return the materials efficiently.

Technology Resources for Teachers

Number Worlds includes several pieces of integrated technology for teachers designed to increase efficiency and effectiveness of instruction and assessment.

eAssess

- Daily Records Use *eAssess* to record daily formal and informal assessments.
- Report Cards Use *eAssess* to print student and class reports to determine grades.
- Parent-Teacher Conferences Use *eAssess* to print student reports to discuss with parents.

Technology Resources for Students

Number Worlds provides engaging technology resources to enrich, apply, and extend learning.

eMathTools are electronic tools that students can use to solve problems, test solutions, explore concepts, or demonstrate understanding.

Building Blocks activities are designed to reinforce key concepts and develop mathematics understanding.

Using Technology

A. Determine rules for computer use and communicate them to students. Rules should include
- sharing available computers. Some teachers have a computer sign-up chart for each computer. Some teachers have the students track these themselves.
- computer time. You might limit the amount of time students can be at the computer or allow students to work in pairs. Some teachers have students work until they complete an activity. Others allow students to continue on with additional activities.

B. Train students on your rules for proper use of computers, including how to turn computers on, load programs, and shut down the computer. Some teachers manage computers themselves; others have an aide or student in charge of computer management.

C. Using the suggestions for eMathTools and Building Blocks, make sure the computers are on and the programs are loaded and that students know how to access the software.

D. At the beginning of each week, demonstrate and discuss any new Building Blocks or eMath Tools computer activities.

E. Make sure all children work on the assigned computer activities individually at least twice per week for about 15 minutes each time.

F. When the children have finished the assigned activities, they should always get the chance to play and learn with the "free explore" activity. This might be individual but is also an excellent opportunity for children to explore cooperatively, posing problems for each other, solving problems together, or just learning through play.

G. Remember that preparation and followup are as necessary for computer activities as they are for any other activities. Do not omit critical whole group discussion sessions following computer work. Help children communicate their solution strategies and reflect on what they've learned.

H. Make sure children make sense of the mathematics.

For more information about Technology, see Appendix C.

Constructing Numbers to 999

Week at a Glance

This week, students begin **Number Worlds,** Level F, Number Sense. Students should explore different ways to visualize and represent quantities and understand that these are equivalent representations for the same number. Students will also explore the fundamentals of regrouping.

Math Background

Number sense is difficult to define but easy to recognize. Students with good number sense can move seamlessly between the real world of quantities and the mathematical world of numbers and numerical expressions.

—*Robbie Case*

How Students Learn

Students' knowledge of numbers and quantity becomes more integrated as their experiences in mathematics continue. Students begin to link numbers to quantities and realize that questions about numbers can be answered with or without the use of concrete objects. Teach the number sense units with a focus on *quantity*, not *numbers*.

Teaching for Understanding

Observe closely while evaluating the assigned tasks this week to see whether students can demonstrate the following understandings.

Benchmark after Lesson 2: Students can identify how many hundreds, tens, and ones are in a number or a combination of two numbers.

Benchmark after Lesson 3: Students can make "trades" when constructing numbers to 999.

Benchmark after Lesson 4: Students can make a connection between the base-ten number system and money.

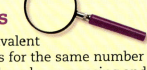

Skills Focus

- Recognize equivalent representations for the same number and generate them by composing and decomposing quantities.
- Separate and regroup double-digit and triple-digit numbers into hundreds, tens, and ones.
- Gain experience with regrouping.

Math at Home

Give one copy of the Letter to Home, page A1, to each student. Encourage students to share and complete the activity with their caregivers.

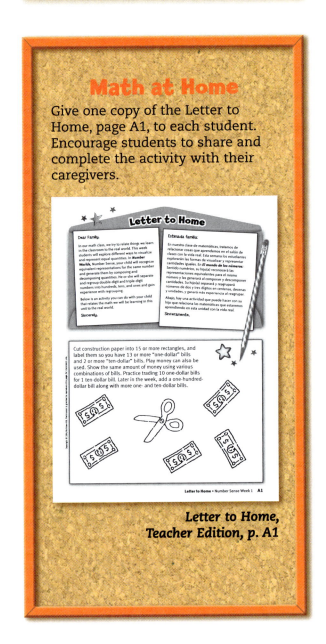

Letter to Home, Teacher Edition, p. A1

Week 1 Planner — Constructing Numbers to 999

PACING	LESSON	LEARNING GOALS	NCTM	MATERIALS	TECHNOLOGY
DAY 1	pages 2–3	Students identify the numbers of tens and ones in a two-digit number.	• Number and Operations • Problem Solving • Communication • Representation	• Base-Ten Blocks* • Number Construction Mat, p. B1 • Place Value Mat, p. B2 • Centimeter ruler	**Building Blocks** Number Snapshots **e MathTools** Base Ten Blocks
DAY 2	pages 4–5	Students identify the numbers of hundreds and tens in a three-digit number.	• Number and Operations • Problem Solving • Communication • Connections • Representation	• Base-Ten Blocks* • Number Construction Mat, p. B1 • Place Value Mat, p. B2 • Number Cards (0–9) • Number 1–6 Cube	**Building Blocks** Number Snapshots **e MathTools** Base Ten Blocks
DAY 3	pages 6–7	Students use Base-Ten Blocks to construct numbers to 999.	• Number and Operations • Problem Solving • Communication • Connections • Representation	• Base-Ten Blocks* • Number Construction Mat, p. B1 • Place Value Mat, p. B2 • Number Cards (1–25)	**Building Blocks** Number Snapshots **e MathTools** Base Ten Blocks
DAY 4	pages 8–9	Students connect the basics of the base-ten number system to money.	• Number and Operations • Problem Solving • Communication • Connections • Representation	• Models of pennies, dimes, dollar bills, and ten-dollar bills* • Base-Ten Money Table, p. B3	**Building Blocks** Coin Combos **e MathTools** Coins and Money
DAY 5	**Review and Assess** pages 10–11	Students review skills learned this week and complete the weekly assessment.	• Number and Operations • Problem Solving • Communication • Connections • Representation	Materials will be selected from Lessons 1–4.	**Building Blocks** **e MathTools** Review previous activities

Math Vocabulary

unit block The Base-Ten Block that represents 1 in the base-ten number system

rod A single block that is equal in size and value to 10 unit blocks

flat A single block equal in size and value to 10 rods, or 100 unit blocks

English Learners

SPANISH COGNATES

English	Spanish
constructing	construyendo
numbers	números
units	unidades
visualizing	visualizando

ALTERNATE VOCABULARY

trade To exchange 10 blocks of a smaller size for 1 block of the next size

Lesson 1

Objective

Students identify the numbers of tens and ones in a two-digit number.

Materials

Program Materials
- Number Construction Mat, p. B1
- Place Value Mat, p. B2

Additional Materials
- Centimeter ruler
- Base-Ten Blocks

Access Vocabulary

sample answers Examples, possible answers
standard form Common numeral form

Creating Context

Acquiring the academic language necessary to be successful in mathematics is challenging for English Learners. Often, the instructions given in a textbook are abbreviated and do not give sufficient direction. For example, the instructions for Problems 4 through 7 say to, "Tell how many tens and ones there are in each number." Some English Learners may interpret this as a "repeat after me" instruction. Provide examples to clarify the instructions.

1 Warm Up 5

Concept Building COMPUTING

Instruct students to take a handful of unit blocks and to group the blocks into sets of 10. Students should skip count the groups by tens and then count the remaining units to name the number their blocks represent. Have students join their blocks with those of another student and regroup the blocks into sets of 10. They should skip count by tens and count on to name the number represented by both students' blocks.

2 Engage 30

Skill Building

Give each student a Number Construction Mat, a Place Value Mat, and Base-Ten Blocks.

Modeling UNDERSTANDING

Instruct students to model the quantity 27 on their mats. Then have students clear their mats and model 27 in a different way.

- **What is the same about these models?** They each represent 27.

Point out that both models are correct.

Strategy Building

Trading COMPUTING

- Have students model 16 on their mats. Without clearing their mats, have students add 8 more unit blocks. Guide students as they trade to get 2 rods in the tens column and 4 unit blocks in the ones column.
- Have students transfer the blocks onto the Number Construction Mat and name the number as 24.
- Have students model and trade the following: 12, and then 9 more; 24, and then 6 more; and 19, and then 5 more.

Monitoring Student Progress

If . . . a student is trading too early,

Then . . . encourage the student to label the circles from 1 (beginning at the bottom) to 10 on the Place Value Mat.

 Building Blocks For additional practice representing double-digit numbers, students should complete **Building Blocks** Number Snapshots.

MathTools Use the Base Ten Blocks tool to demonstrate and explore double-digit numbers.

Using Student Pages

Have students complete **Workbook,** pages 2–3, on their own.

Lesson 1

Key Idea

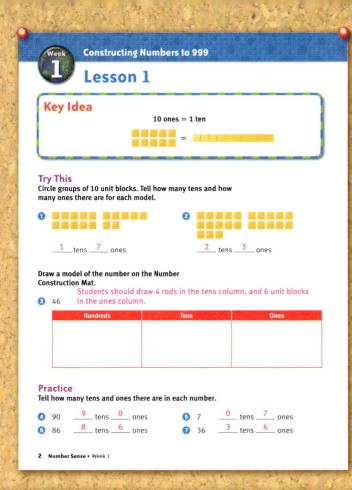

10 ones = 1 ten

Try This

Circle groups of 10 unit blocks. Tell how many tens and how many ones there are for each model.

❶ ___1___ tens ___7___ ones

❷ ___2___ tens ___3___ ones

Draw a model of the number on the Number Construction Mat.

Students should draw 4 rods in the tens column, and 6 unit blocks in the ones column.

❸ 46

Hundreds	Tens	Ones

Practice

Tell how many tens and ones there are in each number.

❹ 90   ___9___ tens ___0___ ones   ❺ 7   ___0___ tens ___7___ ones

❻ 86   ___8___ tens ___6___ ones   ❼ 36   ___3___ tens ___6___ ones

---

Make trades. Write each number in standard form.

❽ 36

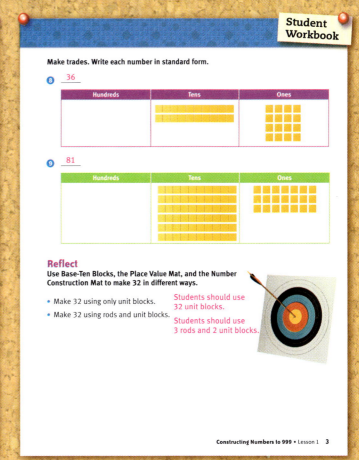

Hundreds	Tens	Ones

❾ 81

Hundreds	Tens	Ones

Reflect

Use Base-Ten Blocks, the Place Value Mat, and the Number Construction Mat to make 32 in different ways.

- Make 32 using only unit blocks.   Students should use 32 unit blocks.
- Make 32 using rods and unit blocks.   Students should use 3 rods and 2 unit blocks.

---

3 Reflect   10

Extended Response   REASONING

Review students' answers to the Reflect prompt at the bottom of student page 3.

- **How did your two models look different?** Each was made of different blocks.
- **How were they alike?** Both represent the same number.

Real-World Application   APPLYING

In many situations, trades are made for equivalent quantities. In the metric system, ten of one unit can be traded for one of the next larger unit. Because the metric system is another base-ten system, the trades are comparable to Base-Ten Blocks. Display a metric ruler.

- **On this ruler, are there 10 smaller units that can be traded for one larger unit?** Every 10 millimeters equals 1 centimeter.

4 Assess

Informal Assessment

Use the Student Assessment Record, *Assessment,* page 100, to record informal observations.

UNDERSTANDING	COMPUTING
Modeling	**Trading**
Did the student	Did the student
❏ make important observations?	❏ respond accurately?
❏ extend or generalize learning?	❏ respond quickly?
❏ provide insightful answers?	❏ respond with confidence?
❏ pose insightful questions?	❏ self-correct?

Lesson 2

Objective

Students identify the numbers of hundreds and tens in a three-digit number.

Materials

Program Materials
• Number Construction Mat, p. B1
• Place Value Mat, p. B2

Additional Materials
• Number Cards (0–9), two sets for each student: one blue set, one yellow set
• Number 1–6 Cube
• Base-Ten Blocks

Access Vocabulary

draw a model To create a picture of an example

Creating Context

Understanding place value is necessary for success in mathematics. This lesson uses mats designed to clearly separate the amounts in each place value and easily identify what trades are needed. This is an excellent strategy for English Learners, especially those at early English proficiency levels who may not be able to verbally explain their understanding. On the Place Value Mat, students can manipulate the objects and point to the answer.

 1 Warm Up 5

Concept Building COMPUTING

Practice skip-counting by tens aloud. Begin with 10 and count to 100. Then begin with 15 and count to 105. Students need to be able to skip count by tens and hundreds, beginning with any number greater than 0.

 2 Engage 30

Skip Building

Instruct students to place the blue set of Number Cards above the tens column on the Number Construction Mat. Place the yellow set above the hundreds column. Have students place their yellow 0 card faceup in the ones space because for this activity, the ones value will always be 0.

Game Show Contestant COMPUTING

Each student will pretend to be a contestant on a game show. The first student to reach 500 points is the winner. Each time you announce a round, each student should roll the Number Cube. This number times ten will be the student's score for that round.

• Have students place rods in the tens column on the Place Value Mat for their Round 1 score.
• Students should also place a blue card to show their score on their Number Construction Mat.
• With each score, the student places that number of rods on the Place Value Mat, makes trades as needed, and places the appropriate Number Card in the tens column.
• After a student's score reaches 100, the yellow cards should be used in the hundreds column.
• Continue with rounds until one student reaches a score of 500.

Monitoring Student Progress

If . . . a student is struggling to make trades at the appropriate times,	**Then . . .** encourage the student to record the result of each roll so you can review the final score with him or her.

Building Blocks For additional practice constructing three-digit numbers, students should complete **Building Blocks** Number Snapshots.

MathTools Use the Base Ten Blocks tool to demonstrate and explore three-digit numbers.

Using Student Pages

Have students complete **Workbook,** pages 4–5, on their own.

Lesson 2

Key Idea

10 tens = 1 hundred

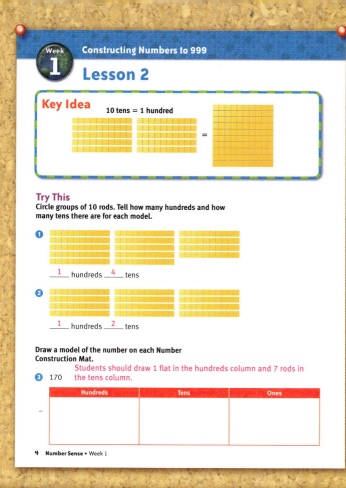

Try This

Circle groups of 10 rods. Tell how many hundreds and how many tens there are for each model.

1

_1_ hundreds _4_ tens

2

_1_ hundreds _2_ tens

Draw a model of the number on each Number Construction Mat.

3 170   Students should draw 1 flat in the hundreds column and 7 rods in the tens column.

Hundreds	Tens	Ones

Practice

Tell how many hundreds and tens there are in each number.

4 280   _2_ hundreds _8_ tens _0_ ones

5 340   _3_ hundreds _4_ tens _0_ ones

6 150   _1_ hundreds _5_ tens _0_ ones

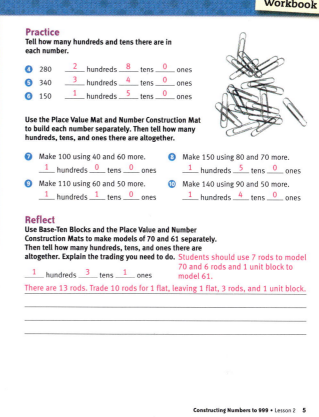

Use the Place Value Mat and Number Construction Mat to build each number separately. Then tell how many hundreds, tens, and ones there are altogether.

7 Make 100 using 40 and 60 more.

_1_ hundreds _0_ tens _0_ ones

8 Make 150 using 80 and 70 more.

_1_ hundreds _5_ tens _0_ ones

9 Make 110 using 60 and 50 more.

_1_ hundreds _1_ tens _0_ ones

10 Make 140 using 90 and 50 more.

_1_ hundreds _4_ tens _0_ ones

Reflect

Use Base-Ten Blocks and the Place Value and Number Construction Mats to make models of 70 and 61 separately. Then tell how many hundreds, tens, and ones there are altogether. Explain the trading you need to do.   Students should use 7 rods to model 70 and 6 rods and 1 unit block to model 61.

_1_ hundreds _3_ tens _1_ ones

There are 13 rods. Trade 10 rods for 1 flat, leaving 1 flat, 3 rods, and 1 unit block.

_____

_____

_____

_____

---

3 Reflect   10

Extended Response   REASONING

Review students' answers to the Reflect prompt at the bottom of student page 5. Invite students to share their explanations of the trade.

- **Explain the trades you made.** Answers will vary.
- **How can you check your answer?** Do the problem again.

Real-World Application   APPLYING

Have students design a sports stadium scoreboard that will allow for two team's scores to reach 999. Students should draw a picture of the scoreboard with team names and a different score for each team.

4 Assess

Informal Assessment

Use the Student Assessment Record, **Assessment,** page 100, to record informal observations.

COMPUTING

Game Show Contestant

Did the student
- ❏ respond accurately?
- ❏ respond quickly?
- ❏ respond with confidence?
- ❏ self-correct?

Lesson 3

Objective

Students use Base-Ten Blocks to construct numbers to 999.

Materials

Program Materials
- Number Construction Mat, p. B1
- Place Value Mat, p. B2

Additional Materials
- Number Cards (1–25), 1 set for each group
- Base-Ten Blocks

Access Vocabulary

explain the trade To describe how and why you made the exchange

Creating Context

English Learners often encounter English words that proficient speakers know well. Many of these words can be confusing because they are spelled the same or almost the same as other concept words. In this lesson, we work with tens and ones. The number *1* means a single item, yet *ones* refers to single digit numbers fewer than ten. It may seem counter-intuitive to use a plural form of a word that represents a singular word. Ask English Learners to draw a labeled illustration to show the difference between *1* and *ones*.

1 Warm Up — 5

Concept Building COMPUTING

As a class, skip count by tens to 100.

■ **How many tens are in 80?** 8

Name other multiples of 10, and ask students to tell how many tens are in the number.

2 Engage — 30

Skill Building

Give each student a Place Value Mat and Number Construction Mat, as well as Base-Ten Blocks. Select a Number Card. Read the number aloud. Students should build that number on their Place Value Mat. Select another card. Read it aloud. Students build that number onto the first number. Make trades when needed. Repeat selecting a card and building the number until there are ten rods in the tens column.

■ **What do you trade for ten rods?** 1 flat

Quick Builders Game COMPUTING

Organize students into groups of five. The object of the game is to be the first one in the group to build a number greater than 100.

- Students take turns choosing a Number Card and building that number on their Place Value Mat. Each student chooses a different card and builds that number.
- After all players have built their first number, play goes around the group again with players taking turns selecting a Number Card and building onto the existing number on their mats.
- Play continues until one player reaches a number greater than 100. This player is the winner.

Monitoring Student Progress

If . . . a student is struggling to understand the build-on concept,

Then . . . give him or her a set of Base-Ten Blocks and the Place Value Mat so he or she can practice with an adult outside of class.

Building Blocks For additional practice combining numbers to 999, students should complete *Building Blocks* Number Snapshots.

MathTools Use the Base Ten Blocks tool to demonstrate and explore combining numbers to 999.

Using Student Pages

Have students complete *Workbook,* pages 6–7, on their own.

Lesson 3

Key Idea
Sometimes when you trade 10 ones for 1 ten, you end up with more than 10 tens.

Then you have to trade 10 tens for 1 hundred.

Try This
Trade ones and tens of each number shown. Write each number in standard form.

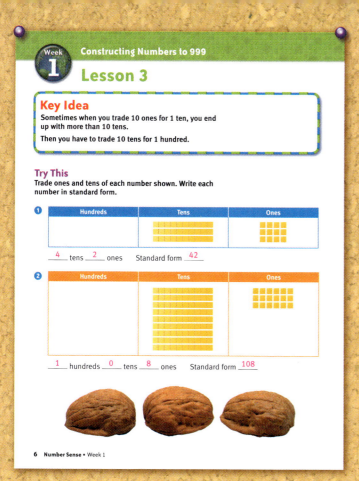

1

Hundreds	Tens	Ones

__4__ tens __2__ ones   Standard form __42__

2

Hundreds	Tens	Ones

__1__ hundreds __0__ tens __8__ ones   Standard form __108__

Practice
Use the Place Value Mat and Number Construction Mat to make each number and make trades. Tell how many there are altogether. Write each number in standard form.

3 72 and 18   __9__ tens __0__ ones   __90__

4 9 and 19   __2__ tens __8__ ones   __28__

5 85 and 22   __1__ hundreds __0__ tens __7__ ones   __107__

Show two numbers that can be joined to make each number. Name the two numbers.   Answers will vary. Sample answers are given.

6 103   __74 and 29__   **7** 110   __55 and 55__

8 171   __129 and 42__   **9** 405   __315 and 90__

10 202   __64 and 138__   **11** 211   __126 and 85__

Reflect
When you have 10 tens, what will be the trade?

10 tens will trade for 1 hundred

Use Base-Ten Blocks to show 90 + 10. Explain the trade.
10 rods trade for 1 flat

Students should model 9 rods and 1 rod and trade for 1 flat.

Use Base-Ten Blocks to show 95 + 5. Explain all trades.
10 unit blocks trade for 1 rod. 10 rods trade for 1 flat.

Students should model 9 rods and 5 unit blocks, and 5 unit blocks, and trade for 1 flat.

---

3 Reflect   10

Extended Response   REASONING

Review students' answers to the Reflect prompt at the bottom of student page 7. Display the sizes of the Base-Ten Blocks in order from least to greatest. Guide students to see the pattern that ten of a lesser-sized block trade for one of the next-sized block.

- **How were the two trades similar?** Both trades resulted in 1 flat.
- **How were the two trades different?** The first trade used only rods. The second trade used unit blocks and rods.

Real-World Application   APPLYING

The fourth-grade classes are going on a field trip. Each bus has 100 seats and must be full before the next bus can begin loading. The classes board the buses in the following order. Class A–28 students, Class B–24 students, Class C–31 students, Class D–29 students, Class E–28 students. Students may use Base-Ten Blocks to solve this problem.

- **Which class will be split between Bus 1 and Bus 2?** Class D will be split.
- **How many students will be on Bus 2?** Bus 2 will have 40 students.

4 Assess

Informal Assessment

Use the Student Assessment Record, *Assessment*, page 100, to record informal observations.

COMPUTING

Quick Builders Game
Did the student
❑ respond accurately?
❑ respond quickly?
❑ respond with confidence?
❑ self-correct?

Lesson 4

Objective

Students connect the basics of the base-ten number system to money.

Materials

Program Materials
Base-Ten Money Table, p. B3

Additional Materials
Models of pennies, dimes, one-dollar bills, and ten-dollar bills

Access Vocabulary

penny One cent, represents a one in the base-ten number system

dime 10 cents, represents a ten in the base-ten number system

one-dollar bill 100 cents, represents a hundred in the base-ten number system

ten-dollar bill 1,000 cents, represents a thousand in the base-ten number system

Creating Context

When a sentence describes more than one object, the noun is plural. In some languages there is no inflected ending. In English an *s* is usually added to the end of the word, but some words also require spelling changes. Help English Learners make a chart of the types of U.S. currency and their plurals.

1 Warm Up 5

Concept Building COMPUTING

Show the class a penny, nickel, dime, quarter, one-dollar bill, and ten-dollar bill. Explain that in this lesson, only the penny, dime, one-dollar bill, and ten-dollar bill will be used.

■ **What do the coins and bills being used in this lesson have in common with the quantities in the previous three lessons?** They are a base-ten number system. Students should say something about the role 10 plays in how different denominations are related to each other.

2 Engage 30

Skill Building

Give each student models for pennies, dimes, one-dollar bills, and ten-dollar bills. Students also need a **Base-Ten Money Table.**

Number "Cents" UNDERSTANDING

In the first row of the **Base-Ten Money Table,** under Amount column, instruct students to write 43¢. Instruct students to use their models to make 43¢ and to record their answers in their tables.

- Discuss all combinations of coins that students describe.
- Repeat the activity using different amounts of money.

Strategy Building

How Much Money? REASONING

Give students the following problems. Allow time to solve.

■ **I have $1.50. I have one bill. I have five coins. What bills and coins do I have?** 1 one-dollar bill, five dimes

■ **I have $2.20. I have one bill. I have more pennies than dimes. What bills and coins do I have?** 1 one-dollar bill, 10 dimes, 20 pennies

Monitoring Student Progress

If . . . a student is doing well solving the problems,

Then . . . encourage the student to write a problem to share with the class.

Building Blocks For additional practice using money to combine numbers to 1,000, students should complete **Building Blocks** Coin Combos.

MathTools Use the Coins and Money tool to demonstrate and explore using money to combine numbers to 1,000.

Using Student Pages

Have students complete **Workbook,** pages 8–9, on their own.

Lesson 4

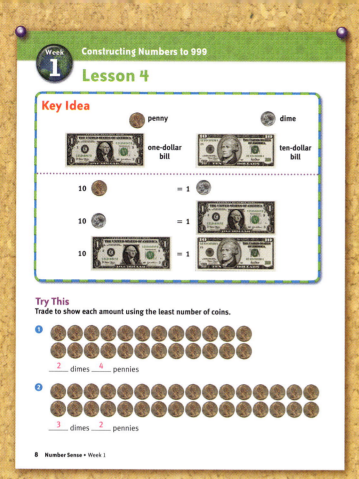

Key Idea

penny   dime

one-dollar bill   ten-dollar bill

10 = 1

10 = 1

10 = 1

Try This

Trade to show each amount using the least number of coins.

①

__2__ dimes __4__ pennies

②

__3__ dimes __2__ pennies

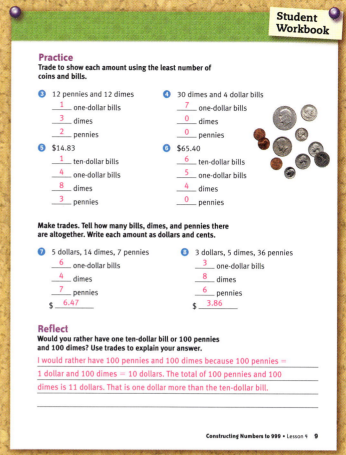

Practice
Trade to show each amount using the least number of coins and bills.

③ 12 pennies and 12 dimes
__1__ one-dollar bills
__3__ dimes
__2__ pennies

④ 30 dimes and 4 dollar bills
__7__ one-dollar bills
__0__ dimes
__0__ pennies

⑤ $14.83
__1__ ten-dollar bills
__4__ one-dollar bills
__8__ dimes
__3__ pennies

⑥ $65.40
__6__ ten-dollar bills
__5__ one-dollar bills
__4__ dimes
__0__ pennies

Make trades. Tell how many bills, dimes, and pennies there are altogether. Write each amount as dollars and cents.

⑦ 5 dollars, 14 dimes, 7 pennies
__6__ one-dollar bills
__4__ dimes
__7__ pennies
$ __6.47__

⑧ 3 dollars, 5 dimes, 36 pennies
__3__ one-dollar bills
__8__ dimes
__6__ pennies
$ __3.86__

Reflect
Would you rather have one ten-dollar bill or 100 pennies and 100 dimes? Use trades to explain your answer.

I would rather have 100 pennies and 100 dimes because 100 pennies =
1 dollar and 100 dimes = 10 dollars. The total of 100 pennies and 100
dimes is 11 dollars. That is one dollar more than the ten-dollar bill.

---

3 Reflect   10

Extended Response APPLYING

Review students' answers to the Reflect prompt at the bottom of student page 9.

Select two students to act out the exchange of 100 pennies and 100 dimes. One student is the banker. The other student is the customer.

■ **How would you explain your reasoning to someone who doesn't understand this concept?** Answers will vary.

Real-World Application APPLYING

Organize students into pairs. Make a note card with an amount of money written on it for each pair of students. Each student in the pair should gather a group of coins and bills equaling the amount on the note card, but each student should use a different arrangement of coins and bills.

4 Assess

Informal Assessment

Use the Student Assessment Record, **Assessment,** page 100, to record informal observations.

UNDERSTANDING	REASONING
Number "Cents"	**How Much Money?**
Did the student	Did the student
❏ make important observations?	❏ provide a clear explanation?
❏ extend or generalize learning?	❏ communicate reasons and strategies?
❏ provide insightful answers?	❏ choose appropriate strategies?
❏ pose insightful questions?	❏ argue logically?

Lesson 5 Review

Objective

Students review skills learned this week and complete the weekly assessment.

Materials

Review materials will be selected from those used in previous activities.

Creating Context

In English the letter *c* has several pronunciations. At the beginning of words, it can have a hard /k/ or soft /s/ sound. When the letter *c* comes before *a, o,* or *u,* it makes the hard sound as in *constructing, cat,* and *cut.* When the letter *c* comes before *e* or *i,* it makes the soft sound as in *cent, cipher,* and *cycle.* Have students make a two-column chart labeled *Hard c* and *Soft c.* Write examples you find during the week.

1 Warm Up 5

Concept Building COMPUTING

As a class skip count by tens to 200, or higher.

- **How many tens are in 160?** 16

Name other multiples of 10 and ask students to tell how many tens are in the number.

2 Engage 20

Skill Building

Free-Choice Activity

For the last day of the week, allow students to choose an activity from the previous lessons. Some activities they may choose are the following:

- **Modeling**
- **Trading**
- **Game Show Contestant**
- **Quick Builders Game**
- **Number "Cents"**
- **How Much Money?**

Make a note of the activities students select. Do they prefer easy or challenging activities? If you believe your students would benefit from extra practice on specific skills, choose an activity for them.

Monitoring Student Progress

If . . . students are not participating,

Then . . . continue the game until all students have participated.

Using Student Pages

Have students complete **Workbook,** pages 10–11, on their own.

3 Reflect 10

Extended Response APPLYING

Review students' answers to the Reflect prompts at the bottom of student pages 10–11.

Discuss these answers with the group to reinforce Week 1 concepts.

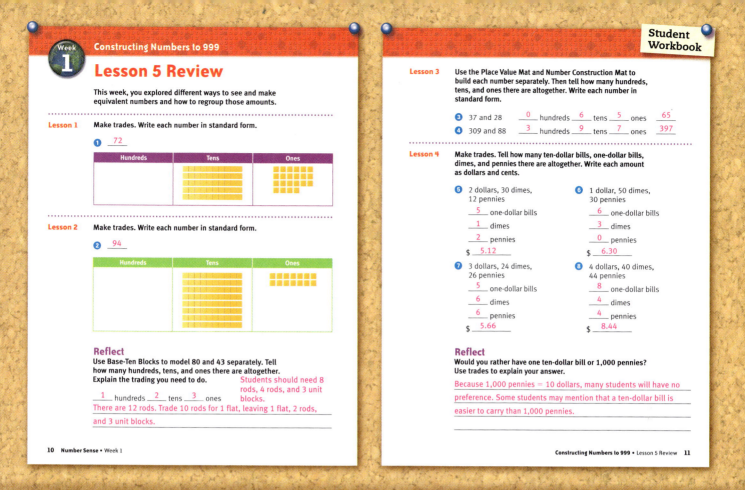

Week 1 Constructing Numbers to 999

Lesson 5 Review

This week, you explored different ways to see and make equivalent numbers and how to regroup those amounts.

Lesson 1 Make trades. Write each number in standard form.

① 72

Hundreds	Tens	Ones

Lesson 2 Make trades. Write each number in standard form.

② 94

Hundreds	Tens	Ones

Reflect

Use Base-Ten Blocks to model 80 and 43 separately. Tell how many hundreds, tens, and ones there are altogether. Explain the trading you need to do.

__1__ hundreds __2__ tens __3__ ones

Students should need 8 rods, 4 rods, and 3 unit blocks. There are 12 rods. Trade 10 rods for 1 flat, leaving 1 flat, 2 rods, and 3 unit blocks.

10 Number Sense • Week 1

Lesson 3 Use the Place Value Mat and Number Construction Mat to build each number separately. Then tell how many hundreds, tens, and ones there are altogether. Write each number in standard form.

③ 37 and 28 __0__ hundreds __6__ tens __5__ ones __65__

④ 309 and 88 __3__ hundreds __9__ tens __7__ ones __397__

Lesson 4 Make trades. Tell how many ten-dollar bills, one-dollar bills, dimes, and pennies there are altogether. Write each amount as dollars and cents.

⑤ 2 dollars, 30 dimes, 12 pennies
__5__ one-dollar bills
__1__ dimes
__2__ pennies
$ __5.12__

⑥ 1 dollar, 50 dimes, 30 pennies
__6__ one-dollar bills
__3__ dimes
__0__ pennies
$ __6.30__

⑦ 3 dollars, 24 dimes, 26 pennies
__5__ one-dollar bills
__6__ dimes
__6__ pennies
$ __5.66__

⑧ 4 dollars, 40 dimes, 44 pennies
__8__ one-dollar bills
__4__ dimes
__4__ pennies
$ __8.44__

Reflect

Would you rather have one ten-dollar bill or 1,000 pennies? Use trades to explain your answer.

Because 1,000 pennies = 10 dollars, many students will have no preference. Some students may mention that a ten-dollar bill is easier to carry than 1,000 pennies.

Constructing Numbers to 999 • Lesson 5 Review 11

---

4 Assess 10 ▶

A Gather Evidence

Formal Assessment

Have students complete the Weekly test on **Assessment,** pages 24–25. Record scores on the Student Assessment Record, **Assessment,** page 100.

B Summarize Findings

Determine whether students have Minimal, Basic, or Secure understanding of the concepts presented in Week 1.

C Differentiate Instruction

Based on your observations, use these teaching strategies next week to follow up.

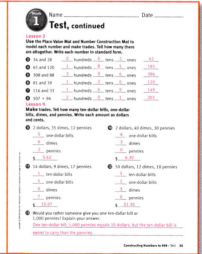

Assessment, pp. 24–25

Minimal Understanding
- Repeat the Warm-Up and Engage activities to develop Week 1 concepts.
- Use **Building Blocks** Number Snapshots and **eMathTools** Base Ten Blocks to develop and reinforce Week 1 concepts.

Basic Understanding
- Repeat Engage activities to reinforce Week 1 concepts.
- Use **Building Blocks** Coin Combos and **eMathTools** Coins and Money to reinforce Week 1 concepts.

Secure Understanding
Use **Building Blocks** and **eMathTools** to reinforce Week 1 concepts.

Week 2 — Place Value to 9,999

Week at a Glance

This week, students continue with *Number Worlds,* Level F, Number Sense, by investigating the meaning of place value. Students should continue to compose and decompose numbers as well as gain an understanding of the role that regrouping plays in place value.

Math Background

By the age of 7 or 8, a student's central conceptual structure for number becomes more complex, permitting him or her to make sense of two distinct quantitative dimensions such as tens and ones, and dollars and cents in money. With this more complex structure, students are able to understand place value, to solve double-digit arithmetic problems in their heads, and to tell which of two double-digit numbers is bigger or smaller.

—*Sharon Griffin*

How Students Learn

After students begin to think using a more elaborate and integrated central conceptual structure, they are able to handle quantities in a more coordinated and organized manner. This level of development, which usually occurs by age 9 or 10, enables students to understand the complete number system and to perform mental computations that involve regrouping.

Teaching for Understanding

Observe closely while evaluating the assigned tasks this week to see whether students can demonstrate the following understandings.

Benchmark after Lesson 2: Students can identify the value of any digit in a three-digit number.

Benchmark after Lesson 3: Students can write any number up to 9,999 in expanded form.

Benchmark after Lesson 4: Students can identify a four-digit number written in a variety of forms.

Skills Focus

- Understand the place-value structure of the base-ten number system for numbers through the thousands place.
- Identify the value of each digit in any four-digit number.
- Use expanded notation to represent numbers.

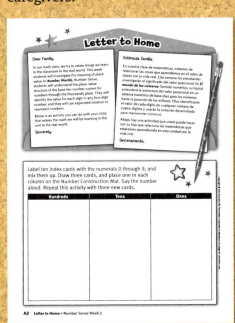

Math at Home

Give one copy of the Letter to Home, page A2, to each student. Encourage students to share and complete the activity with their caregivers.

Letter to Home,
Teacher Edition, p. A2

PACING	LESSON	LEARNING GOALS	NCTM	MATERIALS	TECHNOLOGY
DAY 1	pages 12–13	Students find all possible groupings for quantities of 10 and 100.	• Number and Operations • Problem Solving • Communication	Base-Ten Blocks*	**Building Blocks** Number Snapshots **e MathTools** Base Ten Blocks
DAY 2	pages 14–15	Students write numbers to 999 in expanded form.	• Number and Operations • Communication • Representation	• Number Cards (0–9) • Number Construction Mat, p. B1 • Base-Ten Blocks*	**Building Blocks** Number Snapshots **e MathTools** Base Ten Blocks
DAY 3	pages 16–17	Students write numbers to 9,999 in expanded form.	• Number and Operations • Communication • Representation	• Number Cards (0–9) • Number Cards (1, 10, 100, and 1,000) • 3-Digit Window, p. B4 • 4-Digit Window, p. B5	**Building Blocks** Number Snapshots
DAY 4	pages 18–19	Students identify and write numbers in standard form, expanded form, and word form.	• Number and Operations • Communication • Connections • Representation	Blank check forms, p. B6	**Building Blocks** Number Snapshots
DAY 5	Review and Assess pages 20–21	Students review skills learned this week and complete the weekly assessment.	• Number and Operations • Problem Solving • Communication • Connections • Representation	Materials will be selected from Lessons 1–4.	**Building Blocks** **e MathTools** Review previous activities

Math Vocabulary

expanded form A way of writing numbers that shows the sum of the values of each place value

standard form A way of writing numbers using the digits 0–9

word form A way of writing numbers using words

English Learners

SPANISH COGNATES

English	Spanish
value	valor
count	contar
model	modelar
different	diferente

ALTERNATE VOCABULARY

skip count by tens To count by tens, skipping all of the counting numbers in the interval between tens

* Available from SRA

Lesson 1

Objective

Students find all possible groupings for quantities of 10 and 100.

Materials

Additional Materials
Base-Ten Blocks

Access Vocabulary

skip count by tens To count by tens, skipping all of the counting numbers in the interval between tens

Creating Context

Some languages sound out all vowel sounds. When English Learners transfer their literacy skills from their primary language to English, they encounter non-transferable sounds and grammar skills. In English, some words end with a silent *e*, which means that the *e* at the end of the word is not pronounced. In this lesson, the word *value* is used often and has a silent *e* at the end. Help English Learners understand why the *e* is silent and how the word is pronounced.

1 Warm Up 5

Concept Building COMPUTING

Place 12 unit blocks on the overhead projector or draw them on the board.

- **What is one way to arrange these blocks into two groups?** Answers can be any of the following: 1 and 11, 2 and 10, 3 and 9, 4 and 8, 5 and 7, 6 and 6.

Write one arrangement on the board and ask for another arrangement. Repeat this until students have named all possible answers.

2 Engage 30

Strategy Building

Organize students into groups of two. Each group needs 20 rods and 18 flats.

Valuing Models UNDERSTANDING

Students should work together to list all possible ways that 20 rods can be arranged into two groups, such as 19 and 1, 18 and 2, and so on. Students should model each arrangement and then sketch each model on paper.

- **How many different ways did you arrange the 20 rods?** 10

Students should now work together to list all possible ways that 18 flats can be arranged into two groups. Students should sketch each model on paper.

- **How many different ways did you arrange the 18 flats?** 9

Discuss each arrangement as it is presented to the class.

Monitoring Student Progress

If . . . a student is missing some of the arrangements,

Then . . . suggest an organized way to make a list. Begin with 1 and its match, then 2 and its match, and so on until the first number used is the same as one of the numbers used previously as a match.

Building Blocks For additional practice composing multiples of ten, students should complete **Building Blocks** Number Snapshots.

MathTools Use the Base Ten Blocks tool to demonstrate and explore multiples of ten.

Using Student Pages

Have students complete **Workbook**, pages 12–13, on their own.

Lesson 1

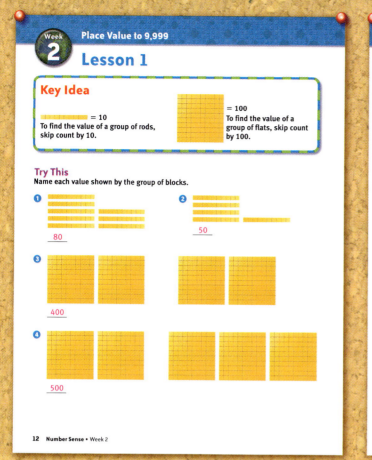

Key Idea

▬▬▬ = 10
To find the value of a group of rods, skip count by 10.

⬛ = 100
To find the value of a group of flats, skip count by 100.

Try This
Name each value shown by the group of blocks.

1. 80

2. 50

3. 400

4. 500

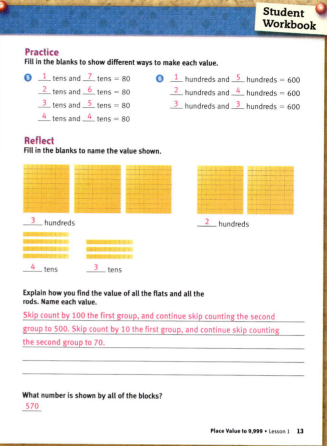

Practice
Fill in the blanks to show different ways to make each value.

5.
  1 tens and 7 tens = 80
  2 tens and 6 tens = 80
  3 tens and 5 tens = 80
  4 tens and 4 tens = 80

6.
  1 hundreds and 5 hundreds = 600
  2 hundreds and 4 hundreds = 600
  3 hundreds and 3 hundreds = 600

Reflect
Fill in the blanks to name the value shown.

3 hundreds   2 hundreds

4 tens   3 tens

Explain how you find the value of all the flats and all the rods. Name each value.

Skip count by 100 the first group, and continue skip counting the second group to 500. Skip count by 10 the first group, and continue skip counting the second group to 70.

What number is shown by all of the blocks?
570

3 Reflect   10

Extended Response   REASONING

Review students' answers to the Reflect prompt at the bottom of student page 13.

Discuss the difference between the place value of a number and the value of the number.

- **Explain how you found the value of all the flats and rods in the Reflect problem.** Answers will vary.

- **Is there more than one way to solve this problem?** Answers will vary.

Real-World Application   APPLYING

Often times, the location of a building is described in general terms using place value. For example, the library is on the 900 block of Main Street. This means that the address of the library could be any number between 900 and 999.

- **Describe the location of these addresses in general terms: 425 Washington Street; 789 Lincoln Avenue.** 400 block of Washington Street; 700 block of Lincoln Avenue

4 Assess

Informal Assessment

Use the Student Assessment Record, **Assessment**, page 100, to record informal observations.

UNDERSTANDING

Valuing Models
Did the student
❏ make important observations?
❏ extend or generalize learning?
❏ provide insightful answers?
❏ pose insightful questions?

Place Value to 9,999 • Lesson 1   12–13

Lesson 2

Objective

Students write numbers to 999 in expanded form.

Materials

Program Materials
Number Construction Mat, p. B1

Additional Materials
- Number Cards (0–9)
- Base-Ten Blocks

Access Vocabulary

digit A single numeral
expanded form A way of writing numbers that shows the sum of the values of each place value

Creating Context

It is important for English Learners to be able to write the names of the basic numbers in English. These serve as the basic root words for the ordinal numbers and the place values in the base-ten system. Complete the chart below to show the similarities and differences between numbers, ordinals, and place values.

NUMBERS	ORDINALS	PLACE VALUES
One	First	Ones
Ten	Tenth	Tens
Hundred	Hundredth	Hundreds

---

1 Warm Up — 5

Concept Building COMPUTING

When multiplying by 10 and multiplying by 100, students can use a "shortcut." For multiplying whole numbers by 10, place one zero at the right end of the number. For multiplying whole numbers by 100, place two zeros at the right end of the number. As a class, verbally practice the following problems.

$7 \times 100 = 700$ $14 \times 10 = 140$ $12 \times 100 = 1,200$
$56 \times 10 = 560$ $100 \times 10 = 1,000$

---

2 Engage — 30

Skill Building

Give each student a Number Construction Mat and Number Cards. Choose cards numbered 2, 4, and 5. Arrange the cards to form the number 425.

- **How many hundreds, tens, and ones does this number have? What is this number?** 4 hundreds, 2 tens, and 5 ones; 425
- **Can you use the same three digits to make a number that has a value less than 425? Name this number.** yes; 245 or 254

Composing Numbers UNDERSTANDING

Instruct students to place a 4 in the hundreds column and a 9 in the tens column on their mats.

- **What is the value of 4 in this number?** 400
- **What is the value of the number in the tens place?** 90; 9 tens or 9×10

Strategy Building

Number Challenge COMPUTING

Organize students into pairs. Students should take turns describing a number and showing that number on the Number Construction Mat. Instruct students to write each number and then to write the value of each digit in that number.

Monitoring Student Progress

If . . . a student is struggling to name the value of a digit,

Then . . . encourage the student to use Base-Ten Blocks so he or she can physically count each block.

Building Blocks For additional practice writing numbers in expanded form, students should complete **Building Blocks** Number Snapshots.

MathTools Use the Base Ten Blocks tool to demonstrate and explore numbers in expanded form.

Using Student Pages

Have students complete **Workbook,** pages 14–15, on their own.

Week 2 — Place Value to 9,999

Lesson 2

Key Idea

368 ← Written in standard form.

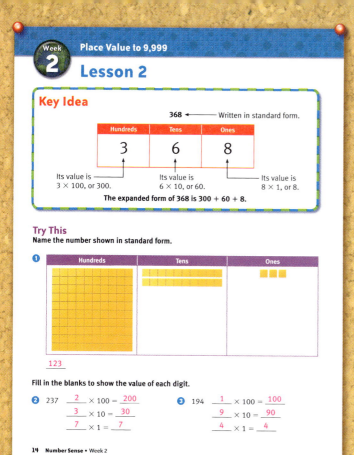

Hundreds	Tens	Ones
3	6	8

Its value is 3 × 100, or 300.

Its value is 6 × 10, or 60.

Its value is 8 × 1, or 8.

The expanded form of 368 is 300 + 60 + 8.

Try This

Name the number shown in standard form.

1

Hundreds	Tens	Ones

123

Fill in the blanks to show the value of each digit.

2 237
$\underline{2} \times 100 = \underline{200}$
$\underline{3} \times 10 = \underline{30}$
$\underline{7} \times 1 = \underline{7}$

3 194
$\underline{1} \times 100 = \underline{100}$
$\underline{9} \times 10 = \underline{90}$
$\underline{4} \times 1 = \underline{4}$

Practice

Combine the numbers to form a single quantity.

4 400 + 10 + 9   419

5 70 + 5   75

6 900 + 30   930

7 200 + 1   201

Tell how many hundreds, tens, and ones are in each number.

8 248
$\underline{2}$ hundreds $\underline{4}$ tens $\underline{8}$ ones

9 325
$\underline{3}$ hundreds $\underline{2}$ tens $\underline{5}$ ones

10 304
$\underline{3}$ hundreds $\underline{0}$ tens $\underline{4}$ ones

11 560
$\underline{5}$ hundreds $\underline{6}$ tens $\underline{0}$ ones

12 56
$\underline{5}$ tens $\underline{6}$ ones

13 792
$\underline{7}$ hundreds $\underline{9}$ tens $\underline{2}$ ones

14 227
$\underline{2}$ hundreds $\underline{2}$ tens $\underline{7}$ ones

15 385
$\underline{3}$ hundreds $\underline{8}$ tens $\underline{5}$ ones

Reflect

Explain how the number 502 differs from the number 532 in terms of place value.

532 has a 3 in the tens place. 532 is 30 more than 502.

---

3 Reflect   10

Extended Response   REASONING

Review students' answers to the Reflect prompt at the bottom of student page 15.

- **How does place value affect a number?** Answers will vary; Possible answer: The number is multiplied by the place value.

Real-World Application   APPLYING

As you learned in Week 1, you can use money in place of Base-Ten Blocks to model dollars and cents. List five different amounts less than $100. Write these amounts in terms of ten-dollar bills and one-dollar bills. Be sure to include the dollar and cent symbols.

4 Assess

Informal Assessment

Use the Student Assessment Record, **Assessment,** page 100, to record informal observations.

UNDERSTANDING	COMPUTING
Composing Numbers	**Number Challenge**
Did the student	Did the student
❏ make important observations?	❏ respond accurately?
❏ extend or generalize learning?	❏ respond quickly?
❏ provide insightful answers?	❏ respond with confidence?
❏ pose insightful questions?	❏ self-correct?

Lesson 3

Objective

Students write numbers to 9,999 in expanded form.

Materials

Program Materials
- 3-Digit Window, p. B4
- 4-Digit Window, p. B5

Additional Materials
- Number Cards (0–9)
- Number Cards (1, 10, 100, 1,000)

Access Vocabulary

expanded form A way of writing numbers that shows the sum of the values of each place value

standard form A way of writing numbers using the digits 0–9

comma A mark that separates the thousands and hundreds column

Creating Context

In this lesson, students learn that in the United States, a comma is placed between the thousands digit and the hundreds digit when writing numbers in standard form. In some countries, a period is used instead of a comma. This may cause confusion for English Learners who have attended school outside the United States. Be sure to explain this to students and their parents.

 1 **Warm Up** 5

Concept Building COMPUTING

The expanded form of a number shows the value of each digit according to its place value. Name the number of thousands, hundreds, tens, and ones in each of the following numbers. Then use the place value location to name the value of each digit. For example, 68 is 6 tens and 8 ones. The 6 is worth 60 and the 8 is worth 8.

5,137: 5,000, 100, 30, 7   1,803: 1,000, 800, 0, 3

 2 **Engage** 30

Strategy Building

Provide each student with Number Cards, a 3-Digit Window, and a 4-Digit Window.

Random Composition COMPUTING

- **Shuffle your cards and hold them facedown.**
- **Turn over one card and write that number in the hundreds column of the 3-Digit Window.**
- **Turn over a second card and write that number in the tens column.**
- **Turn over a third card and write that number in the ones column.**
- **Name your three-digit number.**

The number in the 3-Digit Window is written in standard form. Instruct students to write the value of each digit of their number in the window below the digit.

- **To write a number in expanded form, use the values of each digit with addition signs between them.**

Have students repeat selections to form several more three-digit numbers. Then have students use the 4-Digit Window to form four-digit numbers.

Monitoring Student Progress

If . . . a student needs more guidance composing the number,

Then . . . have him or her select each card and place it faceup above the place-value column where it will be written.

Building Blocks For additional practice combining numbers to 9,999, students should complete **Building Blocks** Number Snapshots.

Using Student Pages

Have students complete **Workbook,** pages 16–17, on their own.

Week 2 — Place Value to 9,999

Lesson 3

Key Idea

3,000 +   100 +   70 +   9

3 thousands + 1 hundred + 7 tens + 9 ones

Thousands	Hundreds	Tens	Ones
3	1	7	9

This number in standard form is 3,179. The comma is placed between the thousands digit and the hundreds digit.

Try This

Write each number in expanded form and in standard form.

1. 2,000 + 500 + 90 + 4
   __2__ thousands + __5__ hundreds + __9__ tens + __4__ ones = __2,594__

2. 1,000 + 700 + 30 + 2
   __1__ thousands + __7__ hundreds + __3__ tens + __2__ ones = __1,732__

Write the value of each digit.

3. 1,962
   __1,000__ + __900__ + __60__ + __2__

4. 6,128
   __6,000__ + __100__ + __20__ + __8__

Practice

Write each number in standard form.

5. 1,000 + 900 + 30 + 7   __1,937__
6. 4,000 + 200 + 60 + 1   __4,261__
7. 5,000 + 600 + 40   __5,640__
8. 3,000 + 70 + 8   __3,078__

Write the expanded form of each number.

9. 2,345   __2,000 + 300 + 40 + 5__
10. 4,756   __4,000 + 700 + 50 + 6__
11. 1,312   __1,000 + 300 + 10 + 2__

Write the place of each underlined digit. Then write the value of each underlined digit.

12. 1,6<u>1</u>4   place __hundreds__   value __600__
13. 6,7<u>1</u>0   place __tens__   value __10__
14. <u>5</u>,692   place __thousands__   value __5,000__
15. 2,07<u>8</u>   place __ones__   value __8__

Reflect

Write the expanded form and the standard form of a number that has seven thousands, three hundreds, eight tens, and nine ones.

Expanded form: __7,000 + 300 + 80 + 9__

Standard form: __7,389__

Tell what digit is in the thousands place, hundreds place, tens place, and ones place.

The digit in the thousands place is 7. The digit in the hundreds place is 3. The digit in the tens place is 8. The digit in the ones place is 9.

---

3 — Reflect   10

Extended Response — REASONING

Review students' answers to the Reflect prompt at the bottom of student page 17.

- **How does the expanded form of a number translate directly into the standard form when you put addition signs between the named numbers?** The expanded numbers can be added together.

Real-World Application — APPLYING

Cashiers use expanded form when totaling their cash drawers.

- **How would a cashier fill out a form asking for the number of 100s, 10s, and 1s for $234?** two 100s, three 10s, four 1s

4 — Assess

Informal Assessment

Use the Student Assessment Record, **Assessment**, page 100, to record informal observations.

COMPUTING

Random Composition

Did the student
- ❏ respond accurately?
- ❏ respond quickly?
- ❏ respond with confidence?
- ❏ self-correct?

Lesson 4

Objective

Students identify and write numbers in standard form, expanded form, and word form.

Materials

Additional Materials

Blank check forms, p. B6

Access Vocabulary

hyphen A dash placed between compound words such as twenty-nine

expanded form A way of writing numbers that shows the sum of the values of each place value

standard form A way of writing numbers using the digits 0–9

word form A way of writing numbers using words

Creating Context

In English, digraphs are sounds represented by a combination of letters. In this lesson, there are many words that have the digraph *th-*, which makes the /th/ sound. When students' primary languages do not have the /th/ sound, students are likely to approximate the sound with /d/ or /t/. Find the words *three, thousand, thirty,* and *length* in this lesson. Help English Learners practice this pronunciation.

1 Warm Up   5

Concept Building   COMPUTING

Select students to go to the board. Assign each student a word or words to write on the board from the following list: *zero, one, two, three, four, five, six, seven, eight, nine, hundred, thousand.* After each word is written, review the word and have the class give a "thumbs up" if the word is spelled correctly. Then have the student write the number that the word describes.

2 Engage   30

Strategy Building

Organize students into groups of eight. Arrange desks to form a circle. Students need Blank Check Forms, page B6.

Pay to the Order?   UNDERSTANDING

One application of writing numbers in words is on personal checks. When a person writes a personal check for an amount, that amount has to be written in standard form and also in word form. Instruct students to write checks to three classmates for an amount up to $900.

- **What place value word did you use when writing the amounts in word form?** hundred

Monitoring Student Progress

If . . . a student needs more practice writing standard form given the word form,	**Then . . .** create an audiotape with 20 numbers read in word form. The student can listen at home and practice writing the numbers.

Building **B**locks For additional practice writing numbers in standard form given the word form, students should complete **Building Blocks** Number Snapshots.

Using Student Pages

Have students complete **Workbook,** pages 18–19, on their own.

Lesson 4

Key Idea

You can write numbers in word form. The names of some place values are used when writing numbers in word form.

541	five hundred forty-one
2,729	two thousand, seven hundred twenty-nine

Notice that the words that name the tens and ones place digits have a hyphen. The words *hundred* and *thousand* follow the name of the digit in those places. A comma is placed after the word *thousand*.

Try This

Fill in the blanks to write each number in word form.

1. 683   __six__ hundred eighty-__three__
2. 945   __nine__ hundred forty-__five__
3. 1,262   __one__ thousand, __two__ hundred sixty-two
4. 4,167   __four__ thousand, __one__ hundred __sixty__-seven
5. 11,304   __eleven__ thousand, __three__ hundred __four__

Practice

Write each number in standard form.

6. two hundred seventy-nine   __279__
7. five thousand, eight hundred thirty-one   __5,831__
8. three thousand, fifty-four   __3,054__

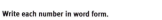

Write each number in word form.

9. 362   __three hundred sixty-two__
10. 1,501   __one thousand, five hundred one__

Write the length of each animal in word form.

11. A whale shark is 41 feet long.
__forty-one feet__
12. A stick insect is 15 inches long.
__fifteen inches__
13. An anaconda snake is 26 feet long.
__twenty-six feet__

Write the weight of each animal in standard form.

14. A blue whale weighs two thousand, ninety pounds.
__2,090 pounds__
15. A saltwater crocodile weighs one thousand, one hundred fifty pounds.
__1,150 pounds__
16. An ostrich weighs three hundred forty-five pounds.
__345 pounds__

Reflect

The table below shows the numbers of animals that are listed as endangered or threatened species. Write each number in word form.

U.S. Endangered Species	389	three hundred eighty-nine
Foreign Endangered Species	516	five hundred sixteen
U.S. Threatened Species	129	one hundred twenty-nine
Foreign Threatened Species	44	forty-four
Grand Total	1,078	one thousand, seventy-eight

---

3 Reflect   10

Extended Response   REASONING

Review students' answers to the Reflect prompt at the bottom of student page 19.

Select students to write their word forms on the board so that other students can compare their answers.

- **When should you use a comma?** after the thousands place
- **When should you use a hyphen?** between the tens and ones places

Real-World Application   APPLYING

Your home address could be a one-, two-, three-, or four-digit number.

Write your address in standard form, expanded form, and word form.

- **Where do you see addresses written in word form?** Answers will vary; on a house; on a front door

4 Assess

Informal Assessment

Use the Student Assessment Record, *Assessment*, page 100, to record informal observations.

UNDERSTANDING

Pay to the Order?

Did the student
- ❏ make important observations?
- ❏ extend or generalize learning?
- ❏ provide insightful answers?
- ❏ pose insightful questions?

Lesson 5 Review

Objective

Students review skills learned this week and complete the weekly assessment.

Materials

Program Materials
Review materials will be selected from those used in previous activities.

Creating Context

An excellent strategy to use with English Learners is to have students work in small groups to solve problems or to work on projects. In this way, English Learners can practice verbal skills in a low-stress environment. When students within a group speak the same primary language, they can discuss new concepts and check for understanding.

1 Warm Up 5

Concept Building UNDERSTANDING

Name the number of thousands, hundreds, tens, and ones in each of the following numbers.

9999: nine thousands, nine hundreds, nine tens, and nine ones

1,706: one thousand, seven hundreds, and six ones

- **What number would you have if you added 1 to 9,999?** 10,000
- **What number is in the thousands place?** 0

2 Engage 20

Skill Building

Free-Choice Activity

For the last day of the week, allow students to choose an activity from previous lessons. Some activities they may choose are the following:

- **Valuing Models**
- **Composing Numbers**
- **Number Challenge**
- **Random Composition**
- **Pay to the Order?**

Make a note of the activities students select. Do they prefer easy or challenging activities? If you believe your students would benefit from extra practice on specific skills, choose an activity for them.

Monitoring Student Progress

If . . . students are not participating,

Then . . . continue the activity until all students have participated.

Using Student Pages

Have students complete **Workbook,** pages 20–21, on their own.

3 Reflect 10

Extended Response APPLYING

Review students' answers to the Reflect prompts at the bottom of student pages 20–21.

Discuss these answers with the group to reinforce Week 2 concepts.

Lesson 5 Review

This week, you learned the meaning of place value and how to compose and decompose numbers.

Lesson 1 Fill in the blanks to show different ways to make each value.

1
_4_ tens and _1_ tens = 50
_3_ tens and _2_ tens = 50

2
_7_ tens and _6_ tens = 130
_9_ tens and _4_ tens = 130
_8_ tens and _5_ tens = 130

Lesson 2 Fill in the blanks to show the value of each digit.

3 815
_8_ hundreds = _800_
_1_ tens = _10_
_5_ ones = _5_

4 3,768
_3_ thousands = _3,000_
_7_ hundreds = _700_
_6_ tens = _60_
_8_ ones = _8_

Reflect

Explain how the expanded form of 709 differs from the expanded form of 749.

The expanded form of 709 is 700 + 9. Because there is no number in the tens place, there are only two numbers in the expanded form. The expanded form of 749 is 700 + 40 + 9. Because there is a number in the tens place, there are three numbers in the expanded form.

Lesson 3 Write each number in standard form.

5 4,000 + 600 + 10 + 8   _4,618_

6 600 + 20 + 1   _621_

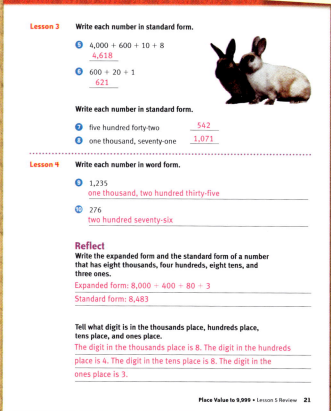

Write each number in standard form.

7 five hundred forty-two   _542_

8 one thousand, seventy-one   _1,071_

Lesson 4 Write each number in word form.

9 1,235
one thousand, two hundred thirty-five

10 276
two hundred seventy-six

Reflect

Write the expanded form and the standard form of a number that has eight thousands, four hundreds, eight tens, and three ones.

Expanded form: 8,000 + 400 + 80 + 3

Standard form: 8,483

Tell what digit is in the thousands place, hundreds place, tens place, and ones place.

The digit in the thousands place is 8. The digit in the hundreds place is 4. The digit in the tens place is 8. The digit in the ones place is 3.

Assess   10

A Gather Evidence

Formal Assessment

Have students complete the Weekly test on **Assessment,** pages 26–27. Record scores on the Student Assessment Record, **Assessment,** page 100.

B Summarize Findings

Determine whether students have Minimal, Basic, or Secure understanding of the concepts presented in Week 2.

C Differentiate Instruction

Based on your observations, use these teaching strategies next week to follow up.

Minimal Understanding
- Repeat the Warm-Up and Engage activities to develop Week 2 concepts.
- Use **Building Blocks** Number Snapshots and **eMathTools** Base Ten Blocks to develop and reinforce Week 2 concepts.

Basic Understanding
- Repeat Engage activities in subsequent weeks.
- Use **Building Blocks** Number Snapshots and **eMathTools** Base Ten Blocks to reinforce Week 2 concepts.

Secure Understanding
Use **Building Blocks** Number Snapshots to reinforce Week 2 concepts.

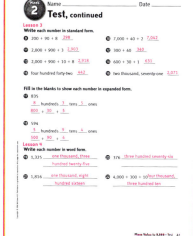

Assessment, pp. 26–27

Place Value and Number Sense

Week at a Glance

This week, students continue with **Number Worlds,** Level F, Number Sense, by developing visual strategies for the skills learned in the previous three weeks. Students see relationships related to place value that should help them develop their mental computational strategies.

Math Background

Visualizing numerals is one of the basic cognitive processes necessary to understanding math. For example, when we see the numeral "3," we know that it represents the concept of three of something— three pennies, three apples, three horses, three dots. While imaging numerals is important to mathematical computation, another aspect of imagery is equally as important: concept imagery.

—Tuley, Kimberly and Nanci Bell. 1997. *On Cloud Nine, Visualizing and Verbalizing for Math.* Gander Publishing.

How Students Learn

Students should use models and other strategies to gain a level of familiarity and understanding about our numeration system. Make a variety of models available to students. When students can move from one type of model to another, it signals that they are internalizing the content and are ready to begin working without the use of models.

Teaching for Understanding

Observe closely while evaluating the assigned tasks this week to see whether students can demonstrate the following understandings.

Benchmark after Lesson 2: Students can add and subtract using multiples of 10.

Benchmark after Lesson 3: Students can use a 100 Chart to complete addition and subtraction sentences.

Benchmark after Lesson 4: Students can use mental math strategies to add and subtract double-digit numbers.

Skills Focus

- Gain experience with numbers 1–100 in a ten-by-ten arrangement
- Understand that in a ten-by-ten arrangement, horizontal movement affects the digit in the ones place, and vertical movement affects the digit in the tens place
- Practice successive mental addition computations

Math at Home

Give one copy of the Letter to Home, page A3, to each student. Encourage students to share and complete the activity with their caregivers.

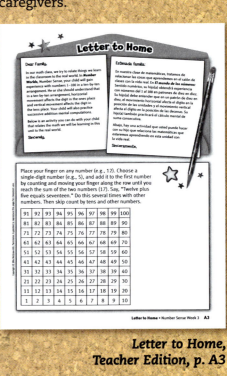

Letter to Home, Teacher Edition, p. A3

Week 3 Planner   Place Value and Number Sense

PACING	LESSON	LEARNING GOALS	NCTM	MATERIALS	TECHNOLOGY
DAY 1	pages 22–23	Students identify properties of the base-ten number system.	• Number and Operations • Communication • Connections • Representation	Neighborhood Number Line	**Building Blocks** School Supply Shop **e MathTools** 100s Table
DAY 2	pages 24–25	Students add and subtract using multiples of ten.	• Number and Operations • Problem Solving • Communication • Connections • Representation	• Number 1–6 Cube • Neighborhood Number Line • Ten-and-One Cube • 15 Monster Cards, p. B7	**Building Blocks** Math-O-Scope
DAY 3	pages 26–27	Students understand the meaning of horizontal and vertical movements in a 100 Chart.	• Number and Operations • Problem Solving • Communication • Connections • Representation	• Hotel Game Board • Room Service Hotel Activity Sheet, p. B8	**Building Blocks** Math-O-Scope **e MathTools** 100s Table
DAY 4	pages 28–29	Students find sums and differences using mental computations.	• Number and Operations • Problem Solving • Communication	Number Cards (0–5)	**Building Blocks** Math-O-Scope **e MathTools** 100s Table
DAY 5	Review and Assess pages 30–31	Students review skills learned this week and complete the weekly assessment.	• Number and Operations • Problem Solving • Communication • Connections • Representation	Materials will be selected from Lessons 1–4.	**Building Blocks** **e MathTools** Review previous activities

Math Vocabulary

skip count A method of counting by which each number stated increases or decreases by the same amount (a number greater than 1)

mental computation Performing math operations in your mind without paper and pencil or a calculator

English Learners

SPANISH COGNATES

English	Spanish
value	valor
count	contar
numeral	número
sense	sentido

ALTERNATE VOCABULARY

counting on Continuing to count from a given number

Lesson 1

Objective

Students identify properties of the base-ten number system.

Materials

Program Materials

Neighborhood Number Line

Access Vocabulary

skip count A method of counting by which each number stated increases by the same amount (a number greater than 1)

missing numbers The numbers that should be present according to a pattern but are blank and need to be written in

Creating Context

In English there are many words that have more than one meaning, such as *block*. This can confuse some English Learners when using math terms. An excellent strategy to use with English Learners is to preview the lesson to identify any potentially confusing words or phrases. Then you can introduce those words or explain them while teaching the lesson.

1 Warm Up 5

Concept Building COMPUTING

As a class, practice counting on. Select a double-digit number, such as 23, that does not increase the digit in the tens column when you count on 5. Then select another double-digit number, such as 38, that does increase the digit in the tens column when you count on 5.

- **What differences did you notice when you counted on from the two numbers?** When we counted on from 23, we stayed in the 20s. When we counted on from 38, we moved into the 40s.

2 Engage 30

Skill Building

Organize students in a semicircle. Tell students that in this activity they will use a neighborhood to explore the base-ten number system.

Making a Neighborhood UNDERSTANDING

Display the first block (1–10) of the **Neighborhood Number Line**.

- **The streets in this neighborhood can have only 10 houses on them.**
- **What house numbers should the next street have?** 11–20

Place the second block of the Neighborhood Number Line with houses numbered 11–20 directly above the first Neighborhood Number Line.

- **If this neighborhood has eight streets, what will be the greatest house number in the neighborhood?** 80

Place six more blocks and let students verify their answers. Then place the remaining two blocks.

- **Describe any patterns that you notice in the numbering of the houses in the neighborhood.** Encourage students to recognize that horizontal travel results in an increase or decrease of 1, and that vertical travel results in an increase or decrease of 10.

Monitoring Student Progress

If . . . students do not notice any patterns,

Then . . . guide them to look at the numbers as they appear horizontally and as they appear vertically.

Building Blocks For additional practice using a ten-by-ten arrangement, students should complete **Building Blocks** School Supply Shop.

MathTools Use the 100s Table tool to demonstrate and explore using a ten-by-ten arrangement.

Using Student Pages

Have students complete **Workbook,** pages 22–23, on their own.

Lesson 1

Key Idea
When you count 1, 2, 3, and so on, each number is 1 more than the number before it.

When you skip count by 10, each number is 10 more than the number before it.

10, 20, 30, 40, 50, …

12, 22, 32, 42, 52, …

Try This
Skip count by 10. Write the next three numbers.

① 20, 30, 40, __50__, __60__, __70__
② 45, 55, 65, __75__, __85__, __95__
③ 33, 43, 53, __63__, __73__, __83__
④ 78, 88, 98, __108__, __118__, __128__

Skip count backward by 10. Write the next three numbers.

⑤ 100, 90, 80, __70__, __60__, __50__
⑥ 95, 85, 75, __65__, __55__, __45__
⑦ 81, 71, 61, __51__, __41__, __31__
⑧ 164, 154, 144, __134__, __124__, __114__

Practice
Name the number described.

⑨ 7 and 3 more ones __10__
⑩ 40 and 2 more tens __60__
⑪ 48 and 4 more ones __52__
⑫ 130 and 8 more ones __138__

Count from 1 to 100 in the chart below.
Write in the missing numbers.

91	92	93	94	**95**	96	97	98	99	100
81	82	83	84	85	86	87	88	89	**90**
71	72	73	74	75	**76**	77	78	79	80
61	**62**	63	64	65	66	67	68	69	70
51	52	53	54	55	56	**57**	58	59	60
41	42	**43**	44	45	46	47	48	49	50
31	32	33	34	35	36	37	38	39	**40**
21	22	23	24	**25**	26	27	28	29	30
11	12	13	**14**	15	16	17	18	19	**20**
1	**2**	**3**	**4**	5	6	7	8	9	10

Reflect
In the table above, begin with 3 and skip count to the number above it in the next row. By what are you skip counting? Explain your answer. Show a set of numbers from the chart that is another example of skip counting.

10; Because each row has ten numbers, when you move up a row, you land on

13. Sample example: 5, 15, 25, 35, 45, 55, 65, 75, 85, 95

---

3 Reflect 10

Extended Response REASONING
Review students' answers to the Reflect prompt at the bottom of student page 23.

- **Why does the ten-by-ten arrangement work best for skip counting?** You can skip count by tens simply by moving your finger up or down the chart.

Discuss why nine columns or rows or eleven columns or rows is not the best arrangement.

Real-World Application APPLYING
Some hotel rooms are numbered similar to a 100 Chart. Rooms on the first floor are numbered 1–10.

- **What rooms are directly above and directly below Room 22?** Above is Room 32, and below is Room 12.

---

4 Assess

Informal Assessment
Use the Student Assessment Record, *Assessment*, page 100, to record informal observations.

UNDERSTANDING

Making a Neighborhood
Did the student
❑ make important observations?
❑ extend or generalize learning?
❑ provide insightful answers?
❑ pose insightful questions?

Lesson 2

Objective

Students add and subtract using multiples of ten.

Materials

Program Materials
- Neighborhood Number Line
- Ten-and-One Cube
- Number 1–6 Cube
- 15 Monster Cards, p. B7

Access Vocabulary

skip count A method of counting by which each number stated increases or decreases by the same amount (a number greater than 1)

counting on Continuing to count from a given number

difference The answer to a subtraction problem

Creating Context

The number chart in this lesson is an excellent visual cue for English Learners to support their understanding. Students at the early levels of English proficiency need practice with counting in English to develop automaticity. They may already have a conceptual understanding of counting but may not have the English fluency to say the number names quickly.

1 Warm Up   5

Concept Building   REASONING

Write the problems below on the board. Instruct students to choose the problems in which the missing number is greater than 10.

$$24 + ? = 32$$
$$56 + ? = 68$$
$$13 + ? = 24$$
$$75 + ? = 85$$

second and third problems

Ask several students to explain their thinking.

2 Engage   30

Strategy Building

Use the Neighborhood Number Lines from Lesson 1. Randomly number fifteen Monster Cards with numbers from 1 to 100. Keep the Monster Cards in order and in your hands. Organize the class into two teams. Team 1 begins play on Square 1. Team 2 begins play on Square 51.

Monsters in the Neighborhood   COMPUTING

Explain that the goal of the game is to find monsters that are hidden in some of these houses.

- Team members take turns rolling the Ten-and-One Die. When a player rolls a 1, he or she may move 1 house right or left on the same street. When a player rolls a 10, he or she may move up or back 1 street. Each player must decide to add or subtract the number on the die based on whether he or she wants to move forward or backward.
- Before searching for monsters, the player should verbally report where he or she is now, how many spaces forward or backward he or she will move, and where he or she will end up. After a player makes the move, he or she announces the number of the house. The teacher then checks whether there is a Monster Card with that house number on it.
- When a monster is found, the team gets that monster card. The team that finds the most monsters wins.

Monitoring Student Progress

| If . . . students are at differing levels, | Then . . . assign students who are ready to work with greater numbers to be on Team 2. |

Building Blocks For additional practice completing an addition or subtraction equation, students should complete **Building Blocks** Math-O-Scope.

Using Student Pages

Have students complete **Workbook,** pages 24–25, on their own.

Place Value and Number Sense

Lesson 2

Key Idea
The chart below is useful for counting by 1s and skip counting by 10s.

91	92	93	94	95	96	97	98	99	100
81	82	83	84	85	86	87	88	89	90
71	72	73	74	75	76	77	78	79	80
61	62	63	64	65	66	67	68	69	70
51	52	53	54	55	56	57	58	59	60
41	42	43	44	45	46	47	48	49	50
31	32	33	34	35	36	37	38	39	40
21	22	23	24	25	26	27	28	29	30
11	12	13	14	15	16	17	18	19	20
1	2	3	4	5	6	7	8	9	10

Try This
Fill in the blanks for each pattern of numbers.

1. 4, __14__, 24, 34, __44__, __54__, __64__
2. 99, 89, __79__, __69__, __59__, __49__, __39__, 29
3. 7, __8__, 9, __10__, __11__, __12__, 13
4. 100, 99, __98__, __97__, __96__, __95__, __94__, __93__

Complete each equation.

5. $19 + \underline{10} = 29$
6. $67 + \underline{1} = 68$
7. $30 - \underline{1} = 29$
8. $78 - \underline{10} = 68$

Practice
Use the chart on the previous page to find each sum or difference.

9. $56 + 30 = \underline{86}$
10. $29 + 50 = \underline{79}$
11. $77 + 20 = \underline{97}$
12. $84 - 40 = \underline{44}$
13. $31 - 20 = \underline{11}$
14. $75 - 60 = \underline{15}$
15. $24 + 6 = \underline{30}$
16. $56 + 5 = \underline{61}$
17. $89 - 40 = \underline{49}$
18. $98 - 60 = \underline{38}$
19. $65 + 9 = \underline{74}$
20. $90 - 70 = \underline{20}$

Reflect
Use the chart to help you write four equations for 36. Write two using addition and two using subtraction.

Sample answers:

$34 + 2 = 36$	$16 + 20 = 36$
$86 - 50 = 36$	$40 - 4 = 36$

---

3 Reflect   10

Extended Response   REASONING
Review students' answers to the Reflect prompt at the bottom of student page 25.

■ **Is the answer affected if the horizontal movement is made before the vertical movement?**

Students who understand the commutative property of addition should know that the order of the movements does not affect the answer.

■ **How can you check your answer?** Solve the problem both ways to see if you get the same answer.

Real-World Application   APPLYING
The 100 Chart looks similar to some game boards. Students can create a game using the 100 Chart and a 1–6 Number Cube. Students should write the basic rules for the game and then play the game with two classmates. Instruct students to write a sentence that describes the math skills their games practice.

4 Assess

Informal Assessment
Use the Student Assessment Record, **Assessment,** page 100, to record informal observations.

COMPUTING

Monsters in the Neighborhood
Did the student
❏ respond accurately?
❏ respond quickly?
❏ respond with confidence?
❏ self-correct?

Place Value and Number Sense • Lesson 2   **24–25**

Lesson 3

Objective

Students understand the meaning of horizontal and vertical movements in a 100 Chart.

Materials

Program Materials
- Hotel Game Board
- The Room Service Hotel activity sheet, p. B8

Access Vocabulary

mental computation Performing math operations in your mind without paper and pencil or a calculator

trace your finger Touching the row of numbers with your finger and moving your finger along the row or column

row A straight line of objects or numbers stretching horizontally (side to side)

columns A straight line of objects or numbers stretching vertically (up and down)

Creating Context

Making concepts less abstract can be helpful to English Learners who may not understand a verbal explanation without some kind of visual or hands-on model. The chart in this lesson can be very helpful not only for counting practice, but also for demonstrating the operations of subtraction and addition.

1 Warm Up 5

Concept Building COMPUTING

Write the problems below on the board. Instruct students to identify the problems in which a horizontal move on the 100 Chart is made and the problems in which a vertical move is made. *first and third problems, horizontal; second and fourth problems, vertical*

$$64 + ? = 68 \qquad 52 + ? = 62$$
$$71 + ? = 79 \qquad 25 + ? = 35$$

Discuss how to determine the movement by looking at place value.

2 Engage 30

Skill Building

Show the Hotel Game Board. Tell students that they are going to pretend to be the waiters at a hotel that has room service, and they will take turns delivering room service orders. Explain the rules found on The Room Service Hotel activity sheet, page B8.

Room Service Delivery COMPUTING

- **When it is your turn to deliver room service orders, stand up and announce the room number. The class will then give you directions on how to travel to the room. After you arrive at the room, state the number sentence that explains the "delivery."**

For example, if the "waiter" has to travel from the kitchen to room 24, the class may direct him or her to walk up two flights of stairs and then across four rooms to room 24. Students will fill out their delivery slips as described in the activity sheet.

Monitoring Student Progress

If . . . students determine the equations easily,	Then . . . randomly select four room numbers and instruct students to make a plan for making the deliveries to all four rooms. They should write an equation to show the path in which the deliveries will be made.

Building Blocks For additional practice adding and subtracting using a ten-by-ten arrangement, students should complete **Building Blocks** Math-O-Scope.

MathTools Use the 100s Table tool to demonstrate and explore using a ten-by-ten arrangement.

Using Student Pages

Have students complete **Workbook,** pages 26–27, on their own.

Lesson 3

Key Idea
The chart below is useful for practicing mental computation.

91	92	93	94	95	96	97	98	99	100
81	82	83	84	85	86	87	88	89	90
71	72	73	74	75	76	77	78	79	80
61	62	63	64	65	66	67	68	69	70
51	52	53	54	55	56	57	58	59	60
41	42	43	44	45	46	47	48	49	50
31	32	33	34	35	36	37	38	39	40
21	22	23	24	25	26	27	28	29	30
11	12	13	14	15	16	17	18	19	20
1	2	3	4	5	6	7	8	9	10

Try This
Use the chart, and trace your finger along the rows and columns. Name what number and operation each movement models.

1. up 3 rows and right 2 columns — add 32
2. up 5 rows and right 1 column — add 51
3. down 2 rows and left 4 columns — subtract 24
4. up 3 rows — add 30

Practice
Use the chart above, and trace your finger along the rows and columns. Name the number where you stop.

5. from 4, go up 5 rows and right 3 columns — 57
6. from 56, go down 2 rows and left 5 columns — 31

Use the chart on the previous page, and trace your finger along the rows and columns. Complete each addition sentence.

7. $44 + \underline{42} = 86$
8. $26 + \underline{5} = 31$
9. $73 + \underline{21} = 94$

Use the chart on the previous page, and trace your finger along the rows and columns. Complete each subtraction sentence.

10. $67 - \underline{23} = 44$
11. $84 - \underline{21} = 63$
12. $45 - \underline{13} = 32$

Use the chart on the previous page, and trace your finger along the rows and columns. Write each equation.

13. from 16, go up 4 rows and right 3 columns
$16 + 43 = 59$

14. from 89, go down 6 rows
$89 - 60 = 29$

15. from 50, go down 1 row and left 6 columns
$50 - 16 = 34$

16. from 21, go up 7 rows and right 6 columns
$21 + 76 = 97$

Reflect
Explain how you use the chart to add or subtract a number greater than 10. Give an example of a movement for a number greater than 10.

In the chart each move from one row to the row above it is the same as adding 10. Each move from one row to the row below it is the same as subtracting 10. Each move to the right is adding 1 for each box, and each move to the left is subtracting 1 for each box. To add 15, move up 1 row and right 5 boxes. To subtract 35, move down 3 rows and left 5 boxes.

---

3 Reflect   10

Extended Response   APPLYING
Review students' answers to the Reflect prompt at the bottom of student page 27.

■ **How can you use the 100 Chart to write equations that have missing addends and equations that have missing sums?** For missing addends, you can choose the first addend and the sum and use the chart to find the missing addend. For missing sums, you can start at the first addend and move your finger the amount of the second addend to find the sum.

Real-World Application   APPLYING
Many times in the real world, people need to use mental math to add more than two numbers. Discuss that the first two numbers should be added and then the sum should be added to the third number.

■ **What is the sum of 14 + 6 + 32?** 52

■ **Write an addition problem with three addends that are less than 30. Find the sum. Then trade with a classmate, and check each other's work.**

4 Assess

Informal Assessment
Use the Student Assessment Record, *Assessment,* page 100, to record informal observations.

COMPUTING

Room Service Delivery
Did the student
❑ respond accurately?
❑ respond quickly?
❑ respond with confidence?
❑ self-correct?

Lesson 4

Objective

Students find sums and differences using mental computations.

Materials

Additional Materials
Number Cards (0–5)

Access Vocabulary

skip count A method counting by which each number stated increases or decreases by the same amount (a number greater than 1).

mental computation Performing math operations in your mind without paper and pencil or a calculator.

Creating Context

In English there are many words that have more than one meaning. This can confuse some English Learners when working with math terms. An excellent strategy to use with English Learners is to preview the lesson to identify any potentially confusing words or phrases. You can introduce the terms or explain them while teaching the lesson. Help English Learners find and list the words *counting on, skip, difference, find,* and *steps.* Make a three-column chart with the confusing word in column one, the math meaning in column two, and the common meaning in column three.

1 Warm Up   5

Concept Building COMPUTING

In this lesson you will practice strategies for doing mental addition and subtraction of double-digit numbers. A good strategy is to take the first number and skip count the number of tens in the second number. Then count on the number of ones in the second number. Give the following problems to the class. Model this strategy using the first problem.

$30 + 28 = 58$	$45 + 21 = 66$
$40 + 28 = 68$	$49 + 20 = 69$

2 Engage   30

Skill Building

This activity will require good listening skills. Students may benefit from having a 100 Chart and scratch paper. Hand out Number Cards to students. Arrange students in a circle. The activity ends when everyone in the circle has had a turn.

Keep It Under 100 COMPUTING

- **This is a mental math game. I will begin with a number. The person on my right will pick two Number Cards to use to make a double-digit number. The person will announce the number and then add it to or subtract it from the previous number.**

- **The person next in the circle takes a turn. Make a double-digit number, announce it, and then add or subtract it. The object of the game is to make it all the way around the circle and keep the number under 100.**

Students will need to listen to the other students carefully and think about the players yet to get a turn before deciding whether to add or subtract their double-digit numbers.

Monitoring Student Progress

If . . . students cannot decide whether to add or subtract,

Then . . . suggest that they do the opposite operation than the one used by the previous player.

Building Blocks For additional practice with mental computation, students should complete **Building Blocks** Math-O-Scope.

e MathTools Use the 100s Table tool to demonstrate and explore using a ten-by-ten arrangement.

Using Student Pages

Have students complete **Workbook,** pages 28–29, on their own

Week 3 — Place Value and Number Sense

Lesson 4

Key Idea

You can find sums and differences mentally even when you do not have the 1–100 chart. Use the same methods of skip counting and counting on or back.

$48 + 34$ Think of 34 as 3 tens and 4 ones.

From 48, skip count by tens 3 times: 58, 68, 78.

From 78, count on 4: 79, 80, 81, 82.

$48 + 34 = 82$

Try This

Use skip counting to find each sum.

1. $38 + 50 = \underline{88}$
2. $17 + 80 = \underline{97}$
3. $25 + 30 = \underline{55}$
4. $73 + 10 = \underline{83}$
5. $64 + 20 = \underline{84}$
6. $42 + 50 = \underline{92}$

Use skip counting to find each difference.

7. $71 - 30 = \underline{41}$
8. $56 - 40 = \underline{16}$
9. $60 - 10 = \underline{50}$
10. $92 - 80 = \underline{12}$

Practice

Name how many tens and how many ones to add. Then find each sum.

11. $50 + 36 =$ add $\underline{3}$ tens $\underline{6}$ ones sum $\underline{86}$
12. $20 + 57 =$ add $\underline{5}$ tens $\underline{7}$ ones sum $\underline{77}$

Name how many tens and how many ones to subtract. Then find each difference.

13. $76 - 43 =$ subtract $\underline{4}$ tens $\underline{3}$ ones difference $\underline{33}$
14. $64 - 32 =$ subtract $\underline{3}$ tens $\underline{2}$ ones difference $\underline{32}$

Use mental math to find each sum or difference.

15. $95 - 32 = \underline{63}$
16. $48 + 51 = \underline{99}$
17. $62 + 34 = \underline{96}$
18. $79 - 55 = \underline{24}$

Reflect

Use mental math. Explain each of your steps.

$34 + 20 - 13 + 22 = \underline{63}$

From 34, skip count by tens 2 times.

44, 54

From 54, skip count by tens backward once. Then count back 3.

$44 \rightarrow 43, 42, 41$

From 41, skip count by tens 2 times. Then count on 2.

$51, 61 \rightarrow 62, 63$

---

3 — Reflect 10

Extended Response REASONING

Review students' answers to the Reflect prompt at the bottom of student page 29.

- **Is there more than one strategy you can use to solve this type of problem?** Answers will vary.
- **How can you check your answer?** use a calculator, use pencil and paper, use a number chart

Real-World Application APPLYING

- **Suppose you received several birthday cards that had money inside. Find the total amount you received. Do not use paper and pencil or a calculator.** $35, $23, $12, $3; total = $73

---

4 — Assess

Informal Assessment

Use the Student Assessment Record, **Assessment**, page 100, to record informal observations.

COMPUTING

Keep It Under 100

Did the student
- ☐ respond accurately?
- ☐ respond quickly?
- ☐ respond with confidence?
- ☐ self-correct?

Lesson 5 Review

Objective

Students review skills learned this week and complete the weekly assessment.

Materials

Review materials will be selected from those used in previous activities.

Creating Context

Be sure that English Learners understand that *use mental math* means "to figure out the answer without paper and pencil or calculator." An excellent mental strategy is to think aloud or verbalize each step to figure out an answer.

2 Engage 20

Skill Building

Free-Choice Activity

For the last day of the week, allow students to choose an activity from the previous lessons. Some activities they may choose are the following:

- **Making a Neighborhood**
- **Monsters in the Neighborhood**
- **Room Service Delivery**
- **Keep it Under 100**

Make a note of the activities students select. Do they prefer easy or challenging activities? If you believe your students would benefit from extra practice on specific skills, choose an activity for them.

Monitoring Student Progress

If . . . students are not participating,

Then . . . continue the activity until all students have participated.

Using Student Pages

Have students complete **Workbook,** pages 30–31, on their own.

3 Reflect 10

Extended Response APPLYING

Review students' answers to the Reflect prompts at the bottom of student pages 30–31.

Discuss these answers with the group to reinforce Week 3 concepts.

1 Warm Up 5

Concept Building UNDERSTANDING

Give the following problems to the class. All answers should be computed mentally.

20 + 15 = 35   55 − 30 = 25
38 + 10 = 48   60 − 45 = 15

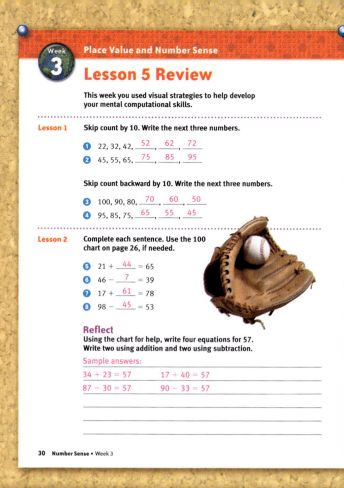

Week 3 Place Value and Number Sense

Lesson 5 Review

This week you used visual strategies to help develop your mental computational skills.

Lesson 1 Skip count by 10. Write the next three numbers.

1. 22, 32, 42, _52_, _62_, _72_
2. 45, 55, 65, _75_, _85_, _95_

Skip count backward by 10. Write the next three numbers.

3. 100, 90, 80, _70_, _60_, _50_
4. 95, 85, 75, _65_, _55_, _45_

Lesson 2 Complete each sentence. Use the 100 chart on page 26, if needed.

5. 21 + _44_ = 65
6. 46 − _7_ = 39
7. 17 + _61_ = 78
8. 98 − _45_ = 53

Reflect

Using the chart for help, write four equations for 57. Write two using addition and two using subtraction.

Sample answers:

| 34 + 23 = 57 | 17 + 40 = 57 |
| 87 − 30 = 57 | 90 − 33 = 57 |

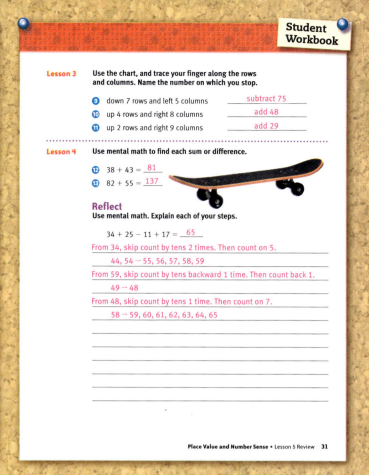

Lesson 3 Use the chart, and trace your finger along the rows and columns. Name the number on which you stop.

9. down 7 rows and left 5 columns ___subtract 75___
10. up 4 rows and right 8 columns ___add 48___
11. up 2 rows and right 9 columns ___add 29___

Lesson 4 Use mental math to find each sum or difference.

12. 38 + 43 = _81_
13. 82 + 55 = _137_

Reflect

Use mental math. Explain each of your steps.

34 + 25 − 11 + 17 = _65_

From 34, skip count by tens 2 times. Then count on 5.
   44, 54 → 55, 56, 57, 58, 59

From 59, skip count by tens backward 1 time. Then count back 1.
   49 → 48

From 48, skip count by tens 1 time. Then count on 7.
   58 → 59, 60, 61, 62, 63, 64, 65

---

4 Assess   10

A Gather Evidence

Formal Assessment

Have students complete the Weekly test on **Assessment,** pages 28–29. Record progress on the Student Assessment Record, **Assessment,** page 100.

B Summarize Findings

Determine whether students have Minimal, Basic, or Secure understanding of the concepts presented in Week 3.

C Differentiate Instruction

Based on your observations, use these teaching strategies next week to follow up.

Minimal Understanding
- Repeat the Warm-Up and Engage activities to develop Week 3 concepts.
- Use **Building Blocks** School Supply Shop and **eMathTools** 100s Table to develop and reinforce Week 3 concepts.

Basic Understanding
- Repeat Engage activities to reinforce Week 3 concepts.
- Use **Building Blocks** Math-O-Scope and **eMathTools** 100s Table to reinforce Week 3 concepts.

Secure Understanding
Use **Building Blocks** Math-O-Scope and **eMathTools** 100s Table to reinforce Week 3 concepts.

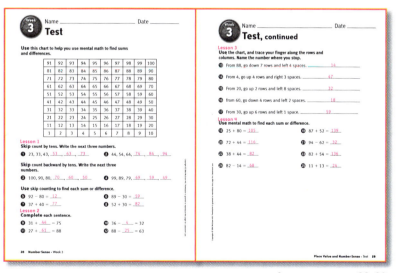

Assessment, pp. 28–29

Place Value and Number Sense • Lesson 5 Review 30–31

Comparing Numbers to 9,999

Week at a Glance

This week, students continue with **Number Worlds,** Level F, Number Sense, by examining the relationships among numbers, including halves and quarters.

Math Background

Equivalence is one of the key mathematical concepts for students to understand early in their mathematical experiences. Equivalence provides a foundation that students use in comparing and ordering numbers. Instructional activities that employ the use of models and manipulatives are valuable as students learn about equivalence, the concept of inequality, and relative value.

How Students Learn

Encountering a variety of instructional activities and strategies for understanding the relationship among numbers helps students grasp the key ideas of equivalence and inequality. It is important that students can compare and order whole numbers before the topics of fractions and decimals are introduced. Students deal informally with equivalence in real-world situations such as money, time, and sharing. Whenever possible, relate lessons to these types of experiences.

Teaching for Understanding

Observe closely while evaluating the assigned tasks this week to see whether students can demonstrate the following understandings.

Benchmark after Lesson 2: Students can use $<$, $>$, and $=$ to relate two numbers correctly.

Benchmark after Lesson 3: Students can visually identify half of a whole and name and compare whole numbers and mixed numbers comprised of halves.

Benchmark after Lesson 4: Students can visually identify a quarter of a whole and two quarters as a half and name and compare whole numbers and mixed numbers comprised of halves and quarters.

Skills Focus

- Compare and order whole numbers to 9,999
- Compare and order money amounts containing combinations of coins and bills
- Develop an informal understanding of the fractions $\frac{1}{2}$ and $\frac{1}{4}$

Math at Home

Give one copy of the Letter to Home, page A4, to each student. Encourage students to share and complete the activity with their caregivers.

Letter to Home, Teacher Edition, p. A4

Week 4 Planner — Comparing Numbers to 9,999

PACING	LESSON	LEARNING GOALS	NCTM	MATERIALS	TECHNOLOGY
DAY 1	pages 32–33	Students compare and order numbers to 9,999.	• Number and Operations • Problem Solving • Communication • Connections	• Base-Ten Blocks* • Number Cards (0–9) • Number Construction Mat, p. B1	**Building Blocks** Party Time **MathTools** Base Ten Blocks
DAY 2	pages 34–35	Students compare and order money amounts.	• Number and Operations • Problem Solving • Communication • Connections • Representation	• Models of pennies, dimes, one-dollar bills, and ten-dollar bills* • Construction paper • Markers	**Building Blocks** Party Time **MathTools** Coins and Money
DAY 3	pages 36–37	Students use models of wholes and halves to name and compare numbers.	• Number and Operations • Problem Solving • Communication • Connections • Representation	Fraction circles*	**Building Blocks** Pizza Pizzazz **MathTools** Fractions
DAY 4	pages 38–39	Students use models of wholes, halves, and quarters to name and compare numbers.	• Number and Operations • Problem Solving • Communication • Connections • Representation	• Fraction circles* • Plain paper • Construction paper	**Building Blocks** Pizza Pizzazz **MathTools** Fractions
DAY 5	Review and Assess pages 40–41	Students review skills learned this week and complete the weekly assessment.	• Number and Operations • Problem Solving • Communication • Connections • Representation	Materials will be selected from Lessons 1–4.	**Building Blocks** **MathTools** Review previous activities

Math Vocabulary

greater than A symbol, >, that shows the relationship of two numbers

less than A symbol, <, that shows the relationship of two numbers

equal to A symbol, =, that shows the relationship of two numbers. The numbers have the same value.

greatest In a group of numbers, the number farthest to the right on a number line

least In a group of numbers, the number farthest to the left on a number line

English Learners

SPANISH COGNATES

English	Spanish
compare	comparar
symbol	símbolo
pair	par
order	orden

ALTERNATE VOCABULARY

how many wholes? The word *wholes*, spelled with an initial silent *w*, means objects that are complete. *Holes*, by contrast, are empty spaces.

* Available from SRA

Lesson 1

Objective

Students compare and order numbers to 9,999.

Materials

Program Materials

Number Construction Mat, p. B1

Additional Materials
- Base-Ten Blocks
- Number Cards (0–9)

Access Vocabulary

pair Two numbers or two objects that go together

explain your answer To describe why your answer works or why you chose it

Creating Context

Review with students the relation symbols and the meanings of *greater than* and *less than*. Brainstorm some number comparisons that are familiar to students such as their ages compared to brothers and sisters or other children. Remind students to describe the number relationship in a complete sentence: I am _____ years old. I am older than _____. _____ is _____ years old. _____ is greater than _____. _____ is less than _____.

1 Warm Up 5

Concept Building UNDERSTANDING

Have one student stand and say a number between 0 and 10. Select a second student to stand and say a number that is greater than the first number. Then select a third student to stand and say a number that is greater than the second number. Write the numbers on the board as students say them. Use the same activity to name a lesser number each time. Introduce the greater-than and less-than symbols, and demonstrate their use with the numbers on the board.

2 Engage 30

Skill Building

Provide students with sets of Number Cards. Instruct students to shuffle the cards and turn the cards facedown on their desks.

Number Shuffle UNDERSTANDING

Students draw three cards from their pile of cards.

- **Use these numbers to list as many two- and three-digit numbers as you can.**
- **Write the numbers in order from least to greatest.**
- **Write the less-than symbol between each number to show that each number is less than the number that follows.**

Review and discuss each number produced.

Strategy Building

Ultimate Number Shuffle COMPUTING

Students draw three cards from their pile of cards.

- **Write the greatest number you can make with these three numbers.**
- **Write the least number you can make with these three numbers.**

Monitoring Student Progress

| If . . . students are still dependent on being able to model the numbers, | Then . . . provide Base-Ten Blocks and the Number Construction Mat. |

Building Blocks For additional practice comparing numbers to 9,999, students should complete **Building Blocks** Party Time.

MathTools Use the Base Ten Blocks tool to demonstrate and explore comparing numbers to 9,999.

Using Student Pages

Have students complete **Workbook,** pages 32–33, on their own.

Lesson 1

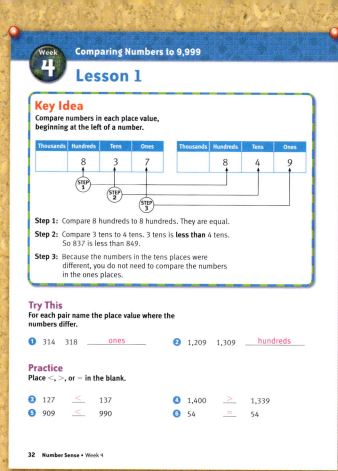

Key Idea

Compare numbers in each place value, beginning at the left of a number.

Thousands	Hundreds	Tens	Ones		Thousands	Hundreds	Tens	Ones
	8	3	7			8	4	9

STEP 1 · STEP 2 · STEP 3

Step 1: Compare 8 hundreds to 8 hundreds. They are equal.

Step 2: Compare 3 tens to 4 tens. 3 tens is **less than** 4 tens. So 837 is less than 849.

Step 3: Because the numbers in the tens places were different, you do not need to compare the numbers in the ones places.

Try This

For each pair name the place value where the numbers differ.

❶ 314   318   _____ones_____

❷ 1,209   1,309   _____hundreds_____

Practice

Place <, >, or = in the blank.

❸ 127   <   137

❹ 1,400   >   1,339

❺ 909   <   990

❻ 54   =   54

Fill in the blank with a digit to make each sentence true.

❼ 434 < 4__1
   4, 5, 6, 7, 8, or 9

❽ 1,000 > __75
   0, 1, 2, 3, 4, 5, 6, 7, 8, or 9

❾ 6__ > 62
   3, 4, 5, 6, 7, 8, or 9

❿ 1__7 = __87
   8; 1

⓫ Write these numbers in order from least to greatest: 393, 313, 339.
   313, 339, 393

⓬ Write these numbers in order from least to greatest: 800, 788, 810.
   788, 800, 810

Reflect

Write the greatest number possible using the digits 1, 4, 6, and 8. Explain your answer.

8,641; the greatest number is in the thousands place, the next greatest number is in the hundreds place, the next greatest number is in the tens place, and the least number is in the ones place.

Write the least number possible using the digits 7, 5, and 2. Explain your answer.

257; the least number is in the hundreds place, the next least number is in tens place, and the greatest number is in the ones place.

---

3 Reflect   10

Extended Response   REASONING

Review students' answers to the Reflect prompt at the bottom of student page 33.

■ **How did you decide where to place each digit?**
For the greatest number, place the largest digit in the highest place value. For the least number, place the smallest digit in the highest place value.

Real-World Application   APPLYING

In some cities and towns, addresses on main streets that go north, south, east, or west are numbered from the "center" of town to the outlying areas. The greater the number of the address, the farther the building is from the "center" of town.

■ **If one building's address is 1452, and another building on the same street is 347, which building is closer to the "center" of town?** 347

---

4 Assess

Informal Assessment

Use the Student Assessment Record, *Assessment,* page 100, to record informal observations.

UNDERSTANDING	COMPUTING
Number Shuffle	**Ultimate**
Did the student	**Number Shuffle**
❏ make important observations?	Did the student
❏ extend or generalize learning?	❏ respond accurately?
❏ provide insightful answers?	❏ respond quickly?
❏ pose insightful questions?	❏ respond with confidence?
	❏ self-correct?

Lesson 2

Objective

Students compare and order money amounts.

Materials

Additional Materials
- Models of pennies, dimes, one-dollar bills, and ten-dollar bills
- Construction paper
- Markers

Access Vocabulary

in order from greatest to the least In order, starting with the largest number and counting down to the smallest number

piggy bank A small, pig-shaped ceramic container for coins

hair dryer A small, electrical appliance used to dry hair

Creating Context

The use of manipulatives increases comprehension with English Learners. Students at the early levels of English proficiency can also use manipulatives to demonstrate their understanding of the math concepts involved. In this lesson, several problems deal with comparing amounts of money. Use play money to have students show the amounts in the lesson, and compare which amounts are greatest and which are least.

 Warm Up 5

Concept Building UNDERSTANDING

Write these amounts on the board: $1.51 and $1.15.

- **Which is more?** $1.51
- **How do you know?** Answers will vary: The number 5 in the tens place of the first number is greater than the 1 in the tens place of the second number.
- **Can you show each amount using play money?** Allow students to do this.

 Engage 30

Strategy Building

Organize students into groups of five. Provide groups with construction paper and a marker.

Who Has the Most? COMPUTING

Instruct students to write (without others seeing) any whole-dollar amount less than $50 on a sheet of construction paper. Explain to students that this activity begins as a small-group activity and ends as a whole-class effort. Teamwork will be important.

- **In your group, hold up your amount so that everyone can see. Then arrange yourselves in a line so that the student with the greatest amount is first in line, then the next greatest amount, and so on.**

Next, arrange the amounts of the entire class. The first students in each small group compare their amounts and begin to form a line at the front of the classroom, the greatest number being first. Then, the next students in each group compare their amounts. Continue adding more in the arrangement until all students are standing in line. The line will go around the room from the greatest amount to the least amount. Students with equal numbers should stand next to each other.

Monitoring Student Progress

If . . . while building the line from greatest to least amounts, a student misplaces himself or herself in line,

Then . . . let the student self-correct the position or have other class members tell that student he or she is out of order.

Building Blocks For additional practice comparing numbers to 9,999, students should complete **Building Blocks** Party Time.

MathTools Use the Coins and Money tool to demonstrate and explore comparing numbers to 9,999.

Using Student Pages

Have students complete **Workbook,** pages 34–35, on their own.

Lesson 2

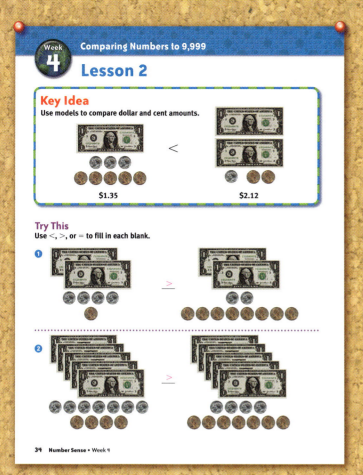

Key Idea

Use models to compare dollar and cent amounts.

$1.35 < $2.12

Try This

Use <, >, or = to fill in each blank.

1 ____ > ____

2 ____ > ____

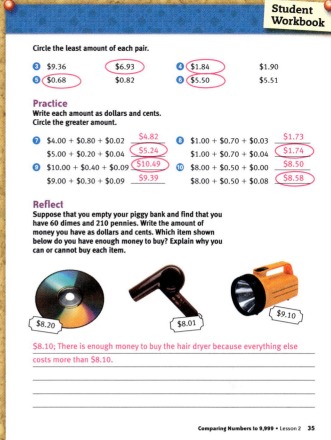

Circle the least amount of each pair.

3 $9.36   ($6.93)   **4** ($1.84) $1.90

5 ($0.68) $0.82   **6** ($5.50) $5.51

Practice

Write each amount as dollars and cents.
Circle the greater amount.

7 $4.00 + $0.80 + $0.02 $4.82
$5.00 + $0.20 + $0.04   ($5.24)

8 $1.00 + $0.70 + $0.03 $1.73
$1.00 + $0.70 + $0.04   ($1.74)

9 $10.00 + $0.40 + $0.09   ($10.49)
$9.00 + $0.30 + $0.09 $9.39

10 $8.00 + $0.50 + $0.00 $8.50
$8.00 + $0.50 + $0.08   ($8.58)

Reflect

Suppose that you empty your piggy bank and find that you have 60 dimes and 210 pennies. Write the amount of money you have as dollars and cents. Which item shown below do you have enough money to buy? Explain why you can or cannot buy each item.

$8.20 $8.01 $9.10

$8.10; There is enough money to buy the hair dryer because everything else costs more than $8.10.

_____
_____
_____

---

3 Reflect   10

Extended Response   REASONING

Review students' answers to the Reflect prompt at the bottom of student page 35.

■ **How did you figure out how much money you had?** Answers will vary.

Discuss and compare methods for totaling the money.

Real-World Application   APPLYING

Bring in sale papers from several area discount stores. Ask students to list five different things they could buy with $10. Instruct students to record the price of each item. Then list the items in order of price from the least expensive to the most expensive.

4 Assess

Informal Assessment

Use the Student Assessment Record, **Assessment,** page 100, to record informal observations

COMPUTING

Who Has the Most?
Did the student
❏ respond accurately?
❏ respond quickly?
❏ respond with confidence?
❏ self-correct?

Lesson 3

Objective

Students use models of wholes and halves to name and compare numbers.

Materials

Additional Materials
Fraction circles

Access Vocabulary

one-half A section of a whole in which two equivalent sections equal one whole. The fraction for one-half is written $\frac{1}{2}$.

name the number To give the numeric answer or total

half the given number Half of the original number

Creating Context

In English, when we talk about more than one object, the name of that object becomes plural. Commonly, we add an *s* to the end of the noun. However, some words change in other ways to show that they are plural. The word *half* is part of a group of English words that change in the middle to help pronunciation. The final *f* is changed to *ve* before the *s* is added. List these pairs of words that follow a similar rule: *shelf/shelves; calf/calves; wolf/wolves; elf/elves; loaf/loaves; leaf/leaves; wife/wives.*

1 Warm Up 5

Concept Building COMPUTING

Discuss the idea of half of an amount, such as 4 is half of 8. Model this concept using manipulatives, if necessary. Say an even number between 20 and 80. Select a student to name the number that is half your number. A correct answer gives students the opportunity to choose a number and a student to name half the number. Continue until all students have had an opportunity to participate.

2 Engage 30

Skill Building

Fraction Circles UNDERSTANDING

Students should build on their informal knowledge of rational numbers to explore the concept of $\frac{1}{2}$ of a whole.

Use fraction circles, or draw on the board circles similar to those used on **Workbook,** pages 36–37. Have students practice adding wholes to halves and multiple halves.

Strategy Building

Half of a Whole UNDERSTANDING

Ask the question below. For students who offer suitable answers, invite them to the board to draw a sketch of their item as a whole and as a half.

- **What kinds of things can you divide in half?**
  Possible answers: sandwich, pizza, pie, hour, dollar, inch, measuring cup, price, and musical note

Monitoring Student Progress

| If . . . any student struggles to visualize one-half, | Then . . . provide two 2-cup measuring cups. Fill one measuring cup with 16 ounces of water. Then pour water from one cup into the other until they have equal amounts of water. |

 For additional practice identifying and comparing numbers with $\frac{1}{2}$, students should complete **Building Blocks** Pizza Pizzazz.

e MathTools Use the Fractions tool to demonstrate and explore identifying and comparing numbers with $\frac{1}{2}$.

Using Student Pages

Have students complete **Workbook,** pages 36–37, on their own.

Lesson 3

Key Idea

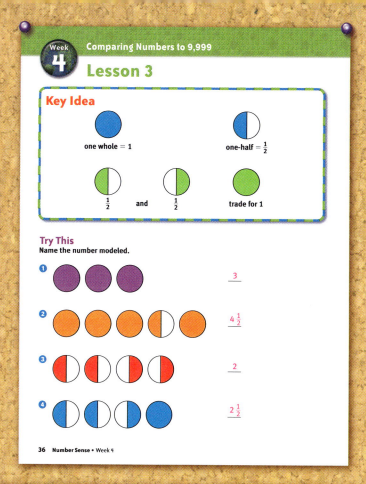

one whole = 1   one-half = $\frac{1}{2}$

$\frac{1}{2}$ and $\frac{1}{2}$ trade for 1

Try This
Name the number modeled.

 1   3

2 $4\frac{1}{2}$

3   2

4 $2\frac{1}{2}$

Practice
Name how many wholes and how many halves there are altogether.

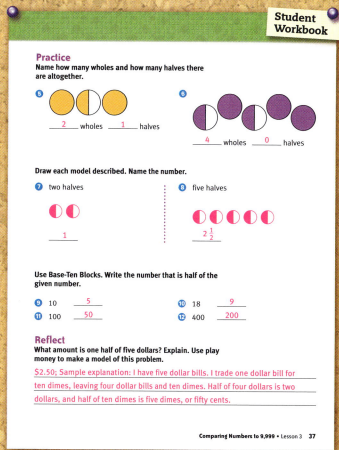

5   2 wholes   1 halves

6   4 wholes   0 halves

Draw each model described. Name the number.

7 two halves

    1

8 five halves

    $2\frac{1}{2}$

Use Base-Ten Blocks. Write the number that is half of the given number.

9 10   5
10 18   9
11 100   50
12 400   200

Reflect
What amount is one half of five dollars? Explain. Use play money to make a model of this problem.

$2.50; Sample explanation: I have five dollar bills. I trade one dollar bill for ten dimes, leaving four dollar bills and ten dimes. Half of four dollars is two dollars, and half of ten dimes is five dimes, or fifty cents.

---

3 Reflect   10

Extended Response   APPLYING

Review students' answers to the Reflect prompt at the bottom of student page 37.

Students may benefit from having model money so they can physically make trades and divide the coins into two equal groups.

- **Was this activity easy or difficult?** Answers will vary.
- **How can you check your answer?** Add the two amounts together.

Real-World Application   APPLYING

Write each of the following amounts on the board: $10.00, $5.00, 50 cents, 10 cents.

Use play money to show amounts that are half the value of these amounts.

4 Assess

Informal Assessment

Use the Student Assessment Record, *Assessment*, page 100, to record informal observations.

UNDERSTANDING	UNDERSTANDING
Fraction Circles	**Half of a Whole**
Did the student	Did the student
❏ make important observations?	❏ make important observations?
❏ extend or generalize learning?	❏ extend or generalize learning?
❏ provide insightful answers?	❏ provide insightful answers?
❏ pose insightful questions?	❏ pose insightful questions?

Comparing Numbers to 9,999 • Lesson 3   36–37

Lesson 4

Objective

Students use models of wholes, halves, and quarters to name and compare numbers.

Materials

Additional Materials
- Fraction circles
- Plain paper
- Construction paper

Access Vocabulary

one-half A section of a whole in which two equivalent sections equal one whole. The fraction for one-half is written $\frac{1}{2}$.

one-quarter A section of a whole in which four equivalent sections equal one whole. The fraction for one-quarter is written $\frac{1}{4}$.

name the number modeled Say the name of the number that is represented by the drawing.

Creating Context

In this lesson, there is an example of an English word with double meaning. Students are asked to count the quarters that are pictured. In everyday speech, English Learners may know the word *quarter* to mean a 25-cent coin. Use four quarters to reinforce that there are four quarters in a dollar, and that we can divide a figure in four equal sections. These sections are also called quarters.

1 Warm Up   5

Concept Building   COMPUTING

Instruct students to draw a square, a rectangle, and a circle on construction papers. Students should then draw lines on their shapes to show four equal sections in each.

- **How much of the whole shape is just one of the sections?** $\frac{1}{4}$, or one quarter
- **How much of the whole shape is two of these sections?** $\frac{1}{2}$ or one-half

2 Engage   30

Skill Building

Half of a Half? UNDERSTANDING

Provide students with a sheet of plain paper. Instruct students to fold the paper so it is sectioned into four equal areas.

- **Did everyone fold his or her paper the same way?** Answer will most likely be no.
- **Even though there are several ways you can fold your paper, what does each student's folded paper show?** The paper shows a whole that has been divided into four sections. It shows four one-quarter sections.

Use fraction circles, or draw on the board circles similar to those in the **Key Idea** box on **Workbook,** page 38. Then draw empty circles, and have students suggest how to shade them to make halves and quarters.

Monitoring Student Progress

If . . . students are grasping the concepts of $\frac{1}{2}$ and $\frac{1}{4}$ quickly,

Then . . . play a game of "What is the question?" Use the following riddles:

- **The answer is 15 minutes.** The question is "What is one-quarter of an hour?"
- **The answer is 6 inches.** The question is "What is one-half of a foot?"
- **The answer is 6 roses.** The question is "What is one-half of a dozen roses?"
- **The answer is 3 roses.** The question is "What is one-quarter of a dozen roses?"

Building Blocks For additional practice identifying and comparing with $\frac{1}{4}$, students should complete **Building Blocks** Pizza Pizzazz.

MathTools Use the Fraction tool to demonstrate and explore identifying and comparing numbers with $\frac{1}{4}$.

Using Student Pages

Have students complete **Workbook,** pages 38–39, on their own.

Lesson 4

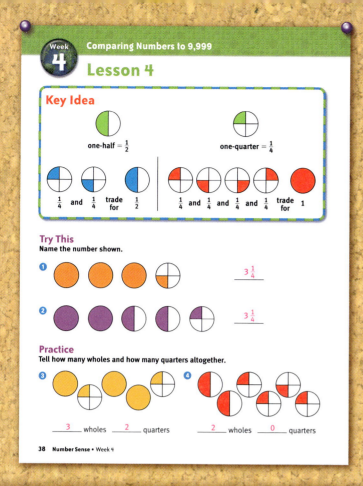

Key Idea

one-half = $\frac{1}{2}$

one-quarter = $\frac{1}{4}$

$\frac{1}{4}$ and $\frac{1}{4}$ trade for $\frac{1}{2}$

$\frac{1}{4}$ and $\frac{1}{4}$ and $\frac{1}{4}$ and $\frac{1}{4}$ trade for 1

Try This
Name the number shown.

1. $3\frac{1}{4}$

2. $3\frac{1}{4}$

Practice
Tell how many wholes and how many quarters altogether.

3.  __3__ wholes __2__ quarters

4. __2__ wholes __0__ quarters

38   Number Sense • Week 4

Draw each model described. Name the number.

5. two quarters
$\frac{1}{2}$

6. one-half and two quarters
1

7. four quarters
1

8. six quarters
$1\frac{1}{2}$

Write the number that is half the given number. Then write the number that is half that number.

9. 12 __6__ __3__
10. 140 __70__ __35__
11. 100 __50__ __25__
12. 1 __$\frac{1}{2}$__ __$\frac{1}{4}$__

Reflect
Use Base-Ten Blocks to explain why 25 is half of 50.

Use 5 rods to model 50. Then arrange the blocks so they are in two equal groups. Put 2 rods in one group and 2 rods in another group. The fifth rod needs to be split in half. Because 1 rod is the same as 10 unit blocks, trade. Now put 5 unit blocks with 2 of the rods and 5 unit blocks with the other 2 rods. 2 rods = 20; 5 unit blocks = 5; 20 + 5 = 25. Check students' drawings.

3 Reflect   10

Extended Response   APPLYING

Review students' answers to the Reflect prompt at the bottom of student page 39.

- **Was the problem easier or more difficult than the reflect question on page 37?** Answers will vary.
- **Is there more than one way to solve this problem?** Answers will vary.

Real-World Application   APPLYING

- **Suppose you are having a party and need to order pizza for you and your guests. You invite 4 people to the party. Plan for each person to eat one-quarter of a pizza. How many pizzas should you order?** Because you are also at the party, buy pizza for 5 people. That is $1\frac{1}{4}$ pizzas. Order two pizzas.

Students can draw pictures to decide how many pizzas to order.

4 Assess

Informal Assessment

Use the Student Assessment Record, *Assessment*, page 100, to record informal observations.

UNDERSTANDING

Half of a Half?

Did the student
- ❑ make important observations?
- ❑ extend or generalize learning?
- ❑ provide insightful answers?
- ❑ pose insightful questions?

Comparing Numbers to 9,999 • Lesson 4   38–39

Lesson 5 Review

Objective

Students review skills learned this week and complete the weekly assessment.

Materials

Review materials will be selected from those used in previous activities.

Creating Context

English Learners may find the game "What is the question?", page 38A, challenging because they must listen to the answer clue and formulate a question to go with it. It may be helpful to provide English Learners with question frames and to have them practice question construction with the critical information left blank.

✓ Unit Assessment

Students should complete the Number Sense Test found on **Assessment,** pages 72–73. Using the key on **Assessment,** page 98, identify incorrect responses. Reteach and review the Warm-Up and Engage activities to reinforce concept understanding.

1 Warm Up   5

Concept Building UNDERSTANDING

Discuss the idea of half of an amount, such as 4 is half of 8. Say an even number between 20 and 80. Select a student to name the number that is half your number. A correct answer gives students the opportunity to choose a number and a student to name half that number. Continue until all students have had an opportunity to participate.

2 Engage   20

Skill Building

Free-Choice Activity

For the last day of the week, allow students to choose an activity from previous lessons. Some activities they may choose are the following:

- **Number Shuffle**
- **Ultimate Number Shuffle**
- **Who Has the Most?**
- **Fraction Circles**
- **Half of a Whole**
- **Half of a Half?**

Make a note of the activities students select. Do they prefer easy or challenging activities? If you believe your students would benefit from extra practice on specific skills, choose an activity for them.

Monitoring Student Progress

If . . . students are not participating,

Then . . . continue the activity until all students have participated.

Using Student Pages

Have students complete **Workbook,** pages 40–41, on their own.

3 Reflect   10

Extended Response APPLYING

Review students' answers to the Reflect prompts at the bottom of student pages 40–41.

Discuss these answers with the group to reinforce Week 4 concepts.

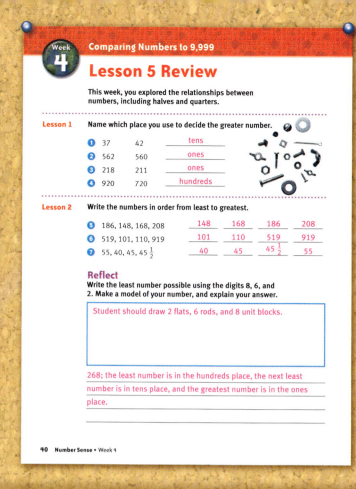

Lesson 5 Review

This week, you explored the relationships between numbers, including halves and quarters.

Lesson 1 Name which place you use to decide the greater number.

1. 37   42   tens
2. 562   560   ones
3. 218   211   ones
4. 920   720   hundreds

Lesson 2 Write the numbers in order from least to greatest.

5. 186, 148, 168, 208   148   168   186   208
6. 519, 101, 110, 919   101   110   519   919
7. 55, 40, 45, 45 $\frac{1}{2}$ 40   45   45 $\frac{1}{2}$ 55

Reflect
Write the least number possible using the digits 8, 6, and 2. Make a model of your number, and explain your answer.

Student should draw 2 flats, 6 rods, and 8 unit blocks.

268; the least number is in the hundreds place, the next least number is in tens place, and the greatest number is in the ones place.

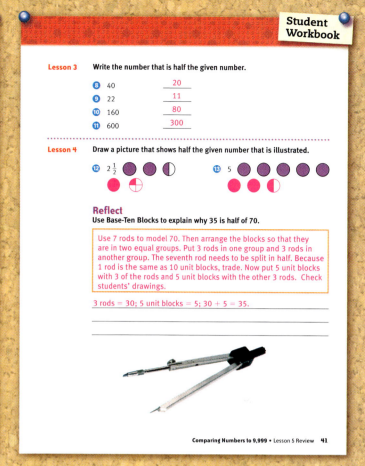

Lesson 3 Write the number that is half the given number.

8. 40   20
9. 22   11
10. 160   80
11. 600   300

Lesson 4 Draw a picture that shows half the given number that is illustrated.

12. 2 $\frac{1}{2}$
13. 5

Reflect
Use Base-Ten Blocks to explain why 35 is half of 70.

Use 7 rods to model 70. Then arrange the blocks so that they are in two equal groups. Put 3 rods in one group and 3 rods in another group. The seventh rod needs to be split in half. Because 1 rod is the same as 10 unit blocks, trade. Now put 5 unit blocks with 3 of the rods and 5 unit blocks with the other 3 rods. Check students' drawings.

3 rods = 30; 5 unit blocks = 5; 30 + 5 = 35.

---

4 Assess   10

A Gather Evidence

Formal Assessment

Have students complete the Weekly test on **Assessment,** pages 30–31. Record scores on the Student Assessment Record, **Assessment,** page 100.

B Summarize Findings

Determine whether students have Minimal, Basic, or Secure understanding of the concepts presented in Week 4.

C Differentiate Instruction

Based on your observations, use these teaching strategies next week to follow up.

Minimal Understanding
- Repeat Engage activities in subsequent weeks.
- Use **Building Blocks** Party Time and **eMathTools** Coins and Money to reinforce Week 4 concepts.

Basic Understanding
- Repeat Engage activities in subsequent weeks.
- Use **Building Blocks** Party Time and **eMathTools** Coins and Money to reinforce Week 4 concepts.

Secure Understanding
Use **Building Blocks** Pizza Pizzazz and **eMathTools** Fractions to reinforce Week 4 concepts.

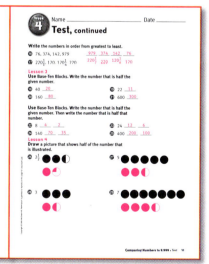

Assessment, pp. 30–31

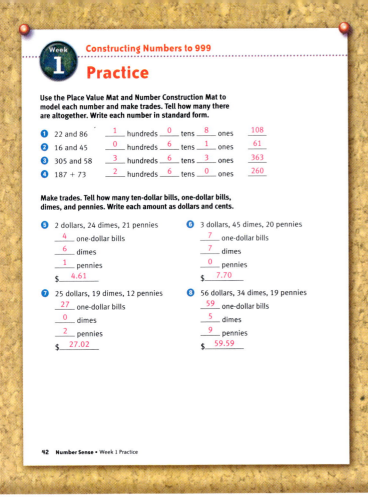

Constructing Numbers to 999

Practice

Use the Place Value Mat and Number Construction Mat to model each number and make trades. Tell how many there are altogether. Write each number in standard form.

1. 22 and 86   __1__ hundreds __0__ tens __8__ ones   108
2. 16 and 45   __0__ hundreds __6__ tens __1__ ones   61
3. 305 and 58   __3__ hundreds __6__ tens __3__ ones   363
4. 187 + 73   __2__ hundreds __6__ tens __0__ ones   260

Make trades. Tell how many ten-dollar bills, one-dollar bills, dimes, and pennies. Write each amount as dollars and cents.

5. 2 dollars, 24 dimes, 21 pennies
   __4__ one-dollar bills
   __6__ dimes
   __1__ pennies
   $ __4.61__

6. 3 dollars, 45 dimes, 20 pennies
   __7__ one-dollar bills
   __7__ dimes
   __0__ pennies
   $ __7.70__

7. 25 dollars, 19 dimes, 12 pennies
   __27__ one-dollar bills
   __0__ dimes
   __2__ pennies
   $ __27.02__

8. 56 dollars, 34 dimes, 19 pennies
   __59__ one-dollar bills
   __5__ dimes
   __9__ pennies
   $ __59.59__

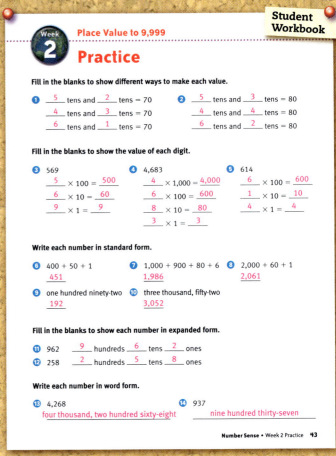

Place Value to 9,999

Practice

Fill in the blanks to show different ways to make each value.

1. __5__ tens and __2__ tens = 70
   __4__ tens and __3__ tens = 70
   __6__ tens and __1__ tens = 70

2. __5__ tens and __3__ tens = 80
   __4__ tens and __4__ tens = 80
   __6__ tens and __2__ tens = 80

Fill in the blanks to show the value of each digit.

3. 569
   __5__ × 100 = __500__
   __6__ × 10 = __60__
   __9__ × 1 = __9__

4. 4,683
   __4__ × 1,000 = __4,000__
   __6__ × 100 = __600__
   __8__ × 10 = __80__
   __3__ × 1 = __3__

5. 614
   __6__ × 100 = __600__
   __1__ × 10 = __10__
   __4__ × 1 = __4__

Write each number in standard form.

6. 400 + 50 + 1
   451

7. 1,000 + 900 + 80 + 6
   1,986

8. 2,000 + 60 + 1
   2,061

9. one hundred ninety-two
   192

10. three thousand, fifty-two
   3,052

Fill in the blanks to show each number in expanded form.

11. 962   __9__ hundreds __6__ tens __2__ ones
12. 258   __2__ hundreds __5__ tens __8__ ones

Write each number in word form.

13. 4,268   four thousand, two hundred sixty-eight
14. 937   nine hundred thirty-seven

Unit 1
Number Sense
Practice Pages

Weeks 1 and 2

1. Assign the Practice pages, **Workbook,** pages 42–43, at the end of Weeks 1 and 2. Students should complete these pages independently.

2. Check student answers using the annotated pages above.

3. If students have difficulty with a Practice page, review the activities completed throughout the week, and have students complete the weekly practice again before they complete the weekly test.

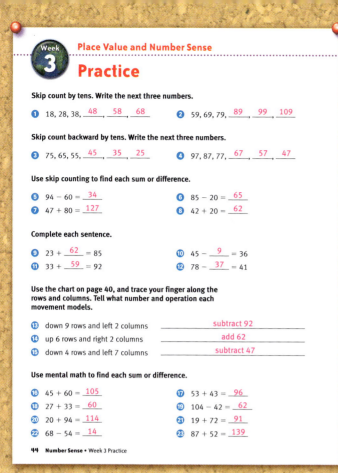

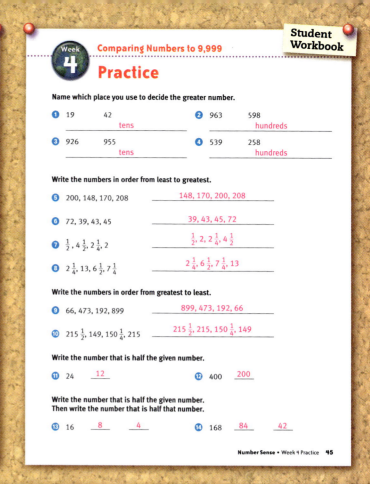

Week 3 — Place Value and Number Sense
Practice

Skip count by tens. Write the next three numbers.

1. 18, 28, 38, __48__, __58__, __68__
2. 59, 69, 79, __89__, __99__, __109__

Skip count backward by tens. Write the next three numbers.

3. 75, 65, 55, __45__, __35__, __25__
4. 97, 87, 77, __67__, __57__, __47__

Use skip counting to find each sum or difference.

5. $94 - 60 = $ __34__
6. $85 - 20 = $ __65__
7. $47 + 80 = $ __127__
8. $42 + 20 = $ __62__

Complete each sentence.

9. $23 + $ __62__ $= 85$
10. $45 - $ __9__ $= 36$
11. $33 + $ __59__ $= 92$
12. $78 - $ __37__ $= 41$

Use the chart on page 40, and trace your finger along the rows and columns. Tell what number and operation each movement models.

13. down 9 rows and left 2 columns — subtract 92
14. up 6 rows and right 2 columns — add 62
15. down 4 rows and left 7 columns — subtract 47

Use mental math to find each sum or difference.

16. $45 + 60 = $ __105__
17. $53 + 43 = $ __96__
18. $27 + 33 = $ __60__
19. $104 - 42 = $ __62__
20. $20 + 94 = $ __114__
21. $19 + 72 = $ __91__
22. $68 - 54 = $ __14__
23. $87 + 52 = $ __139__

Week 4 — Comparing Numbers to 9,999
Practice

Student Workbook

Name which place you use to decide the greater number.

1. 19   42   tens
2. 963   598   hundreds
3. 926   955   tens
4. 539   258   hundreds

Write the numbers in order from least to greatest.

5. 200, 148, 170, 208   148, 170, 200, 208
6. 72, 39, 43, 45   39, 43, 45, 72
7. $\frac{1}{2}$, $4\frac{1}{2}$, $2\frac{1}{4}$, 2 $\frac{1}{2}$, 2, $2\frac{1}{4}$, $4\frac{1}{2}$
8. $2\frac{1}{4}$, 13, $6\frac{1}{2}$, $7\frac{1}{4}$ $2\frac{1}{4}$, $6\frac{1}{2}$, $7\frac{1}{4}$, 13

Write the numbers in order from greatest to least.

9. 66, 473, 192, 899   899, 473, 192, 66
10. $215\frac{1}{2}$, 149, $150\frac{1}{4}$, 215 $215\frac{1}{2}$, 215, $150\frac{1}{4}$, 149

Write the number that is half the given number.

11. 24   12
12. 400   200

Write the number that is half the given number. Then write the number that is half that number.

13. 16   8   4
14. 168   84   42

Unit 1
Number Sense
Practice Pages
Weeks 3 and 4

1. Assign the Practice pages, **Workbook,** pages 44–45, at the end of Weeks 3 and 4. Students should complete these pages independently.

2. Check student answers using the annotated pages above.

3. If students have difficulty with a Practice page, review the activities completed throughout the week, and have students complete the weekly practice again before they complete the weekly test.

Number Sense • Weeks 3 and 4   **44–45**

Exploring Patterns

Week at a Glance

This week, students begin *Number Worlds,* Level F, Number Patterns and Relationships. Students will identify and build growing patterns. They will discuss, generalize, and extend those patterns.

Math Background

Mathematicians repeatedly make the point that one of the primary activities in mathematics is to describe patterns: patterns in nature, patterns we invent, even patterns within other patterns. By examining a wide range of patterns, we notice regularity, variety, and the ways topics interconnect. We also see that certain patterns occur again and again. . . . Some patterns can be represented using rules or functions. Other patterns can be represented both numerically and geometrically and help us link arithmetic and geometry.

—Chapin, Suzanne, and Art Johnson. 2000. *Math Matters: Grades K–6. Understanding the Math You Teach.* Math Solutions Publications. Page 125.

How Students Learn

Students need to be able to identify the rule for a pattern in order to continue the pattern. Patterns can be presented orally or by using movement, numbers, pictures, and objects. Students will benefit from exposure to, and practice with, patterns presented in a variety of ways.

Teaching for Understanding

Observe students closely while evaluating the assigned tasks this week to see whether students can demonstrate the following understandings.

Benchmark after Lesson 2: Students can identify growing patterns and extend those patterns.

Benchmark after Lesson 3: Students can identify a missing set from a pattern.

Benchmark after Lesson 4: Students can create growing patterns with Pattern blocks.

Skills Focus

- Identify same-step and changing-step growing patterns
- Extend patterns and find missing sets in patterns
- Use Pattern blocks to represent growing patterns

Math at Home

Give one copy of the Letter to Home, page A5, to each student. Encourage students to share and complete the activity with their caregivers.

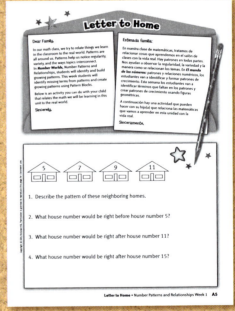

Letter to Home, Teacher Edition, p. A5

Week 1 Planner — Exploring Patterns

PACING	LESSON	LEARNING GOALS	NCTM	MATERIALS	TECHNOLOGY
DAY 1	pages 2–3	Students identify same-step growing patterns and extend those patterns.	• Algebra • Problem Solving • Communication • Representation	Pattern blocks*	**Building Blocks** Pattern Free Explore
DAY 2	pages 4–5	Students identify changing-step growing patterns and extend those patterns.	• Algebra • Problem Solving • Communication • Representation	Pattern blocks*	**MathTools** Shape Tool
DAY 3	pages 6–7	Students identify missing sets from growing patterns.	• Algebra • Problem Solving • Communication • Representation	Pattern blocks*	**Building Blocks** Pattern Free Explore **MathTools** Tessellations
DAY 4	pages 8–9	Students use manipulatives to create growing patterns.	• Algebra • Problem Solving • Communication • Representation	Pattern blocks*	**Building Blocks** Pattern Free Explore **MathTools** Tessellations
DAY 5	**Review and Assess** pages 10–11	Students review skills learned this week and complete the weekly assessment.	• Algebra • Problem Solving • Communication • Representation	Materials will be selected from Lessons 1–4.	**Building Blocks** **MathTools** Review previous activities

Math Vocabulary

same-step pattern A pattern in which the amount of change or growth is the same from each set to the next set

changing-step pattern A pattern in which the amount of growth changes in a regular way from one set to the next

English Learners

SPANISH COGNATES

English	Spanish
illustration	ilustración
patterns	patrones
terms	términos

ALTERNATE VOCABULARY

stack A pile in which flat objects are laid on top of one another

extend To make something longer; to carry it farther in the same pattern

Lesson 1

Objective

Students identify same-step growing patterns and extend those patterns.

Materials

Additional Materials

Pattern blocks

Access Vocabulary

same-step pattern A pattern that changes the same way every step

Creating Context

Have students raise their hands if they can find a pattern in the clothes they are wearing. Ask students to identify and describe the pattern. What other types of patterns can you point to in the classroom? Throughout the day, have students look for patterns and make a list of the patterns they find.

1 Warm Up 5

Concept Building REASONING

- Erin collected seashells on vacation. On Monday she had 5 seashells. On Tuesday she had 10 seashells, and on Wednesday she had 15 seashells. When she counted her seashells on Thursday, she had 20 seashells.
- **Describe the pattern.** Each day she finds 5 more seashells.
- **If the pattern continues, how many seashells will Erin have altogether on Friday?** 25 seashells

2 Engage 30

Skill Building

Use Pattern blocks or centimeter squares cut from paper for **Patterns.** Students should be encouraged to discover patterns on their own and describe them in their own words.

Patterns COMPUTING

Show the following pattern of squares on the board or overhead projector.

- **How many squares are in the first stack?** 3
- **What changes as you go from the first stack to the second stack?** One square is added.
- **What changes as you go from the second stack to the third stack?** One square is added.
- **If the pattern is continued, how many squares will be in the fourth stack? In the fifth stack?** 6 squares; 7 squares
- **This is called a same-step pattern. The pattern grows or changes in the same way every additional step.**

If time allows, create additional same-step growing patterns with the squares, and have students verbally describe the patterns and then extend them.

Monitoring Student Progress

If . . . students are having difficulty extending the square patterns,

Then . . . have them count the number of squares in each set of the pattern without regard to how the squares are arranged.

Building Blocks For additional practice with identifying and extending same-step patterns, students should complete **Building Blocks** Pattern Free Explore.

Using Student Pages

Have students complete **Workbook,** pages 2–3, on their own.

Week 1 — Exploring Patterns

Lesson 1

Key Idea

Same-step patterns are patterns in which the amount of change or growth is the same from one set to the next.

Ask the following questions to help you look for a pattern.

- What is changing from one set to the next?
- What stays the same from one set to the next?
- How would you create the next set?

Try This

Describe what changes in each pattern. Sketch and label the next figure in the pattern.

1

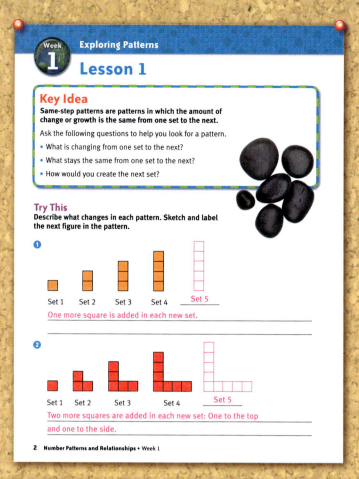

Set 1 Set 2 Set 3 Set 4 Set 5

One more square is added in each new set.

2

Set 1 Set 2 Set 3 Set 4 Set 5

Two more squares are added in each new set: One to the top and one to the side.

Practice

Sketch and label the next two terms in each pattern.

3

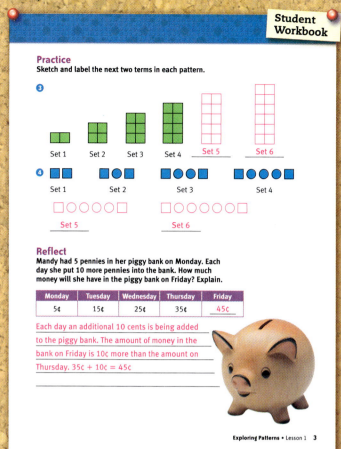

Set 1 Set 2 Set 3 Set 4 Set 5 Set 6

4

Set 1 Set 2 Set 3 Set 4

Set 5 Set 6

Reflect

Mandy had 5 pennies in her piggy bank on Monday. Each day she put 10 more pennies into the bank. How much money will she have in the piggy bank on Friday? Explain.

Monday	Tuesday	Wednesday	Thursday	Friday
5¢	15¢	25¢	35¢	45¢

Each day an additional 10 cents is being added to the piggy bank. The amount of money in the bank on Friday is 10¢ more than the amount on Thursday. 35¢ + 10¢ = 45¢

---

3 Reflect 10

Extended Response REASONING

Review students' answers to the Reflect prompt at the bottom of student page 3.

- **How much money would Mandy have on Saturday?** 55¢
- **How could you model this pattern?** pennies, dimes, Counters

Real-World Application APPLYING

- **Addresses are assigned to houses in a neighborhood according to a same-step growing pattern.**

Show the neighborhood house pattern on the board or overhead projector.

- **What is the pattern in the numbering of the houses?** Each house number increases by 2.
- **What should the next house number be, according to the pattern?** 13

4 Assess

Informal Assessment

Use the Student Assessment Record, *Assessment,* page 100, to record informal observations.

COMPUTING

Patterns

Did the student
- ❏ respond accurately?
- ❏ respond quickly?
- ❏ respond with confidence?
- ❏ self-correct?

Lesson 2

Objective

Students identify changing-step growing patterns and extend those patterns.

Materials

Additional Materials
Pattern blocks

Access Vocabulary

extend To continue the same pattern for another step

changing-step pattern A pattern in which the amount of growth changes in a regular way from one set to the next

Creating Context

A fun game played by children is the game of telephone. Teach English Learners how to play by whispering a funny message in the ear of the first person in a circle, who whispers it to the second person, passing it along to each person until it returns to you. Often the message has significantly changed.

1 Warm Up   5

Concept Building   REASONING

Display the circle pattern on the board or overhead projector.

- **Describe the pattern of circles.** Each set has more circles than the previous set. The change in the number of circles begins with 1 and increases by 1 more with each successive set.

2 Engage   30

Skill Building

Use Pattern blocks or centimeter squares cut from paper for *What Is Changing?* Students should discover patterns on their own and describe them in their own words.

What Is Changing?   COMPUTING

Show the changing square pattern on the board or overhead projector.

- **How many squares are in the first set?** 1
- **What changes as you go from the first set to the second set?** Three squares are added.
- **What changes as you go from the second set to the third set?** Five squares are added.
- **What changes as you go from the third set to the fourth set?** Seven squares are added.
- **Describe the pattern in your own words.** Answers will vary.
- **If the pattern is continued, how many blocks will be in the fifth set?** 25 blocks
- **How did you figure that out?** We first added three squares, then five, and then seven. Nine should be added next, making twenty-five.
- **This is a changing-step pattern. The amount of growth changes in a regular way from one set to the next.**

Monitoring Student Progress

If . . . students have trouble identifying how the pattern changes in a regular way,

Then . . . have them record the pattern numerically: 1, 3, 9, 16.

MathTools Use the Shape Tool to demonstrate and explore patterns.

Using Student Pages

Have students complete **Workbook,** pages 4–5, on their own.

Week 1 — Exploring Patterns

Lesson 2

Key Idea

Changing-step patterns are patterns in which the amount of change or growth changes in a regular and predictable way from one set to the next.

Ask the following questions to help you identify a changing-step pattern:

- What changes from one set to the next?
- What stays the same from one set to the next?
- How would you create the next set?

Try This

Describe the changes in each pattern. Then sketch and label the next set in the pattern.

1

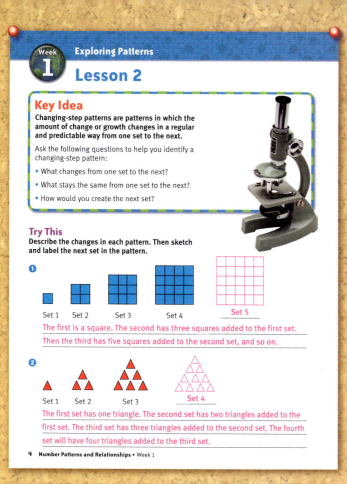

Set 1 Set 2 Set 3 Set 4 Set 5

The first is a square. The second has three squares added to the first set. Then the third has five squares added to the second set, and so on.

2

Set 1 Set 2 Set 3 Set 4

The first set has one triangle. The second set has two triangles added to the first set. The third set has three triangles added to the second set. The fourth set will have four triangles added to the third set.

Practice

Sketch and label the next term in the pattern.

3

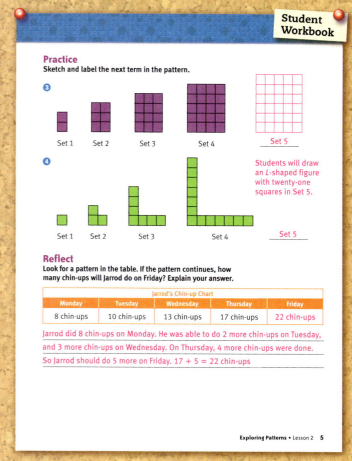

Set 1 Set 2 Set 3 Set 4 Set 5

4

Set 1 Set 2 Set 3 Set 4 Set 5

Students will draw an L-shaped figure with twenty-one squares in Set 5.

Reflect

Look for a pattern in the table. If the pattern continues, how many chin-ups will Jarrod do on Friday? Explain your answer.

Jarrod's Chin-up Chart				
Monday	Tuesday	Wednesday	Thursday	Friday
8 chin-ups	10 chin-ups	13 chin-ups	17 chin-ups	22 chin-ups

Jarrod did 8 chin-ups on Monday. He was able to do 2 more chin-ups on Tuesday, and 3 more chin-ups on Wednesday. On Thursday, 4 more chin-ups were done. So Jarrod should do 5 more on Friday. $17 + 5 = 22$ chin-ups

3 Reflect 10

Extended Response REASONING

Review students' answers to the Reflect prompt at the bottom of student page 5.

- **Describe the pattern.** Answers will vary.
- **How many chin-ups will Jarrod be able to do on Saturday?** 28

Real-World Application APPLYING

A telephone tree is often used to get information quickly to a large group of people. Suppose that one person at the top of the tree calls two people. Then those two people each call two more people beneath them on the tree.

- **How many people are at the top of the tree (set 1)?** 1
- **How many people are at the second level of the tree (set 2)?** 2
- **How many people are at the third level of the tree (set 3)?** 4
- **How many people will be at the fourth level of the tree (set 4)?** 8

4 Assess

Informal Assessment

Use the Student Assessment Record, **Assessment**, page 100, to record informal observations.

COMPUTING

What Is Changing?
Did the student
- ❏ respond accurately?
- ❏ respond quickly?
- ❏ respond with confidence?
- ❏ self-correct?

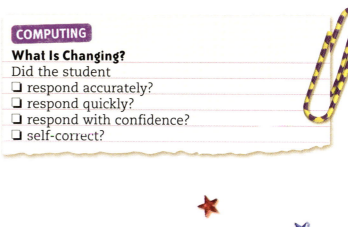

Lesson 3

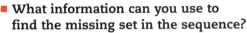

Objective

Students identify missing sets from growing patterns.

Materials

Additional Materials
Pattern blocks

Access Vocabulary

changing-step pattern A pattern in which the amount of growth changes with each step

Creating Context

English Learners may benefit from clarification of some common phrases and words that proficient English speakers probably know. Occasionally words have more than one meaning or are used in potentially puzzling idiomatic expressions. Before or during the lesson, be sure to clarify the words and phrases that may be confusing.

---

② Engage 30

Strategy Building

■ **What information can you use to find the missing set in the sequence?**

Discuss student answers. Guide students to follow this series of questions and steps to determine missing sets:

- How is the pattern growing?
- What changes from one term to the next?
- Is it a same-step or changing-step pattern?
- Sketch the missing set.
- Does it follow the pattern?

The Missing Set **UNDERSTANDING**

Students should work individually or in pairs. They will need Pattern blocks of centimeter squares cut from paper to create growing patterns.

Each pair of students uses squares to create a same-step growing pattern at their desks. After each pair has completed five sets of the pattern, instruct them to remove the fourth set. Rotate the students so that each pair is examining the pattern of another group. The students work together to find the missing set.

Monitoring Student Progress

| **If . . .** students are having difficulty finding the missing set, | **Then . . .** have them write how many shapes are being added in each successive set. |

Building Blocks For additional practice with identifying missing sets from patterns, students should complete **Building Blocks** Pattern Free Explore.

ⓔ MathTools Use Tessellations to demonstrate and explore patterns.

Using Student Pages

Have students complete **Workbook,** pages 6–7, on their own.

---

① Warm Up 5

Concept Building **REASONING**

Display the following pattern on the board or overhead projector. Ask students to figure out how the pattern changes from the first set to the second, and from the fourth set to the fifth.

Set 1   Set 2   Set 3   Set 4   Set 5

■ **How many dots should be in the missing set of the pattern?** 6 dots

Week 1 — Exploring Patterns

Lesson 3

Key Idea
You can use clues from the sets given in a pattern to identify missing sets.

Try This
Look for the changes in each pattern. Then draw and label the missing set.

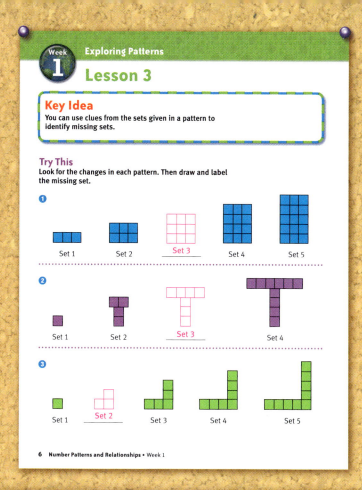

1 Set 1 | Set 2 | Set 3 | Set 4 | Set 5

2 Set 1 | Set 2 | Set 3 | Set 4

3 Set 1 | Set 2 | Set 3 | Set 4 | Set 5

Practice
Look for the changes in each pattern. Tell whether it is a same-step growing pattern or a changing-step growing pattern. Then draw and label the missing set.

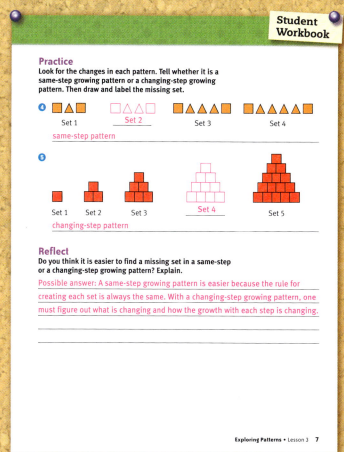

4 Set 1 | Set 2 | Set 3 | Set 4
same-step pattern

5 Set 1 | Set 2 | Set 3 | Set 4 | Set 5
changing-step pattern

Reflect
Do you think it is easier to find a missing set in a same-step or a changing-step growing pattern? Explain.

Possible answer: A same-step growing pattern is easier because the rule for creating each set is always the same. With a changing-step growing pattern, one must figure out what is changing and how the growth with each step is changing.

---

3 Reflect   10

Extended Response   REASONING
Review students' answers to the Reflect prompt at the bottom of student page 7.

■ **What strategies do you use to find missing sets?**
Answers will vary.

Real-World Application   APPLYING

■ **Gina wants to be healthier. She has decided to increase the number of glasses of water she drinks each day. She will begin by drinking 20 ounces of water each day for one week. She expects to take four weeks to get to the final amount she wants to drink each day.**

Create a table on the board with the following data:

Week	1	2	3	4
Ounces of water drank each day	20	28	36	

■ **How many ounces of water does Gina plan to drink each day during week 4?** 44

4 Assess

Informal Assessment
Use the Student Assessment Record **Assessment,** page 100, to record informal observations.

UNDERSTANDING

The Missing Set
Did the student
❏ make important observations?
❏ extend or generalize learning?
❏ provide insightful answers?
❏ pose insightful questions?

Week 1 Exploring Patterns

Lesson 4

Objective

Students use manipulatives to create growing patterns.

Materials

Additional Materials

Pattern blocks

Access Vocabulary

sketch To draw a quick illustration without many details

Creating Context

Not all languages have word contractions. Two of the game choices this week are **What's Missing?** and **What Is Changing?** Write the names of the two games on the board, and point out to English Learners that the contraction *What's* means the same as *What is.* Help students change the name of What Is Changing? by using the contraction. List other phrases, titles of books, and so on that are contractions.

1 Warm Up 5

Concept Building UNDERSTANDING

Explain to students that in this lesson they will have the opportunity to be more creative while building patterns. In addition to Pattern blocks, have available other materials suitable for building patterns. Students may also explore rhythmic patterns.

2 Engage 30

Strategy Building

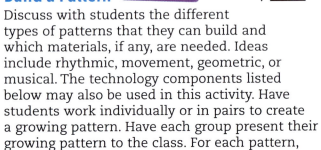

Build a Pattern UNDERSTANDING

Discuss with students the different types of patterns that they can build and which materials, if any, are needed. Ideas include rhythmic, movement, geometric, or musical. The technology components listed below may also be used in this activity. Have students work individually or in pairs to create a growing pattern. Have each group present their growing pattern to the class. For each pattern, discuss with the class the following questions:

- **Is this a same-step or a changing-step growing pattern?** Answers will vary.
- **How is the pattern changing from set to set?** Answers will vary.
- **What will the next set of the pattern look like?** Answers will vary.

Repeat the activity. Have each group of students create a different kind of growing pattern. For example, if the students previously created a same-step pattern, have them create a changing-step pattern this time.

Monitoring Student Progress

| If . . . students are progressing well, | Then . . . challenge them to construct a different growing pattern. |

Building Blocks For additional practice with using manipulatives to create growing patterns, students should complete **Building Blocks** Pattern Free Explore.

eMathTools Use Tessellations to demonstrate and explore patterns.

Using Student Pages

Have students complete **Workbook,** pages 8–9, on their own.

Lesson 4

Key Idea
You can use pattern blocks to create your own growing patterns.

Try This
Draw and label the next set for each pattern. Use the pattern to answer the questions.

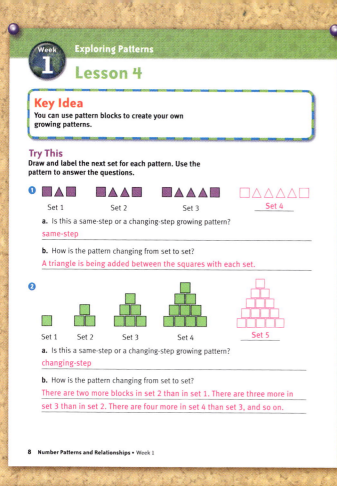

❶

Set 1   Set 2   Set 3   Set 4

a. Is this a same-step or a changing-step growing pattern?
same-step

b. How is the pattern changing from set to set?
A triangle is being added between the squares with each set.

❷

Set 1   Set 2   Set 3   Set 4   Set 5

a. Is this a same-step or a changing-step growing pattern?
changing-step

b. How is the pattern changing from set to set?
There are two more blocks in set 2 than in set 1. There are three more in set 3 than in set 2. There are four more in set 4 than in set 3, and so on.

Practice
Draw your own pattern in the space below. Exchange your pattern with a partner, and have your partner answer each question.

❸ Is this a same-step or a changing-step growing pattern?
Answers will vary.

❹ How is the pattern changing from set to set?
Answers will vary.

❺ What would the next set of the pattern look like? Describe it.
Answers will vary.

Reflect
Use different shapes to create the first few sets of a pattern. Explain how the pattern changes from set to set.

Answers will vary.

---

3 Reflect   10

Extended Response   REASONING

Review students' answers to the Reflect prompt at the bottom of student page 9.

Have students share their centimeter square patterns with the rest of the class. Ask the class to identify the type of pattern and to predict how many squares will be needed to make the next set.

Real-World Application   APPLYING

Manipulatives such as Pattern blocks or centimeter squares can be used to model real-world growing patterns.

■ Janice is trying to improve her miniature golf game. She has a goal to reduce her score three strokes with each game that she plays. Her most recent score was 54. In how many games should she be able to score 45 or less? **3 games**

4 Assess

Informal Assessment

Use the Student Assessment Record, **Assessment,** page 100, to record informal observations.

UNDERSTANDING

Build a Pattern
Did the student
❏ make important observations?
❏ extend or generalize learning?
❏ provide insightful answers?
❏ pose insightful questions?

Lesson 5 Review

Objective

Students review skills learned this week and complete the weekly assessment.

Materials

Review materials will be selected from those used in previous activities.

Creating Context

English Learners may benefit from clarification of some common phrases and words that proficient English speakers probably know. Occasionally words have more than one meaning or are used in potentially puzzling idiomatic expressions. Before or during the lesson, be sure to clarify the words and phrases that may be confusing.

1 Warm Up 5

Concept Building UNDERSTANDING

- **What is a same-step pattern?** a pattern in which the amount of change or growth is the same from each set to the next set
- **What is a changing-step pattern?** a pattern in which the amount of growth changes in a regular way from one set to the next

2 Engage 20

Skill Building

Free-Choice Activity ENGAGING

For the last day of the week, allow students to choose an activity from the previous lessons. Some activities they may choose include the following:

- Patterns
- What Is Changing?
- The Missing Set
- Build a Pattern

Make a note of the activities students select. Do they prefer easy or challenging activities? If you believe your students would benefit from extra practice on specific skills, choose an activity for them.

Monitoring Student Progress

If . . . students are identifying patterns incorrectly,

Then . . . have them extend the pattern to include more sets and to test the pattern from each set to the next.

Using Student Pages

Have students complete **Workbook,** pages 10–11, on their own.

3 Reflect 10

Extended Response REASONING

Review students' answers to the Reflect prompts at the bottom of student pages 10–11.

Discuss the answers with the group to reinforce Week 1 concepts.

Lesson 5 Review

This week you explored patterns. You examined same-step patterns and changing-step patterns.

Lesson 1 Sketch and label the next set in the pattern.

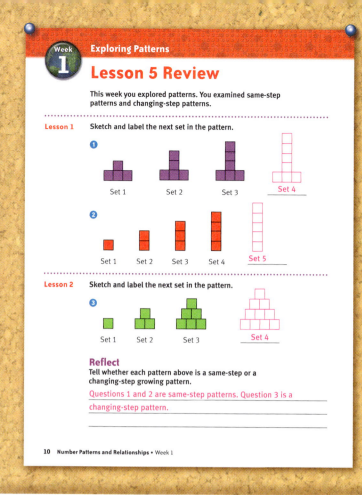

Lesson 2 Sketch and label the next set in the pattern.

Reflect
Tell whether each pattern above is a same-step or a changing-step growing pattern.

Questions 1 and 2 are same-step patterns. Question 3 is a changing-step pattern.

Lesson 3 Sketch and label the missing set in the pattern.

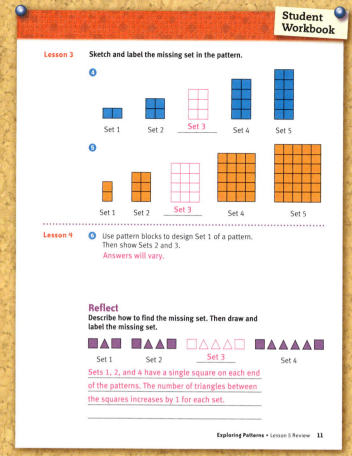

Lesson 4
⑥ Use pattern blocks to design Set 1 of a pattern. Then show Sets 2 and 3.
Answers will vary.

Reflect
Describe how to find the missing set. Then draw and label the missing set.

Sets 1, 2, and 4 have a single square on each end of the patterns. The number of triangles between the squares increases by 1 for each set.

4 Assess   10

A Gather Evidence

Formal Assessment
Have students complete the weekly test, **Assessment,** pages 32–33. Record progress on the Student Assessment Record, **Assessment,** page 100.

B Summarize Findings
Determine whether students have Minimal, Basic, or Secure understanding of the concepts presented in Week 1.

C Differentiate Instruction
Based on your observations, use these teaching strategies next week to follow up.

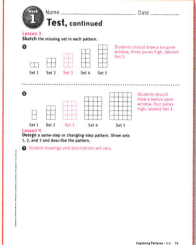

Assessment, pp. 32–33

Minimal Understanding
- Repeat the Warm-Up and Engage activities to develop Week 1 concepts.
- Use **Building Blocks** computer activities and **eMathTools** to develop and reinforce this week's concepts.

Basic Understanding
- Repeat Engage activities in subsequent weeks.
- Use **Building Blocks** computer activities and **eMathTools** to reinforce Week 1 concepts.

Secure Understanding
- Use **Building Blocks** activities and **eMathTools.**
- Use variations of the weekly Warm-Up and Engage activities, using higher numbers and multiple steps.

Exploring Patterns • Lesson 5 Review   10–11

Week 2 Patterns and Relationships

Week at a Glance

This week, students continue with **Number Worlds,** Level F, Number Patterns and Relationships, by creating tables of values to represent the growth in given patterns. Students will use input/output tables to explore patterns and to make conjectures about those patterns.

Math Background

Learning algebra is an important milestone in a student's mathematical development. It opens the door to organized abstract thinking and supplies a tool for logical reasoning. It gives the student the satisfaction of finding simplicity in what appears to be complex and finding generality in a collection of particulars . . . For this reason, the language of algebra is the standard medium for precise communication about numbers and functions . . . It is important for students to master this new language.

—Kaye, Stacey and Mollie MacGregor. February, 1997. *Building Foundations for Algebra.* Mathematics Teaching for the Middle School, NCTM Volume 2.

How Students Learn

Studying patterns and functions is about studying relationships and making connections to sets of data. Many elementary students learn about functions by using an imaginary device called an input/output machine or a function machine. The machine accepts one number and applies a specific rule to produce a related number. Function machines are an effective way to apply pattern rules and to generalize rules for data sets.

Teaching for Understanding

Observe closely while evaluating the assigned tasks this week to see whether students can demonstrate the following understandings.

Benchmark after Lesson 2: Students can use tables of values to represent the growth of same-step and changing-step patterns.

Benchmark after Lesson 3: Students can use input/output tables to create models of real-world patterns.

Benchmark after Lesson 4: Students can use input/output tables to make decisions.

Skills Focus

- Represent the growth of patterns by creating tables of values
- Use input/output tables to create models of real-world patterns
- Use input/output tables to make decisions about real-life scenarios

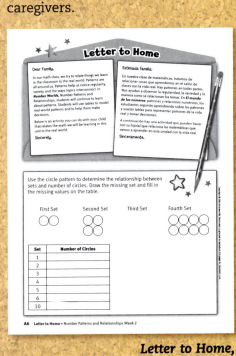

Letter to Home, Teacher Edition, p. A6

Math at Home

Give one copy of the Letter to Home, page A6, to each student. Encourage students to share and complete the activity with their caregivers.

PACING	LESSON	LEARNING GOALS	NCTM	MATERIALS	TECHNOLOGY
DAY 1	pages 12–13	Students create tables of values to represent the growth of same-step patterns.	• Algebra • Problem Solving • Communication • Representation	Pattern blocks*	**MathTools** Shape Tool
DAY 2	pages 14–15	Students create tables of values to represent the growth of changing-step patterns.	• Algebra • Problem Solving • Communication • Representation	Pattern blocks*	**MathTools** Spreadsheet
DAY 3	pages 16–17	Students use input/output tables to create models of real-world patterns.	• Algebra • Problem Solving • Communication • Connections	Straightedge	**Building Blocks** Function Machine **MathTools** Function Machine
DAY 4	pages 18–19	Students use input/output tables to make decisions about real-life scenarios.	• Algebra • Problem Solving • Communication • Connections	Straightedge	**Building Blocks** Function Machine **MathTools** Function Machine
DAY 5	**Review and Assess** pages 20–21	Students review skills learned this week and complete the weekly assessment.	• Algebra • Problem Solving • Communication • Connections • Representation	Materials will be selected from Lessons 1–4.	**Building Blocks** **MathTools** Review previous activities

Math Vocabulary

same-step pattern A pattern in which the amount of change or growth is the same from each set to the next set

changing-step pattern A pattern in which the amount of growth changes regularly from one set to the next

input/output table A table that has a unique output value for each input value

English Learners

SPANISH COGNATES

English	Spanish
figure	figura
function	funciones
pattern	patrón
sequence	secuencia
table	tabla

ALTERNATE VOCABULARY

table A graphic organizer used to display data

* Available from SRA

Lesson 1

Objective

Students create tables of values to represent the growth of same-step patterns.

Materials

Additional Materials

Pattern blocks

Access Vocabulary

value What something is worth; how much someone would pay for something

Creating Context

English Learners may find it confusing that in mathematics we talk about tables and sets, but at home we may set the table. Many words have alternate meanings that are commonly used in daily life. Encourage students to always ask about words they do not understand.

1 Warm Up 5

Concept Building REASONING

Display the Same-Step Squares on the board or overhead projector.

  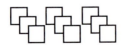

- **How many squares are in the first set? The second set? The third set?** 3, 6, 9
- **Can you think of a rule that describes this pattern?** Each set has three more squares than the set before it.

2 Engage 30

Skill Building

- **One way to represent this pattern is with the squares; another way to represent the same pattern is with a table.**

Display the table on the board or overhead projector.

Set	Number of Squares
1	3
2	6
3	9

Instruct students to look at the numbers in the second column and determine the relationship from one row to the next row. The relationship of these numbers establishes the rule.

Make a Table APPLYING

Students should work in groups to create a same-step pattern with Pattern blocks or squares. Have the groups describe their patterns to the rest of the class and discuss the following:

- **How many blocks are used in each set of the pattern?** Answers will vary.
- **How can you describe the pattern? What rule was used?** Answers will vary.

Work with the class to create on the board a table of values for each pattern.

- **Use the table to extend the pattern. How many blocks will be needed to make the next set of the pattern?** Answers will vary.

Monitoring Student Progress

If . . . students have difficulty using the table to extend a pattern,

Then . . . have them write the rule for the pattern between each term. For example, write "+2" or "+3" between each two terms according to the pattern.

MathTools Use the Shape Tool to demonstrate and explore patterns.

Using Student Pages

Have students complete **Workbook,** pages 12–13, on their own.

Lesson 1

Key Idea
You can use numbers to represent the growth in patterns.

Try This
Use the growing pattern below to answer each question.

❶ Use the pattern to complete the table below.

Set	Number of Squares
1	1
2	2
3	3
4	4
5	5
6	6

❷ What patterns do you notice in the table between the set number and the number of squares?
The set number is the same as the number of squares.

❸ Use the pattern from Problem 2 to predict the number of squares in the tenth set without building or drawing all the sets up to the tenth.
10

Practice
Use the pattern to complete the table.

❹

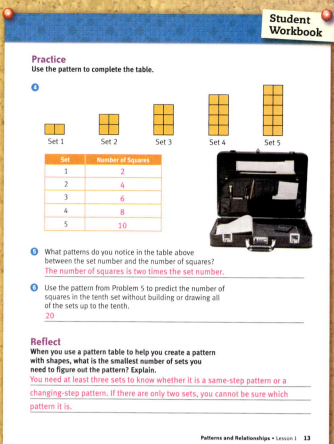

Set	Number of Squares
1	2
2	4
3	6
4	8
5	10

❺ What patterns do you notice in the table above between the set number and the number of squares?
The number of squares is two times the set number.

❻ Use the pattern from Problem 5 to predict the number of squares in the tenth set without building or drawing all of the sets up to the tenth.
20

Reflect
When you use a pattern table to help you create a pattern with shapes, what is the smallest number of sets you need to figure out the pattern? Explain.
You need at least three sets to know whether it is a same-step pattern or a changing-step pattern. If there are only two sets, you cannot be sure which pattern it is.

③ Reflect 10 ▶

Extended Response REASONING
Review students' answers to the Reflect prompt at the bottom of student page 13.

■ **How does the third set determine the pattern?**
Answers will vary.

■ **Can you determine a pattern with only two sets? Why?** No; explanations will vary.

Real-World Application APPLYING
Use pattern blocks or centimeter cubes to model the deer population of a local county. Let each block represent 50 deer. Build stacks of blocks to represent 300, 350, 400, 450, and 500 deer (6, 7, 8, 9, and 10 blocks) over the past 5 decades. Instruct students to create a table of values to model the deer population.

■ **If the pattern continues, how many deer will there be in the county in the next decade?**
550 deer

④ Assess

Informal Assessment
Use the Student Assessment Record, **Assessment**, page 100, to record informal observations.

APPLYING

Make a Table
Did the student
❏ apply learning to new situations?
❏ contribute concepts?
❏ contribute answers?
❏ connect mathematics to the real world?

Patterns and Relationships • Lesson 1 12–13

Lesson 2

Objective

Students create tables of values to represent the growth of changing-step patterns.

Materials

Additional Materials
Pattern blocks

Access Vocabulary

changing-step pattern A pattern in which the amount of growth changes in a regular way from one term to the next

sequence A specific order of numbers

Creating Context

In English we add the suffix *-ship* to the end of words such as *friend*, creating the word *friendship*. *Friendship* means "the condition or relation of being friends." Other examples are *relation*, *relationship*; *author*, *authorship*; *citizen*, *citizenship*; and *owner*, *ownership*. Help English Learners create new words with these roots: *leader*, *sponsor*, *musician*, and *workman*.

1 Warm Up 5

Concept Building COMPUTING

Use Pattern blocks or centimeter squares cut from paper to show a changing-step growing pattern to the class. Ask students to identify how many blocks are used in each set of the pattern. Review **Make a Table** from Lesson 1 to represent the growing pattern.

2 Engage 30

Skill Building

Students should have Pattern blocks or centimeter squares cut from paper.

Represent the Pattern UNDERSTANDING

- **Create a changing-step pattern with the Pattern blocks or squares.**
- **Select several students to describe their patterns to the class.**

For each pattern, discuss the following:

- **How many blocks are used in each set of the pattern?** Answers will vary.
- **How would you describe the pattern? What rule was used?** Answers will vary.

Work with the class to create on the board a table of values for each pattern.

- **According to the table, how many blocks will be needed to make the next set of the pattern?** Answers will vary.

Monitoring Student Progress

If . . . students have difficulty visualizing the described patterns,

Then . . . have students draw their patterns on the board.

ⓔ MathTools Use the Spreadsheet tool to demonstrate and explore data tables.

Using Student Pages

Have students complete **Workbook,** pages 14–15, on their own.

Patterns and Relationships
Lesson 2

Key Idea
You can use numbers to represent the growth in patterns.

Try This
Use the growing pattern below to answer each question.

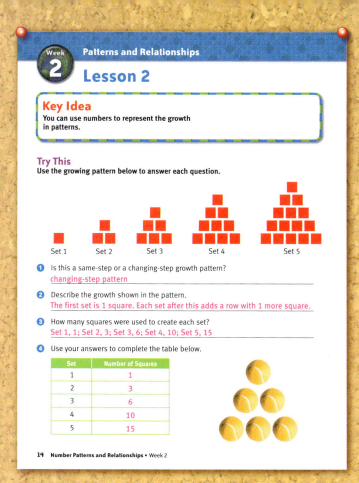

Set 1 Set 2 Set 3 Set 4 Set 5

1 Is this a same-step or a changing-step growth pattern?
changing-step pattern

2 Describe the growth shown in the pattern.
The first set is 1 square. Each set after this adds a row with 1 more square.

3 How many squares were used to create each set?
Set 1, 1; Set 2, 3; Set 3, 6; Set 4, 10; Set 5, 15

4 Use your answers to complete the table below.

Set	Number of Squares
1	1
2	3
3	6
4	10
5	15

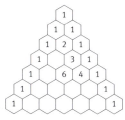

Practice
Use the pattern to complete the table.

5

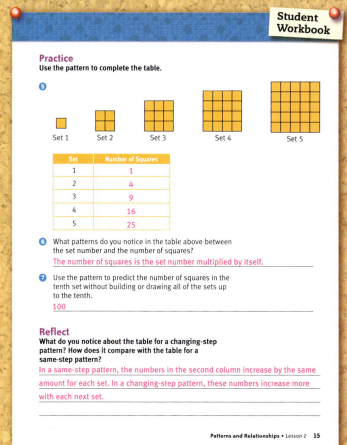

Set 1   Set 2   Set 3   Set 4   Set 5

Set	Number of Squares
1	1
2	4
3	9
4	16
5	25

6 What patterns do you notice in the table above between the set number and the number of squares?
The number of squares is the set number multiplied by itself.

7 Use the pattern to predict the number of squares in the tenth set without building or drawing all of the sets up to the tenth.
100

Reflect
What do you notice about the table for a changing-step pattern? How does it compare with the table for a same-step pattern?
In a same-step pattern, the numbers in the second column increase by the same amount for each set. In a changing-step pattern, these numbers increase more with each next set.

---

3 Reflect 10

Extended Response REASONING

Review students' answers to the Reflect prompt at the bottom of student page 15.

Write the table for a same-step pattern on the board and the table for a changing-step pattern. Discuss with students the differences between the two tables.

Real-World Application APPLYING

Display Pascal's Triangle on the board or overhead projector.

■ **Pascal's Triangle is a triangle built of numbers. This triangle is commonly used in the study of algebra and probability. Look for the pattern so that you can fill in the blank hexagons.** fourth row: 3; fifth row: 4; sixth row: 5, 10, 10, 5; seventh row: 6, 15, 20, 15, 6

---

4 Assess

Informal Assessment

Use the Student Assessment Record, **Assessment,** page 100, to record informal observations.

UNDERSTANDING

Represent the Pattern
Did the student
❏ make important observations?
❏ extend or generalize learning?
❏ provide insightful answers?
❏ pose insightful questions?

Patterns and Relationships • Lesson 2   14–15

Lesson 3

Objective

Students use input/output tables to create models of real-world patterns.

Materials

Additional Materials
Straightedge

Access Vocabulary

input/output table A table that has a certain output value for each input value
input The number put into a function machine
output The number that comes out of a function machine

Creating Context

In this lesson, we see a table that describes how much money Thomas earns from mowing lawns. Discuss with students that *mow* means "to cut grass" and the word *lawn* means "an expanse of grass in the yard of a house." In the United States, most lawns are mowed with a power mower or push mower. In some places outside the United States, grass is cut with a scythe.

1 Warm Up   5

Concept Building   UNDERSTANDING

- Input/output tables are generated by using a rule. Many real-world patterns can be modeled with input/output tables. Suppose that each large pizza at a restaurant has 8 slices.
- This pattern can be modeled by an input/output table in which the input values represent the number of pizzas and the output values represent the number of slices.

Draw a blank table on the board, and fill the cells with the above information.

2 Engage   30

Skill Building

Cereal Servings   UNDERSTANDING

Display the Cereal Servings input/output table on the board or overhead projector.

Cereal Servings	
Input (servings consumed)	**Output (servings left in box)**
0	10
1	9
2	8
3	7
4	6

Use this input/output table to answer the following questions.

- **What is the rule shown in the table?** For each serving consumed, there is one less in the box.
- **How much cereal is left after 5 servings have been consumed?** 5
- **How is this table different from the pizza-slice table we did earlier?** In this table, the output decreases when the input increases.

Monitoring Student Progress

If . . . students are struggling to extend the table in **Cereal Servings,**	**Then . . .** have students write the amount of change between each pair of output values.

 For additional practice using input/output tables, students should complete **Building Blocks** Function Machine.

MathTools Use Function Machine to demonstrate and explore input/output tables.

Using Student Pages

Have students complete *Workbook,* pages 16–17, on their own.

Week 2 — Patterns and Relationships
Lesson 3

Key Idea
The pattern tables are examples of input/output tables.

For each input value (the set number), there is a certain output value (number of squares).

Input/output tables can also be used to model real-world situations.

Try This
Thomas mows lawns in his neighborhood to earn money. He earns $8 for each lawn he mows. Use the input/output table to answer each question.

Input (lawns mowed)	Output (money earned)
1	$8
2	$16
3	$24
4	$32
5	$40

1 How much money would Thomas earn if he mowed 6 lawns? Explain how you found your answer.
Thomas earns $8 for each lawn that he mows. The last row of the table shows earnings of $40 for mowing 5 lawns. If he mows another lawn, he will earn $40 + $8 = $48.

2 Write a mathematical rule for determining the amount of money Thomas will earn for any given number of lawns mowed.
number of lawns mowed × 8 = $ earned

Practice
Complete the table and answer each question.

3 Movie tickets cost $6 each. Complete the input/output table.

Input (number of tickets)	Output (total cost)
1	$6
2	$12
3	$18
4	$24
5	$30

4 How much would it cost to buy 4 movie tickets?
$24

5 Write a mathematical rule for determining the total cost of tickets for any given number of tickets.
number of tickets × 6 = $ total cost

6 For every hour Lisa drives, she uses 2 gallons of gasoline. Her gas tank holds 18 gallons when it is full. Complete the input/output table.

Input (hours of driving)	Output (gas remaining in her tank)
1	16 gallons
2	14 gallons
3	12 gallons
4	10 gallons
5	8 gallons

7 How much gasoline is in Lisa's tank after 5 hours of driving?
8 gallons

8 Write a mathematical rule for determining the amount of gas remaining for any given number of hours driven.
16 − (number of hours driven × 2) = gas remaining

Reflect
What is different about the lawn mowing input/output table and the gasoline input/output table?
In the lawn mowing table, the numbers in the output column are increasing.
In the gasoline table, these numbers are decreasing.

3 Reflect   10

Extended Response   REASONING
Review students' answers to the Reflect prompt at the bottom of student page 17.

- **What are some other examples of growing patterns and shrinking patterns that can be modeled by input/output tables?** Possible answers: growing—distance traveled in a car after so many hours; shrinking—number of sweaters left in a store during a big sale

Real-World Application   APPLYING
- **The world's fastest marine animal is the blue whale. A blue whale can swim at a speed of 30 miles per hour.**
- **Let's create an input/output table for how far a blue whale can swim in different lengths of time, up to 5 hours.**
- **What are the input values?** the time in hours
- **What are the output values?** the distance the whale swam

Have students create the table.

4 Assess

Informal Assessment
Use the Student Assessment Record, *Assessment,* page 100, to record informal observations.

UNDERSTANDING

Cereal Servings
Did the student
- ❏ record important observations?
- ❏ extend or generalize learning?
- ❏ provide insightful answers?
- ❏ pose insightful questions?

Patterns and Relationships • Lesson 3   16–17

Lesson 4

Objective

Students use input/output tables to make decisions about real-life scenarios.

Materials

Additional Materials
Straightedge

Access Vocabulary

input/output table A table that has a certain output value for each input value

pet sit To care for someone's pet while they are away

Creating Context

An excellent strategy for English Learners to use to figure out the meaning of new words is to recognize compound words. These are large words made up of two smaller words that retain their meaning. Work with English Learners to define these examples in this lesson: *input, output, bookstore, bathtub,* and *same-step pattern.*

1 Warm Up   5

Concept Building   UNDERSTANDING

- **When you need to buy a product or service, it is often a good idea to shop around first to find the best deal.**

Discuss with students the concept of renting items, such as a car or an appliance. Explain that sometimes they can use input/output tables to help them find the least expensive option.

2 Engage   30

Strategy Building

Students will need paper, pencils, and straightedges. Write the important details on the board as you explain the scenario.

Festival Planning   UNDERSTANDING

- Imagine that our school wants to have a festival next year. There are two companies that can provide the rides for us, but each charges different prices. Carl's Clowns charges a $500 set-up fee and $100 per ride. Friendly Festivals charges $200 per ride with no set-up fee.
- Make an input/output table for each rental company.
- What should the input and output values be? the number of rides and the total cost

Allow students time to complete tables.

- What is the total cost for renting 3 rides from Carl's Clowns? $800
- What is the total cost for renting 3 rides from Friendly Festivals? $600
- For 3 rides, which company should we choose? Friendly Festivals
- If we want to rent 8 rides, should we still choose Friendly Festivals? Explain your reasoning. No; for 8 rides, Carl's charges $1,300 and Friendly's charges $1,600.

Monitoring Student Progress

| If . . . students are having difficulty creating the input/output tables, | Then . . . help them get started by setting up the table with the appropriate labels and columns. |

 Building Blocks For additional practice using input/output tables to make decisions, students should complete **Building Blocks** Function Machine.

 MathTools Use Function Machine to demonstrate and explore input/output tables.

Using Student Pages

Have students complete **Workbook,** pages 18–19, on their own.

Lesson 4

Key Idea
You can use input/output tables to help you make choices.

Try This
Anna's neighbors have hired her to pet sit their dog for seven days. They have offered two different options for being paid.

- **Option 1:** Anna receives $10 for the first day and an additional $2 per day after the first day.
- **Option 2:** Anna receives $1 for the first day. Every day after the first day she receives an additional amount that is $1 more than the previous day.

❶ Complete the input/output tables for each option.

Option 1		Option 2	
Day	**Total Amount Earned**	**Day**	**Total Amount Earned**
1	$10	1	$1
2	$12	2	$3
3	$14	3	$6
4	$16	4	$10
5	$18	5	$15
6	$20	6	$21
7	$22	7	$28

❷ If Anna chooses Option 1, how much money will she be paid? If Anna chooses Option 2, how much money will she be paid?
$22; $28

Practice
Jim was hired to do yard work for his neighbor. The neighbor expects the work to last 5 days, but it could last 7 days. Payment options are as follows:

- **Option 1:** Jim receives $12 for the first day and $2 per day after the first day.
- **Option 2:** Jim receives $2 for the first two days. Every day thereafter he receives an amount that is $1 more than the previous day.

Complete the tables to find Jim's total earnings for the week for each option.

❸ Complete the table for each option.

Option 1			Option 2		
Day	**Amount Earned for the Day**	**Total Earnings**	**Day**	**Amount Earned for the Day**	**Total Earnings**
1	$12	$12	1	$2	$2
2	$2	$14	2	$2	$4
3	$2	$16	3	$3	$7
4	$2	$18	4	$4	$11
5	$2	$20	5	$5	$16
6	$2	$22	6	$6	$22
7	$2	$24	7	$7	$29

❹ How much will Jim earn for 7 days if he chooses Option 1? $24
❺ How much will Jim earn for 7 days if he chooses Option 2? $29
❻ If the job is for only 5 days, which option will pay better? Option 1

Reflect
What kind of growth pattern is shown by Option 1? What kind of growth pattern is represented by Option 2?
Option 1 is a same-step pattern, and Option 2 is a changing-step pattern.

3 Reflect 10

Extended Response REASONING
Review students' answers to the Reflect prompt at the bottom of student page 19.

- **How can you tell what type of pattern a table represents?** Look at the rules.
- **What rules were used to create each input/output table?** Option 1 is +2 for each step. Option 2 is +2, +3, +4, and so on.

Real-World Application APPLYING
Sometimes you have to decide whether you want to rent an item or whether it makes sense to buy the item.

- **Suppose that Nathan needs to use a power washer to clean his deck and other areas around his house.**
- **He can rent one for $12 per hour, or he can purchase one for $120.**
- **When does it make sense to purchase the power washer instead of renting it?** Possible answer: when the rental cost is close to or exceeds the purchase cost

4 Assess

Informal Assessment
Use the Student Assessment Record, *Assessment*, page 100, to record informal observations.

UNDERSTANDING
Festival Planning
Did the student
- ❑ record important observations?
- ❑ extend or generalize learning?
- ❑ provide insightful answers?
- ❑ pose insightful questions?

Patterns and Relationships • Lesson 4 18–19

Lesson 5 Review

Objective

Students review skills learned this week and complete the weekly assessment.

Materials

Review materials will be selected from those used in previous activities.

Creating Context

English Learners may benefit from clarification of some common phrases and words that proficient English speakers probably know. Occasionally words have more than one meaning or are used in potentially puzzling idiomatic expressions. Before or during the lesson, be sure to clarify the words and phrases that English Learners may find confusing.

1 Warm Up 5

Concept Building UNDERSTANDING

- **What is an input/output table?** a table that has a certain output value for each input value
- **If you are given the input values and the rule, what can you generate?** output values

2 Engage 20

Skill Building

Free-Choice Activity ENGAGING

For the last day of the week, allow students to choose an activity from the previous lessons. Some activities they may choose include the following:

- **Make a Table**
- **Represent the Pattern**
- **Cereal Servings**
- **Festival Planning**

Make a note of the activities children select. Do they prefer easy or challenging activities? If you believe your students would benefit from extra practice on specific skills, choose an activity for them.

Monitoring Student Progress

| **If . . .** students create sloppy input/output tables, | **Then . . .** provide them with lined or graph paper on which to create a table. |

Using Student Pages

Have students complete **Workbook,** pages 20–21, on their own.

3 Reflect 10

Extended Response REASONING

Review students' answers to the Reflect prompts at the bottom of student pages 20–21.

Discuss the answers with the group to reinforce Week 2 concepts.

Week 2 — Patterns and Relationships

Lesson 5 Review

This week you explored patterns and relationships. You looked at how visual patterns can be related to number patterns. You also learned about input/output tables and solving problems.

Lessons 1 and 2 Complete the table for the pattern shown below.

❶

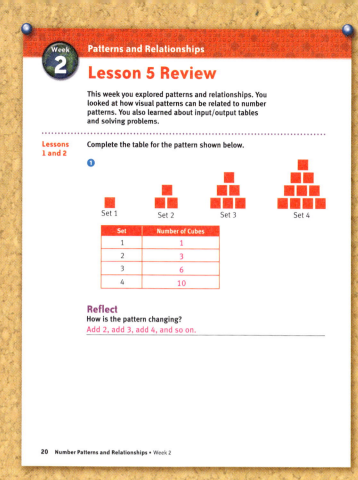

Set	Number of Cubes
1	1
2	3
3	6
4	10

Reflect
How is the pattern changing?
Add 2, add 3, add 4, and so on.

Lesson 3 Complete the input/output table.

❷ The bookstore sells pencils for 15¢ each.

Input (number of pencils)	Output (total cost)
1	15¢
2	30¢
3	45¢
4	60¢

Lesson 4 A bathtub holds 60 gallons of water. When the drain plug is pulled, 12 gallons drain from the tub each minute.

❸ How long does it take for the tub to fully drain?
5 minutes

Input (number of minutes)	Output (water remaining in the tub)
0	60 gallons
1	48 gallons
2	36 gallons
3	24 gallons
4	12 gallons
5	0 gallons

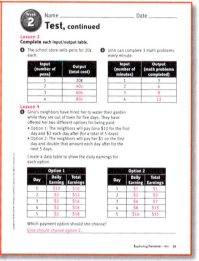

Reflect
How many gallons of water are left in the tub 1 minute after the plug is pulled? How many gallons of water are left in the tub 3 minutes after the plug is pulled? Is this an example of same-step pattern or a changing-step pattern?
48 gallons; 24 gallons; same-step pattern

---

4 Assess 10

A Gather Evidence

Formal Assessment

Have students complete the weekly test, *Assessment*, pages 34–35. Record progress on the Student Assessment Record, *Assessment*, page 100.

B Summarize Findings

Determine whether students have Minimal, Basic, or Secure understanding of the concepts presented in Week 2.

C Differentiate Instruction

Based on your observations, use these teaching strategies next week to follow up.

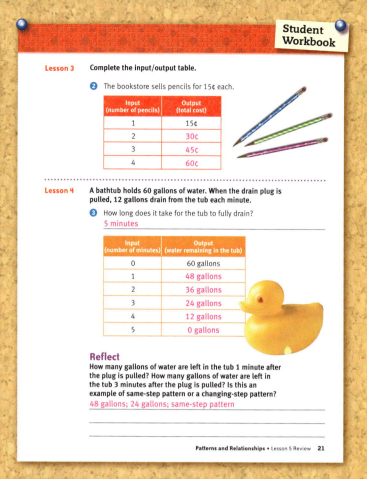

Assessment, pp. 34–35

Minimal Understanding
- Repeat the Warm-Up and Engage activities to develop Week 2 concepts.
- Use *eMathTools* computer activities beginning with Spreadsheet to develop and reinforce this week's concepts.

Basic Understanding
- Repeat Engage activities in subsequent weeks.
- Use *Building Blocks* computer activities beginning with Function Machine to reinforce Week 2 concepts.

Secure Understanding
- Use subsequent *Building Blocks* and *eMathTools* activities.
- Use variations of the weekly Warm-Up and Engage activities, using higher numbers and multiple steps.

Patterns and Graphs

Week at a Glance

This week, students continue with *Number Worlds,* Level F, Number Patterns and Relationships, by investigating how to construct and interpret graphs by using input/output tables. Students will create coordinate line graphs of a given set of values and explore the relationship between two variables.

Math Background

Although *algebra* is a word that until recently was not commonly used in elementary classrooms, the thinking and investigations of many application-based problems and activities include elements of basic algebra. Algebraic reasoning that is often commonplace in elementary classrooms includes describing patterns and representing them in tables, looking for relationships between quantities, making generalizations that work in particular situations, and using graphs to identify a pattern and make predictions.

How Students Learn

Students should analyze the structure of a pattern and how it grows or changes, organize this information systematically, and use their analyses to develop generalizations about the mathematical relationship in the pattern. Students should be encouraged to explain patterns verbally and to make predictions about what will happen if the pattern is continued.

Teaching for Understanding

Observe closely while evaluating the assigned tasks this week to see whether students can demonstrate the following understandings.

Benchmark after Lesson 2: Students can create and interpret coordinate line graphs showing patterns from input/output tables.

Benchmark after Lesson 3: Students can compare the coordinate line graphs of two related patterns.

Benchmark after Lesson 4: Students can match an appropriate story to a graph or create a story for a given graph.

Skills Focus

- Construct and interpret coordinate line graphs
- Compare and contrast the coordinate line graphs of two related patterns
- Match the appropriate story to a given graph
- Create a suitable story for a given graph

Math at Home

Give one copy of the Letter to Home, page A7, to each student. Encourage students to share and complete the activity with their caregivers.

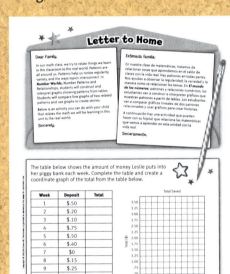

Letter to Home, Teacher Edition, p. A7

PACING	LESSON	LEARNING GOALS	NCTM	MATERIALS	TECHNOLOGY
DAY 1	pages 22–23	Students interpret coordinate line graphs that illustrate a pattern.	• Algebra • Problem Solving • Communication • Representation	No additional materials needed	**e MathTools** Graphing Tool
DAY 2	pages 24–25	Students construct coordinate line graphs from input/output tables.	• Algebra • Problem Solving • Communication • Representation	• Straightedge • Graph paper	**e MathTools** Graphing Tool
DAY 3	pages 26–27	Students compare the coordinate line graphs of two related patterns.	• Algebra • Problem Solving • Communication • Representation	• Straightedge • Graph paper	**e MathTools** Coordinate Grid
DAY 4	pages 28–29	Students match an appropriate story to a graph or create a story for a given graph.	• Algebra • Problem Solving • Communication • Representation	• Payment Options, p. B9 • Straightedge • Graph paper	**Building Blocks** Word Problems with Tools **e MathTools** Coordinate Grid
DAY 5	**Review and Assess** pages 30–31	Students review skills learned this week and complete the weekly assessment.	• Algebra • Problem Solving • Communication • Representation	Materials will be selected from Lessons 1–4.	**Building Blocks** **e MathTools** Review previous activities

Math Vocabulary

input/output table A table that has a certain output value for each input value

coordinate grid A grid used to locate points according to their distances from two number lines that are perpendicular to each other

graph A visual display of a relationship

English Learners

SPANISH COGNATES

English	Spanish
horizontal	horizontal
interpret	interpretar
distance	distancia
patterns	patrones
vertical	vertical

ALTERNATE VOCABULARY

earnings Money paid for a job

match To pair up

Lesson 1

Objective

Students interpret coordinate line graphs that illustrate a pattern.

Materials

No additional materials needed

Access Vocabulary

plot To determine where a line goes on a graph by using two or more points of location

Creating Context

Patterns can be represented by pictures, tables, and graphs. Look at the graphs in this lesson, and have English Learners draw pictures to demonstrate what is represented on these graphs. For English Learners at the beginning level of proficiency, asking them to draw illustrations is an excellent strategy for checking for understanding.

1 Warm Up 5

Concept Building UNDERSTANDING

- **When you are traveling, does the distance from your starting point increase or decrease?** increase

Discuss with students how the total distance traveled increases as the length of time increases.

2 Engage 30

Skill Building

Display the *Distance Traveled* graph on the board or overhead projector.

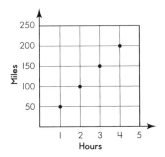

Are We There Yet? UNDERSTANDING

- **This is a line graph. The vertical line on the left side of the graph with the arrow pointing up is called the y-axis. The horizontal line on the bottom of the graph with the arrow pointing to the right is called the x-axis.**
- **Which arrow line, or axis, represents the amount of time spent driving?** the x-axis
- **Which axis represents the number of miles driven?** the y-axis
- **For every hour on the x-axis there is a corresponding data point above it on the grid that relates the time to the distance traveled on the y-axis.**
- **As the amount of time increases, what happens to the total distance traveled?** It also increases.
- **Do you see any patterns in the graph?** For each hour spent driving, you travel 50 miles.

Monitoring Student Progress

| **If . . .** students are unable to describe the pattern in the line graph, | **Then . . .** show how the data points on the graph could be determined from an input/output table. |

MathTools Use the Graphing Tool to demonstrate and explore coordinate line graphs and patterns.

Using Student Pages

Have students complete **Workbook,** pages 22–23, on their own.

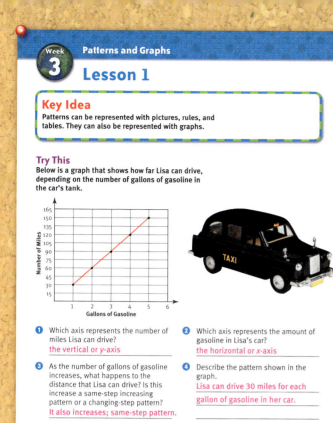

Key Idea
Patterns can be represented with pictures, rules, and tables. They can also be represented with graphs.

Try This
Below is a graph that shows how far Lisa can drive, depending on the number of gallons of gasoline in the car's tank.

① Which axis represents the number of miles Lisa can drive?
the vertical or y-axis

② Which axis represents the amount of gasoline in Lisa's car?
the horizontal or x-axis

③ As the number of gallons of gasoline increases, what happens to the distance that Lisa can drive? Is this increase a same-step increasing pattern or a changing-step pattern?
It also increases; same-step pattern.

④ Describe the pattern shown in the graph.
Lisa can drive 30 miles for each gallon of gasoline in her car.

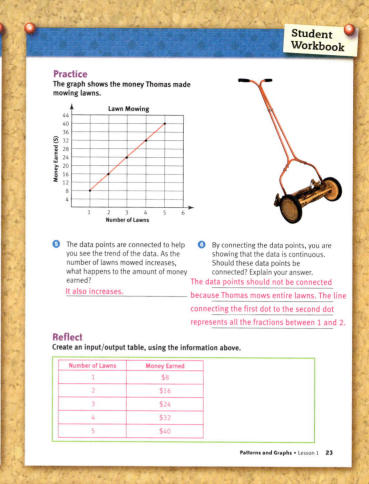

Practice
The graph shows the money Thomas made mowing lawns.

⑤ The data points are connected to help you see the trend of the data. As the number of lawns mowed increases, what happens to the amount of money earned?
It also increases.

⑥ By connecting the data points, you are showing that the data is continuous. Should these data points be connected? Explain your answer.
The data points should not be connected because Thomas mows entire lawns. The line connecting the first dot to the second dot represents all the fractions between 1 and 2.

Reflect
Create an input/output table, using the information above.

Number of Lawns	Money Earned
1	$8
2	$16
3	$24
4	$32
5	$40

Reflect   10

Extended Response   REASONING
Review students' answers to the Reflect prompt at the bottom of student page 23.

- **Was this activity easy or difficult?**
- **How would you explain this to someone who has never done this before?**

Discuss with students the relationship between the paired data points of an input/output table and the plotted points on a coordinate line graph.

Real-World Application   APPLYING

- **Coordinate line graphs can be used by meteorologists to plot the change in the outside temperature. Suppose that the temperature was 68°F at 8:00 this morning and the temperature increased by 2°F each hour. What would the coordinate line graph look like for this situation?**
an increasing pattern that grows 2 degrees per hour

Assess

Informal Assessment
Use the Student Assessment Record, *Assessment,* page 100, to record informal observations.

UNDERSTANDING
Are We There Yet?
Did the student
- ❏ record important observations?
- ❏ extend or generalize learning?
- ❏ provide insightful answers?
- ❏ pose insightful questions?

Lesson 2

Objective

Students construct coordinate line graphs from input/output tables.

Materials

Additional Materials
• Straightedge
• Graph paper

Access Vocabulary

segments Parts of a line

Creating Context

An excellent strategy to use with English Learners is cooperative learning. Working in small groups encourages discussion, and many English Learners need opportunities to practice academic language. Have English Learners work in pairs or small groups to create a story for Problem 1 or 2.

1 Warm Up 5

Concept Building COMPUTING

Create an input/output table on the board with the following paired numbers: (1, 1), (3, 5), (4, 7), and (6, 11). Display a labeled coordinate grid on the board or overhead projector. Show students how you constructed the coordinate grid. Invite student volunteers to come to the board to plot points from the input/output table. Instruct students that the first number corresponds to the *x*-axis and that the second number corresponds to the *y*-axis (*x, y*).

2 Engage 30

Skill Building

Students need a straightedge and graph paper. Assist students in constructing a coordinate grid. Display the Dog Walking input/output table on the board or overhead projector.

Input (dogs walked)	Output (money earned)
1	$5
2	$10
3	$15
4	$20

Walk the Dog APPLYING

■ Enrique walks dogs in his neighborhood after school to earn spending money. The table shows how much he earns for walking different numbers of dogs.

■ Plot points on the coordinate grid, using the data in the input/output table. If you want to see the pattern or trend of the points you plotted, line up a straightedge with the points. You should notice that the points form a line segment.

After students have had time to sketch the graph, pose these questions.

■ Describe the pattern that is shown in the graph. Enrique earns $5 for each dog he walks.

■ How much money would Enrique earn for walking 5 dogs? $25

Monitoring Student Progress

If . . . students are plotting points incorrectly,

Then . . . make sure that they are making the horizontal movement first, using the input value, and then the vertical movement, using the output value.

 MathTools Use the Graphing Tool to demonstrate and explore creating coordinate line graphs from input/output tables.

Using Student Pages

Have students complete *Workbook,* pages 24–25, on their own.

Lesson 2

Key Idea
When creating a graph, be sure to label the axes and give it a title.

Try This
Follow the steps to create a graph of the pattern.

Input (movie tickets)	Output (total cost)
1	$6
2	$12
3	$18
4	$24
5	$30

Step 1 Label the horizontal axis and the vertical axis.

Step 2 Plot a point for each pair of numbers in the table.

Step 3 Give your graph a title.

Movie Ticket Cost

(vertical axis) Total Cost — $6, $12, $18, $24, $30
(horizontal axis) Number of Tickets — 1 2 3 4 5

Practice
Mandy is baby-sitting for her neighbors. Graph the pattern shown in the input/output table.

Input (number of hours)	Output (money earned)
1	$5
2	$10
3	$15
4	$20
5	$25

Baby-sitting Earnings

(vertical axis) Amount Earned — $5, $10, $15, $20, $25
(horizontal axis) Number of Hours — 1 2 3 4 5

1. Describe the pattern shown in the graph.
   Mandy earns $5 for each hour that she spends baby-sitting.

2. How much does Mandy earn for baby-sitting 6 hours?
   $30

Reflect
Can you determine the rule for a pattern by just looking at the graph? Explain and give an example.
Yes; you can look at the graph of Mandy's earnings and see that she earns $5 for each hour that she spends baby-sitting.

---

3 Reflect 10 ▶

Extended Response — REASONING
Review students' answers to the Reflect prompt at the bottom of student page 25.

Discuss with students how you can look at different coordinate line graphs and determine the rule used to create the pattern.

- **How would you explain this to someone who has never done this before?**
- **How can you check your answer?**

Real-World Application — APPLYING
- **Coordinate line graphs can be used to display many different kinds of weather facts. Suppose that it is snowing at an average rate of 2 inches per hour.**

Guide students to set up an input/output table for this pattern. Then use the table to create a coordinate line graph on a coordinate grid. In this case the points can be connected because the data represents an event that is continuous, so something is happening between the points.

4 Assess

Informal Assessment
Use the Student Assessment Record, **Assessment,** page 100, to record informal observations.

APPLYING

Walk the Dog
Did the student
❏ apply learning to new situations?
❏ contribute concepts?
❏ contribute answers?
❏ connect mathematics to the real world?

Lesson 3

Objective

Students compare the coordinate line graphs of two related patterns.

Materials

Additional Materials
- Straightedge
- Graph paper

Access Vocabulary

sold The past tense of *sell*

cafeteria A large kitchen that prepares food to sell to large numbers of people

Creating Context

On some graphs in this lesson, the days of the week are abbreviated. Review with English Learners the meaning of each of the abbreviations. Why do you think Saturday and Sunday are not part of these graphs?

1 Warm Up 5

Concept Building UNDERSTANDING

- **Suppose that you are going to be in charge of a fund-raiser for the fourth-grade class at your school. Your goal is to sell five window stickers each day. How can you track how many stickers you have sold and how many remain in stock?** in two data tables

Discuss with students the idea of managing an inventory of products for a fund-raiser. Include in the discussion the way tables can be used to organize data about sales items.

2 Engage 30

Skill Building

Students will need graph paper and a straightedge. Display the Fund-Raising tables on the board or overhead projector.

Stickers Sold		Stickers Left	
Input (day)	Output (total sold)	Input (day)	Output (total left)
Monday	5	Monday	35
Tuesday	10	Tuesday	30
Wednesday	15	Wednesday	25
Thursday	20	Thursday	20
Friday	25	Friday	15

The Fund-Raiser APPLYING

- **Create two coordinate line graphs, one for each input/output table.**

After students have had time to complete their graphs, ask the following questions:

- **Describe the pattern shown in the Stickers Sold graph.** Five stickers were sold each day of the week.
- **Describe the pattern shown in the Stickers Left graph.** After each day, there are 5 fewer stickers in stock.
- **How are the two graphs related?** The number of stickers that are sold each day determines how many stickers are left in stock.

Monitoring Student Progress

If . . . students understand the concept of inventory and sales,

Then . . . ask them to assess when an order for more stickers should be placed and how many stickers they should order.

 MathTools Use the Coordinate Grid tool to compare coordinate line graphs of related patterns.

Using Student Pages

Have students complete **Workbook,** pages 26–27, on their own.

Week 3 — Patterns and Graphs

Lesson 3

Key Idea
You can use graphs to compare two related patterns.

Try This
Create a graph for each input/output table. Answer each question.

Milk Cartons Sold	
Input (day)	Output (milk sold for the week)
Monday	25 cartons
Tuesday	50 cartons
Wednesday	75 cartons
Thursday	100 cartons
Friday	125 cartons

Milk Cartons Left	
Input (day)	Output (milk cartons left in the cafeteria)
Monday	125 cartons
Tuesday	100 cartons
Wednesday	75 cartons
Thursday	50 cartons
Friday	25 cartons

Milk Cartons Sold

Milk Cartons Left

1 Which arrow line or axis represents the day of the week? Which represents the number of milk cartons left?
The x-axis represents the days of the week. The y-axis represents the number of milk cartons left.

Practice
Use your graphs from Try This to answer each question.

2 Describe the pattern shown in the first graph.
The number of cartons of milk sold increases at a constant rate.

3 Describe the pattern shown in the second graph.
The number of cartons left decreases at a constant rate.

4 Which of the graphs shows a growing pattern?
the first graphs showing the number of cartons sold

5 Are these graphs same-step patterns or changing-step patterns?
same-step patterns

6 What stays the same in the first graph? What changes?
the number of cartons being sold each day; the total number sold

7 What stays the same in the second graph? What changes?
the number of cartons being used each day; the total number left

8 How are the two graphs related?
The number of cartons sold each day determines how many cartons of milk the cafeteria has left.

Reflect
Can you show both patterns on the same graph? Explain.
Yes; you can have two different line graphs on the same set of axes. You can show one pattern with solid line segments and the other pattern with dashed line segments.

---

3 Reflect
10

Extended Response — REASONING

Review students' answers to the Reflect prompt at the bottom of student page 27.

- **How does placing both lines on one graph make it easier to compare the number of sales and the amount of stock remaining?**

If time permits, have students draw both coordinate line graphs from the Fundraising tables on the same coordinate grid.

Real-World Application — APPLYING

- **There are many real-world patterns that involve related growing and shrinking patterns. For example, as the number of people on a bus increases, the number of people waiting at the bus stop decreases.**

Brainstorm with the class to think of other real-world examples of related patterns. For example, as the number of students being served lunch increases, the number of students standing in line decreases.

4 Assess

Informal Assessment

Use the Student Assessment Record, **Assessment,** page 100, to record informal observations.

APPLYING

The Fund-Raiser
Did the student
❏ apply learning to new situations?
❏ contribute concepts?
❏ contribute answers?
❏ connect mathematics to the real world?

Lesson 4

Objective

Students match an appropriate story to a graph or create a story for a given graph.

Materials

Program Materials
Payment Options, p. B9

Additional Materials
- Graph paper
- Straightedge

Access Vocabulary

clog Something blocking the drain
rode Past tense of the word *ride*

Creating Context

Help beginning English Learners create a story for one of the graphs on **Workbook,** page 28, by using story A and story B and by underlining the key details. Work in pairs to replace the key details with new details to match the graph.

1 Warm Up 5

Concept Building

Discuss with students some simple pattern relationships, such as the amount of water in a tub after the plug is pulled or the amount of money earned for working different hours. Talk about what kinds of coordinate line graphs would best model these "stories." Guide students to understand that not all graphs are straight lines. Many graphs are curves or a combination of curves and straight lines.

2 Engage 30

Strategy Building

Display Anna's Options and Jim's Options on the board or overhead projector. Students will need a straightedge and graph paper.

Graph the Options REASONING

- **What do you expect the graph for Anna's Option 1 to look like?** Answers will vary.
- **What do you expect the graph for Anna's Option 2 to look like?** Answers will vary.
- **Use grid paper to create a graph for each of Anna's Options.**
- **Do the graphs look like you expected them to look?** Answers will vary.
- **How can you use the graphs to decide which is the better option?** Answers will vary; students should see that the better option depends on how many days the option is exercised.

Repeat the questions and graphing for Jim's Options.

Monitoring Student Progress

If . . . students are progressing well,	Then . . . tell them a story about how a variable changes with time, such as speed, height, or distance traveled. Then have each group sketch a graph to match the story.

Building Blocks For additional practice with matching appropriate stories to graphs, students should complete **Building Blocks** Word Problems with Tools.

MathTools Use the Coordinate Grid tool to demonstrate and explore coordinate line graphs.

Using Student Pages

Have students complete **Workbook,** pages 28–29, on their own.

Lesson 4

Key Idea
You can use graphs to tell a story or make an informed decision.

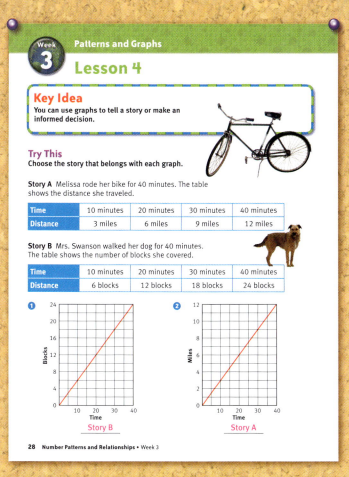

Try This
Choose the story that belongs with each graph.

Story A Melissa rode her bike for 40 minutes. The table shows the distance she traveled.

Time	10 minutes	20 minutes	30 minutes	40 minutes
Distance	3 miles	6 miles	9 miles	12 miles

Story B Mrs. Swanson walked her dog for 40 minutes. The table shows the number of blocks she covered.

Time	10 minutes	20 minutes	30 minutes	40 minutes
Distance	6 blocks	12 blocks	18 blocks	24 blocks

① Story B

② Story A

Practice
Create a graph for the data in the table.

③
Number of Books	1	2	4	6
New Vocabulary Words	2	4	8	12

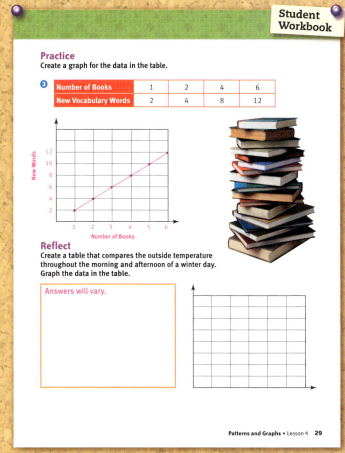

Reflect
Create a table that compares the outside temperature throughout the morning and afternoon of a winter day. Graph the data in the table.

Answers will vary.

3 Reflect 10

Extended Response REASONING
Review students' answers to the Reflect prompt at the bottom of student page 29.

Invite students to come to the board to draw their temperature graphs. Make sure that the axes are labeled and the points plotted correctly.

- **How would you explain this to someone who has never done this before?**
- **How can you check your answer?**

Real-World Application APPLYING
- **Coordinate line graphs can be used to model many real-world situations.**
- **Suppose that the price of a share of stock begins trading at $12 per share at the beginning of the week.**

Draw a line graph on the board showing fluctuations in the price of the stock.

- **Tell the story shown on the line graph, including when the price of the stock rose, fell, and stayed the same.**

4 Assess

Informal Assessment
Use the Student Assessment Record, **Assessment,** page 100, to record informal observations.

REASONING

Graph the Options
Did the student
❏ provide a clear explanation?
❏ communicate reasons and strategies?
❏ choose appropriate strategies?
❏ argue logically?

Patterns and Graphs • Lesson 4 28–29

Lesson 5 Review

Objective

Students review skills learned this week and complete the weekly assessment.

Materials

Review materials will be selected from those used in previous activities.

Creating Context

An excellent strategy for English Learners to use to figure out the meaning of new words is to recognize compound words. These are large words made of two smaller words that retain their meanings. Work with English Learners to define these examples in this lesson; *input, output, sweatshirts,* and *bookstore.*

1 Warm Up 5

Concept Building UNDERSTANDING

■ **What is the difference between same-step patterns and changing-step patterns?**
Same-step patterns change the same way with each step. Changing-step patterns vary the amount of growth from one step to the next.

■ **What type of graph did you make when you used input/output tables this week?** coordinate line graph

2 Engage 20

Skill Building

Free-Choice Activity ENGAGING

For the last day of the week, allow students to choose an activity from the previous lessons. Some activities they may choose include the following:

- **Are We There Yet?**
- **Walk the Dog**
- **The Fund-Raiser**
- **Graph the Options**

Make a note of the activities children select. Do they prefer easy or challenging activities? If you believe your students would benefit from extra practice on specific skills, choose an activity for them.

Monitoring Student Progress

If . . . students choose **Walk the Dog** or **The Bookstore,**

Then . . . have them change the data and the topic so that they can work with other applications of input/output tables.

Using Student Pages

Have students complete **Workbook,** pages 30–31, on their own.

3 Reflect 10

Extended Response REASONING

Review students' answers to the Reflect prompts at the bottom of student pages 30–31.

Discuss the answers with the group to reinforce Week 3 concepts.

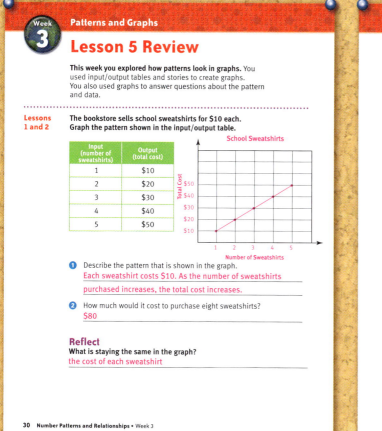

Week 3 — Patterns and Graphs

Lesson 5 Review

This week you explored how patterns look in graphs. You used input/output tables and stories to create graphs. You also used graphs to answer questions about the pattern and data.

Lessons 1 and 2
The bookstore sells school sweatshirts for $10 each. Graph the pattern shown in the input/output table.

Input (number of sweatshirts)	Output (total cost)
1	$10
2	$20
3	$30
4	$40
5	$50

School Sweatshirts (graph)

❶ Describe the pattern that is shown in the graph.
Each sweatshirt costs $10. As the number of sweatshirts purchased increases, the total cost increases.

❷ How much would it cost to purchase eight sweatshirts?
$80

Reflect
What is staying the same in the graph?
the cost of each sweatshirt

30 Number Patterns and Relationships • Week 3

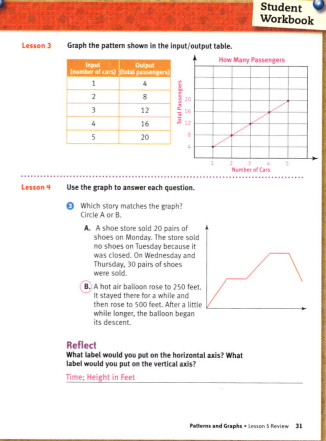

Student Workbook

Lesson 3 Graph the pattern shown in the input/output table.

Input (number of cars)	Output (total passengers)
1	4
2	8
3	12
4	16
5	20

How Many Passengers (graph)

Lesson 4 Use the graph to answer each question.

❸ Which story matches the graph? Circle A or B.

A. A shoe store sold 20 pairs of shoes on Monday. The store sold no shoes on Tuesday because it was closed. On Wednesday and Thursday, 30 pairs of shoes were sold.

(B.) A hot air balloon rose to 250 feet. It stayed there for a while and then rose to 500 feet. After a little while longer, the balloon began its descent.

Reflect
What label would you put on the horizontal axis? What label would you put on the vertical axis?

Time; Height in Feet

Patterns and Graphs • Lesson 5 Review 31

4 Assess 10

A Gather Evidence

Formal Assessment

Have students complete the weekly test on *Assessment,* pages 36–37. Record progress on the Student Assessment Record, *Assessment,* page 100.

B Summarize Findings

Determine whether students have Minimal, Basic, or Secure understanding of the concepts presented in Week 3.

C Differentiate Instruction

Based on your observations, use these teaching strategies next week to follow up.

Assessment, pp. 36–37

Minimal Understanding
- Repeat the Warm-Up and Engage activities to develop Week 3 concepts.
- Use **eMathTools** Graphing Tool to develop and reinforce this week's concepts.

Basic Understanding
- Repeat Engage activities in subsequent weeks.
- Use **eMathTools** Coordinate Grid to reinforce this week's concepts.

Secure Understanding
- Use **Building Blocks** Word Problems with Tools to reinforce this week's concepts.
- Use variations of the weekly Warm-Up and Engage activities, using higher numbers and multiple steps.

Patterns and Graphs • Lesson 5 Review 30–31

Week 4

Variables and Equality

Week at a Glance

This week, students continue with **Number Worlds,** Level F, Number Patterns and Relationships, by exploring variables and equations. Students will find unknown values in equations and use the concept of weight to help them solve problems.

Math Background

Both curriculum developers and teachers should be aware of the need to work in an environment of multiple representations—that is, an environment that allows the representation of a problem and its solution in several ways. Although each representation has its disadvantages, their combined use can cancel out the disadvantages and prove to be an effective tool.

—Friedlander, Alex, and Michal Tabach. 2001. *The Roles of Representations in School Mathematics.* National Council of Teachers of Mathematics. Page 174.

How Students Learn

Visual learners think in terms of pictures and learn best from visual displays such as diagrams, illustrations, videos, and graphic organizers. Visual learners create a vivid mental image that they reference in their mind to retain and recall information. Oftentimes, students have several preferred learning styles, but one style is more dominant. Be aware that many students are predominately visual learners.

Teaching for Understanding

Observe closely while evaluating the assigned tasks this week to see whether students can demonstrate the following understandings.

Benchmark after Lesson 2: Students can find unknown values in equations.

Benchmark after Lesson 3: Students can use the concept of equality to help them solve problems involving weights.

Benchmark after Lesson 4: Students can use the concept of balance scales to help them solve equations.

Skills Focus

- Find unknown values in equations
- Use the concept of equality to solve problems involving weights
- Use the concept of balance scales to solve equations

Math at Home

Give one copy of the Letter to Home, page A8, to each student. Encourage students to share and complete the activity with their caregivers.

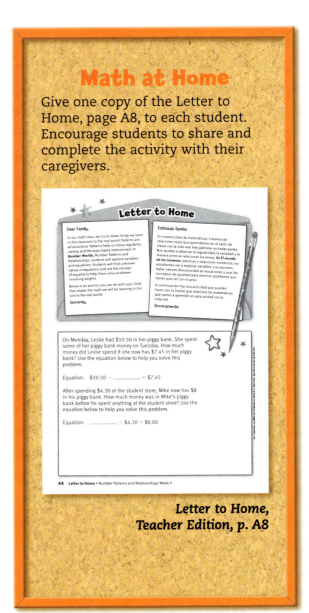

Letter to Home, Teacher Edition, p. A8

Week 4 Planner — Variables and Equality

PACING	LESSON	LEARNING GOALS	NCTM	MATERIALS	TECHNOLOGY
DAY 1	pages 32–33	Students find unknown values in equations.	• Algebra • Problem Solving • Communication • Representation	Counters*	**Building Blocks** Easy as Pie
DAY 2	pages 34–35	Students explore the concept of equality by using weights.	• Algebra • Problem Solving • Communication • Representation	Scale	**Building Blocks** Easy as Pie
DAY 3	pages 36–37	Students solve more challenging problems involving weights, using the concept of equality.	• Algebra • Problem Solving • Communication • Representation	Scale	**Building Blocks** Word Problems with Tools
DAY 4	pages 38–39	Students explore the concept of equality by using balance scales.	• Algebra • Problem Solving • Communication • Representation	• Balance scale • Counters*	**Building Blocks** Eggstremely Equal
DAY 5	**Review and Assess** pages 40–41	Students review skills learned this week and complete the weekly assessment.	• Algebra • Problem Solving • Communication • Representation	Materials will be selected from Lessons 1–4.	**Building Blocks** Review previous activities

Math Vocabulary

equation A number sentence

English Learners

SPANISH COGNATES

English	Spanish
equal	igual
problems	problemas
scales	escalas
values	valores
variables	variables

ALTERNATE VOCABULARY

unknown A missing term from an equation that needs to be filled in to solve the equation

Lesson 1

Objective

Students find unknown values in equations.

Materials

Additional Materials
Counters

Access Vocabulary

equation A number sentence

reel in the kite Wind the string to shorten it, bringing the kite closer to the ground

Creating Context

English Learners may benefit from clarification of some common phrases and words that proficient English speakers probably know. Occasionally words have more than one meaning or are used in potentially puzzling idiomatic expressions. Before or during the lesson, be sure to clarify the words and phrases that may be confusing.

1 Warm Up 5

Concept Building UNDERSTANDING

Discuss with students the concept of equality and equations.

- **What makes two sets of numbers equal?** when they represent the same amount

Write some number sentences on the board, and ask students whether the sentences represent equations.

2 + 3 = 5 yes   7 = 3 + 4 yes
2 + 3 = 1 + 4 yes 5 + 1 = 6 yes
7 = 3 + 5 no   5 + 1 = 6 + 1 no

2 Engage 30

Skill Building

Use a small empty box or container and some Counters for the Skill Building activity.

How Many Counters? COMPUTING

Write a value on a sheet of paper or poster board, and conceal the value by placing the empty box over it. The box will model for students the unknown value in an equation. Use stacks of Counters, and write mathematical symbols on the paper to model the rest of the equation. Then ask students how many Counters need to be placed in the box to balance the equation. The number of Counters should match the value under the box. Use equations similar to those below.

☐ + 4 = 7 3   9 − ☐ = 2 7   ☐ + 2 = 13 11

Monitoring Student Progress

If . . . students do not know how to find the unknown,

Then . . . guide them to understand that each side of the equal sign must contain the same number of Counters to form a true number sentence.

Building Blocks For additional practice finding unknown values in equations, students should complete **Building Blocks** Easy as Pie.

Using Student Pages

Have students complete **Workbook,** pages 32–33, on their own.

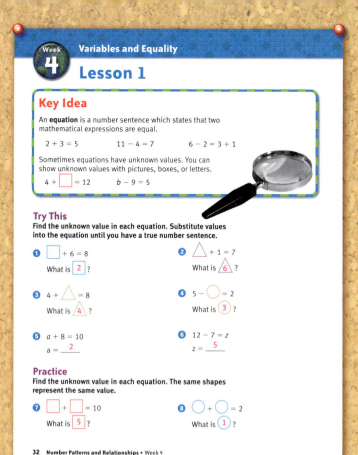
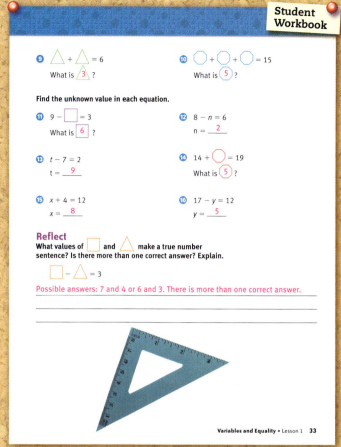

Variables and Equality

Lesson 1

Key Idea

An **equation** is a number sentence which states that two mathematical expressions are equal.

$2 + 3 = 5$ $11 - 4 = 7$ $6 - 2 = 3 + 1$

Sometimes equations have unknown values. You can show unknown values with pictures, boxes, or letters.

$4 + \square = 12$ $b - 9 = 5$

Try This

Find the unknown value in each equation. Substitute values into the equation until you have a true number sentence.

1. $\square + 6 = 8$ What is 2 ?
2. $\triangle + 1 = 7$ What is 6 ?
3. $4 + \triangle = 8$ What is 4 ?
4. $5 - \bigcirc = 2$ What is 3 ?
5. $a + 8 = 10$ $a = $ 2
6. $12 - 7 = z$ $z = $ 5

Practice

Find the unknown value in each equation. The same shapes represent the same value.

7. $\square + \square = 10$ What is 5 ?
8. $\bigcirc + \bigcirc = 2$ What is 1 ?

32 Number Patterns and Relationships • Week 4

9. $\triangle + \triangle = 6$ What is 3 ?
10. $\bigcirc + \bigcirc + \bigcirc = 15$ What is 5 ?

Find the unknown value in each equation.

11. $9 - \square = 3$ What is 6 ?
12. $8 - n = 6$ $n = $ 2
13. $t - 7 = 2$ $t = $ 9
14. $14 + \bigcirc = 19$ What is 5 ?
15. $x + 4 = 12$ $x = $ 8
16. $17 - y = 12$ $y = $ 5

Reflect

What values of $\square$ and $\triangle$ make a true number sentence? Is there more than one correct answer? Explain.

$\square - \triangle = 3$

Possible answers: 7 and 4 or 6 and 3. There is more than one correct answer.

Variables and Equality • Lesson 1   33

---

3 Reflect   10

Extended Response   REASONING

Review students' answers to the Reflect prompt at the bottom of student page 33.

- **Is there more than one way to solve this problem?** Have students share some of their answers to illustrate that there is indeed more than one correct answer.
- **Model another equation that has more than one correct answer.**

Real-World Application   APPLYING

- Equations are very useful tools for modeling and solving real-world problems. Suppose that Jeremy has a kite at the end of 80 feet of string. He begins winding the string to reel in the kite. If he has already pulled in 45 feet of string, how much string is left to wind? Model this situation with an equation, and solve for the unknown value.

  $\square + 45 = 80$ 35 feet $80 - 45 = \square$ 35 feet

4 Assess

Informal Assessment

Use the Student Assessment Record, **Assessment,** page 100, to record informal observations.

COMPUTING

How Many Counters?
Did the student
- ❑ respond accurately?
- ❑ respond quickly?
- ❑ respond with confidence?
- ❑ self-correct?

Lesson 2

Objective

Students explore the concept of equality by using weights.

Materials

Additional Materials

Scale

Access Vocabulary

equality When two numbers or amounts are equal

piggy bank A small ceramic or plastic container in the shape of a pig in which to store change

Creating Context

English Learners may find it confusing to hear homophones such as *way* and *weigh* and know that they have two different meanings and spellings. Practice with English Learners, using context to determine meaning, and encourage them to ask when they are uncertain of the meaning of a word or phrase.

1 Warm Up — 5

Concept Building COMPUTING

■ Suppose that Megan's cat weighs 4 pounds. The cat and her dog weigh 19 pounds altogether. How can you find the weight of Megan's dog?
The dog weighs 15 pounds.

Discuss with students how to set up and solve an equation for this problem.

2 Engage — 30

Skill Building

How Much Does That Weigh? UNDERSTANDING

If you have a scale available, weigh a classroom object such as a textbook. Then weigh the object along with another object, such as a stapler. Have students set up and solve an equation for the weight of the second object, such as the following:

- The book weighs 6 pounds.
- The book and stapler weigh 7 pounds.
- $7 = 6 + \boxed{1}$
- The stapler weighs 1 pound.

Otherwise, you can draw a scale on the board and write the weights of objects on it for this activity. It is important that the numbers be reasonable, but it is not necessary that they be exact weights.

Monitoring Student Progress

If . . . students have a hard time setting up an equation,

Then . . . have them represent the unknown weight by drawing a box. Add to this the known weight, and set the result equal to the known, combined weight.

 Building Blocks For additional practice solving problems involving weights, students should complete **Building Blocks** Easy as Pie.

Using Student Pages

Have students complete **Workbook,** pages 34–35, on their own.

Lesson 2

Key Idea
You can use the idea of weights to help solve equations.

Try This
Answer each question to find the weight of the toy car.

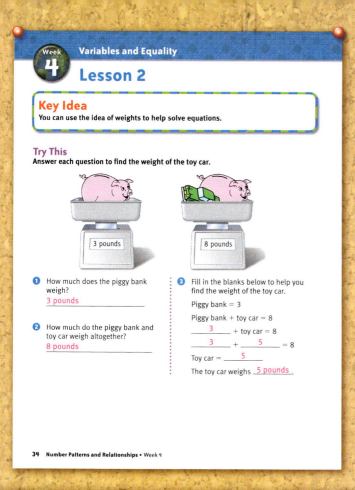

1 How much does the piggy bank weigh?
3 pounds

2 How much do the piggy bank and toy car weigh altogether?
8 pounds

3 Fill in the blanks below to help you find the weight of the toy car.

Piggy bank = 3

Piggy bank + toy car = 8

____3____ + toy car = 8

____3____ + ____5____ = 8

Toy car = ____5____

The toy car weighs __5 pounds__.

Practice
Find each unknown weight. Write a number sentence to show your work.

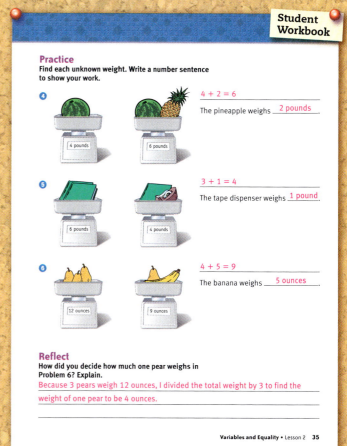

4 4 + 2 = 6
The pineapple weighs __2 pounds__.

5 3 + 1 = 4
The tape dispenser weighs __1 pound__.

6 4 + 5 = 9
The banana weighs __5 ounces__.

Reflect
How did you decide how much one pear weighs in Problem 6? Explain.

Because 3 pears weigh 12 ounces, I divided the total weight by 3 to find the weight of one pear to be 4 ounces.

3 Reflect   10

Extended Response   REASONING

Review students' answers to the Reflect prompt at the bottom of student page 35.

- **How did you solve this problem?**
- **If 2 kittens that are the same size weigh 20 ounces altogether, how much does each kitten weigh?** 10 ounces

Real-World Application   APPLYING

- **Suppose that you purchased a $6 book at the bookstore. While you were there, you also picked out a specialty bookmark. Your final cost was $7.50. How much did the bookmark cost?**
$1.50

4 Assess

Informal Assessment

Use the Student Assessment Record, **Assessment,** page 100, to record informal observations.

UNDERSTANDING

How Much Does That Weigh?
Did the student
- ❏ make important observations?
- ❏ extend or generalize learning?
- ❏ demonstrate understanding?
- ❏ provide insight?

Variables and Equality • Lesson 2   34–35

Lesson 3

Objective

Students solve more challenging problems involving weights, using the concept of equality.

Materials

Additional Materials

Scale

Access Vocabulary

equation A number sentence
scale A device used to determine weight

Creating Context

The pictures in this lesson show a variety of sports equipment. We look at the weight of different types of balls from various sports. Have English Learners point to each picture and name the sport that uses each ball. Say the name, and have students repeat after the teacher.

1 Warm Up 5

Concept Building UNDERSTANDING

■ **Kenneth's calculator weighs 8 ounces. Together his calculator and marker weigh 9 ounces. Kenneth's marker and computer mouse weigh 4 ounces altogether. How much does the computer mouse weigh?**

Discuss with students how to set up and solve two equations for this problem. 8 + marker = 9, so the marker = 1 ounce; 1 + mouse = 4, so the computer mouse weighs 3 ounces.

2 Engage 30

Skill Building

More Weights UNDERSTANDING

If you have a scale available, weigh an object such as an apple. Then weigh the apple together with another object, such as a banana. Finally, weigh the banana and a third object, such as an orange. If a scale is not available, draw a similar situation on the board. Have students set up and solve equations to find the weight of the orange. For example, if the apple weighs 4 ounces, the apple and banana weigh 9 ounces altogether, and the banana and orange weigh 10 ounces altogether, what is the weight of the orange?
Apple = 4, apple + banana = 9, so banana = 5.
Banana + orange = 10, so orange = 5 ounces.

Monitoring Student Progress

If . . . students are having difficulty,

Then . . . explain that they need to break the problem into two parts: solve an equation to find the weight of the banana, and then use that information to find the weight of the orange.

Building Blocks For additional practice solving more challenging problems, students should complete **Building Blocks** Word Problems with Tools.

Using Student Pages

Have students complete **Workbook,** pages 36–37, on their own.

Week 4 — Variables and Equality

Lesson 3

Key Idea
Use reasoning to solve more challenging problems involving weights.

Try This
Answer each question to find the weight of each shape.

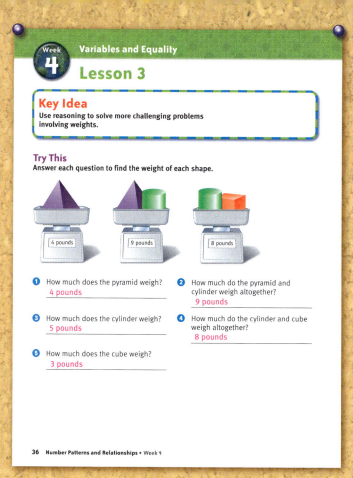

1 How much does the pyramid weigh?
4 pounds

2 How much do the pyramid and cylinder weigh altogether?
9 pounds

3 How much does the cylinder weigh?
5 pounds

4 How much do the cylinder and cube weigh altogether?
8 pounds

5 How much does the cube weigh?
3 pounds

Practice
Find each unknown weight.

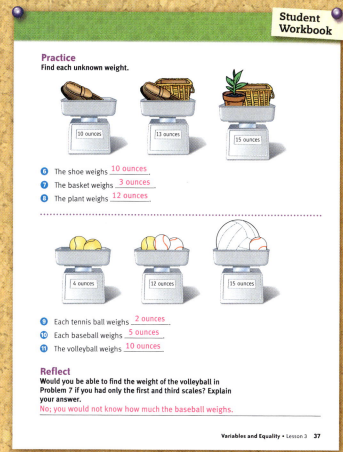

6 The shoe weighs 10 ounces

7 The basket weighs 3 ounces

8 The plant weighs 12 ounces

9 Each tennis ball weighs 2 ounces

10 Each baseball weighs 5 ounces

11 The volleyball weighs 10 ounces

Reflect
Would you be able to find the weight of the volleyball in Problem 7 if you had only the first and third scales? Explain your answer.
No; you would not know how much the baseball weighs.

---

3 Reflect   10

Extended Response   REASONING

Review students' answers to the Reflect prompt at the bottom of student page 37.

- **What was difficult about this activity?**
- **Did this activity help you do anything you could not do before?**

Discuss with students how these more challenging weight problems are two-step problems. All three scales are needed to find the final answer.

Real-World Application   APPLYING

- **Nuts are usually sold by the pound. Suppose that you know that a bag of peanuts weighs 4 pounds, and the bag of peanuts weighed with a bag of walnuts is 6 pounds. The bag of walnuts weighed with a bag of cashews weighs 7 pounds. How much does the bag of cashews weigh?** 5 pounds

4 Assess

Informal Assessment

Use the Student Assessment Record, *Assessment,* page 100, to record informal observations.

UNDERSTANDING

More Weights
Did the student
- ❏ make important observations?
- ❏ extend or generalize learning?
- ❏ demonstrate understanding?
- ❏ provide insight?

Lesson 4

Objective

Students explore the concept of equality by using balance scales.

Materials

Additional Materials
- Balance scale
- Counters

Access Vocabulary

balance scales A scale with two pans that are balanced when both pans are filled with equal weights

Creating Context

An excellent vocabulary strategy for English Learners is to listen and look for words that have the same root. These words are usually related. For example, in this lesson we talk about *equal*, *equality*, and *equivalent*. What is the root they share? What does each word mean? Can you find other words that share this same root?

1 Warm Up   5

Discuss the concepts of equality and balance scales. Use a balance scale for demonstration if one is available.

- **If you put 5 ounces into one side of a balance scale, how many ounces must you put into the other side to have a balanced scale?** 5 ounces

- **If we add one ounce to the left pan of the balance scale, what do we have to do to keep the scale balanced?** Add one ounce to the right pan.

2 Engage   30

Skill Building

Balance the Scale COMPUTING

If you have a balance scale and different objects that you can use to show equivalent weights, obtain those materials for the Skill-Building activity. Otherwise, draw a balance scale on the board. The scale should be balanced with two circles on the left and six squares on the right. Draw a second scale balanced with one circle and one triangle on the left and five squares on the right. Finally draw a third scale balanced with 2 squares on the left and a question mark on the right. Guide students through the process of balancing the third scale with the appropriate shape. 1 triangle

Repeat this activity with new values for the shapes.

Monitoring Student Progress

If . . . students are progressing well,

Then . . . challenge them to work in groups to create their own balance-scale equality problems.

Building Blocks For additional practice exploring equality, students should complete **Building Blocks** Eggstremely Equal.

Using Student Pages

Have students complete **Workbook,** pages 38–39, on their own.

Lesson 4

Key Idea
Balance scales can be used to help solve equations.
When a scale is balanced, both sides are equal.

Try This
Use each balance scale to find two equal weights.

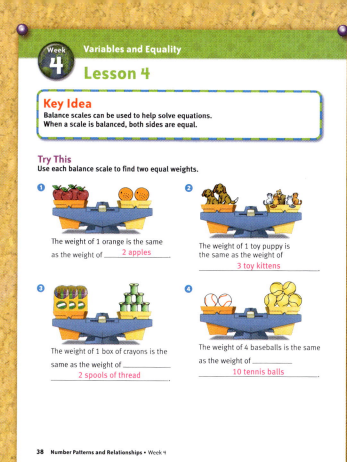

1. The weight of 1 orange is the same as the weight of **2 apples**.

2. The weight of 1 toy puppy is the same as the weight of **3 toy kittens**.

3. The weight of 1 box of crayons is the same as the weight of **2 spools of thread**.

4. The weight of 4 baseballs is the same as the weight of **10 tennis balls**.

Practice
Find each unknown weight. Draw your answer on the scale with the question mark.

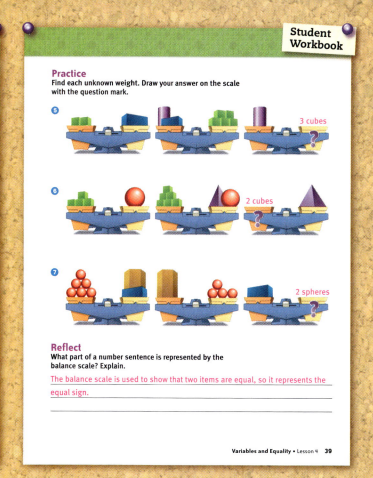

5. **3 cubes**

6. **2 cubes**

7. **2 spheres**

Reflect
What part of a number sentence is represented by the balance scale? Explain.

The balance scale is used to show that two items are equal, so it represents the equal sign.

3 Reflect 10

Extended Response REASONING
Review students' answers to the Reflect prompt at the bottom of student page 39.

- **How is a perfectly balanced scale like an equal sign?** Answers will vary.
- **What symbol would you use if the scale were unbalanced?** less than or greater than symbols

Real-World Application APPLYING
- A teeter-totter, or seesaw, is a popular piece of playground equipment that is a real-world example of a balance scale. If two children who weigh the same amount each sit on a side of the teeter-totter, it will be balanced.
- Suppose that Wendy, who weighs 60 pounds, is sitting on one side of a teeter-totter. Her smaller sister is holding their dog on the other side of the teeter-totter. If Wendy's sister weighs 48 pounds and the teeter-totter is balanced, how much does the dog weigh? 12 pounds

4 Assess

Informal Assessment
Use the Student Assessment Record, *Assessment*, page 100, to record informal observations.

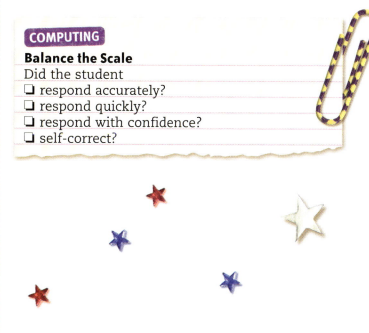

COMPUTING

Balance the Scale
Did the student
- ❏ respond accurately?
- ❏ respond quickly?
- ❏ respond with confidence?
- ❏ self-correct?

Lesson 5 Review

Objective

Students review skills learned this week and complete the weekly assessment.

Materials

Review materials will be selected from those used in previous activities.

Creating Context

Adding a prefix to a word often changes its meaning to the opposite. In this lesson, we have words such as *inequality* and *unknown*. The prefixes *in-* and *un-* both mean "not." Make a two-column chart, collect other words with the prefixes *in-* and *un-,* and have English Learners tell what they mean.

☑ Unit Assessment

Students should complete the Number Patterns and Relationship Test, *Assessment,* pages 74–75. Using the key, *Assessment,* page 98, identify incorrect responses. Reteach and review the Warm-Up and Engage activities to reinforce concept understanding.

① Warm Up 5

Concept Building UNDERSTANDING

■ Which was easier to use to solve equations: the traditional scale or the balanced scale?

■ Use your own words to explain how to solve a problem when you are given three balance scales. Answers will vary.

② Engage 20

Skill Building

Free-Choice Activity ENGAGING

For the last day of the week, allow students to choose an activity from the previous lessons. Some activities they may choose include the following:

- **How Many Cubes?**
- **How Much Does That Weigh?**
- **More Weights**
- **Balance the Scale**

Make a note of the activities children select. Do they prefer easy or challenging activities? If you believe your students would benefit from extra practice on specific skills, choose an activity for them.

Monitoring Student Progress

| If . . . students finish early, | Then . . . allow time for students to select three objects from the classroom that they can use to create their own balance problems. They can trade problems for other students to solve. |

Using Student Pages

Have students complete **Workbook,** pages 40–41, on their own.

③ Reflect 10

Extended Response REASONING

Review students' answers to the Reflect prompts at the bottom of student pages 40–41.

Discuss the answers with the group to reinforce Week 4 concepts.

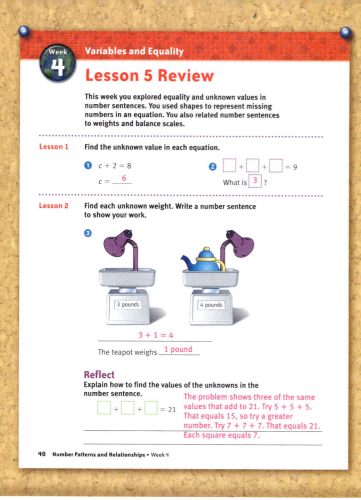

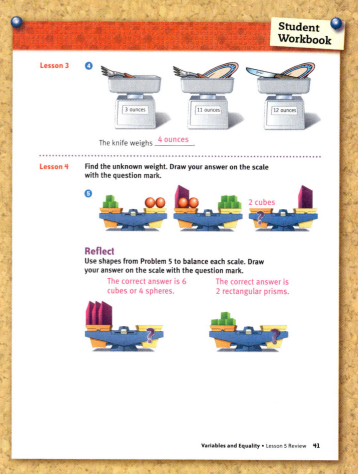

Week 4 — Variables and Equality

Lesson 5 Review

This week you explored equality and unknown values in number sentences. You used shapes to represent missing numbers in an equation. You also related number sentences to weights and balance scales.

Lesson 1 Find the unknown value in each equation.

① $c + 2 = 8$
$c =$ __6__

② ☐ + ☐ + ☐ = 9
What is ☐ 3 ?

Lesson 2 Find each unknown weight. Write a number sentence to show your work.

③

3 pounds   4 pounds

$3 + 1 = 4$

The teapot weighs __1 pound__.

Reflect
Explain how to find the values of the unknowns in the number sentence.

☐ + ☐ + ☐ = 21

The problem shows three of the same values that add to 21. Try $5 + 5 + 5$. That equals 15, so try a greater number. Try $7 + 7 + 7$. That equals 21. Each square equals 7.

40 Number Patterns and Relationships • Week 4

Lesson 3 ④

3 ounces   11 ounces   12 ounces

The knife weighs __4 ounces__.

Lesson 4 Find the unknown weight. Draw your answer on the scale with the question mark.

⑤   2 cubes

Reflect
Use shapes from Problem 5 to balance each scale. Draw your answer on the scale with the question mark.

The correct answer is 6 cubes or 4 spheres.

The correct answer is 2 rectangular prisms.

Variables and Equality • Lesson 5 Review 41

---

4 Assess   10

A Gather Evidence

Minimal Understanding

Have students complete the weekly test on **Assessment,** pages 38–39. Record progress on the Student Assessment Record, **Assessment,** page 100.

B Summarize Findings

Determine whether students have Minimal, Basic, or Secure understanding of the concepts presented in Week 4.

C Differentiate Instruction

Based on your observations, use these teaching strategies next week to follow up.

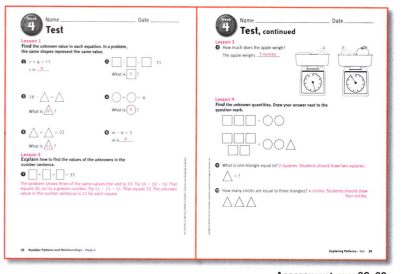

Assessment, pp. 38–39

Minimal Understanding
- Repeat the Warm-Up and Engage activities to develop Week 4 concepts.
- Use **Building Blocks** computer activities beginning with Easy as Pie to develop and reinforce this week's concepts.

Basic Understanding
- Repeat Engage activities in subsequent weeks.
- Use **Building Blocks** Word Problems with Tools to reinforce Week 4 concepts.

Secure Understanding
- Use **Building Blocks** Eggstremely Equal to reinforce Week 4 concepts.
- Use variations of the weekly Warm-Up and Engage activities, using higher numbers and multiple steps.

Variables and Equality • Lesson 5 Review   40–41

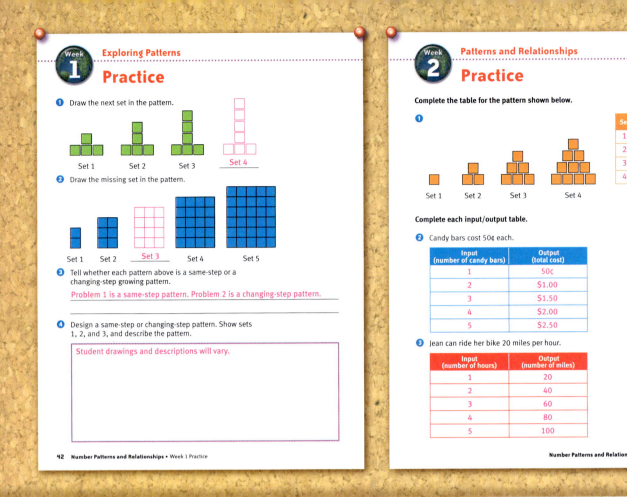

Exploring Patterns

Practice

❶ Draw the next set in the pattern.

Set 1   Set 2   Set 3   Set 4

❷ Draw the missing set in the pattern.

Set 1   Set 2   Set 3   Set 4   Set 5

❸ Tell whether each pattern above is a same-step or a changing-step growing pattern.

Problem 1 is a same-step pattern. Problem 2 is a changing-step pattern.

❹ Design a same-step or changing-step pattern. Show sets 1, 2, and 3, and describe the pattern.

Student drawings and descriptions will vary.

Patterns and Relationships

Practice

Complete the table for the pattern shown below.

❶

Set 1   Set 2   Set 3   Set 4

Set	Number of Squares
1	1
2	3
3	6
4	10

Complete each input/output table.

❷ Candy bars cost 50¢ each.

Input (number of candy bars)	Output (total cost)
1	50¢
2	$1.00
3	$1.50
4	$2.00
5	$2.50

❸ Jean can ride her bike 20 miles per hour.

Input (number of hours)	Output (number of miles)
1	20
2	40
3	60
4	80
5	100

Unit 2
Number Patterns and Relationships Practice Pages

Weeks 1 and 2

1 Assign the Practice pages, *Workbook,* pages 42–43, at the end of Weeks 1 and 2. Students should complete these pages independently.

2 Check student answers using the annotated pages above.

3 If students have difficulty with a Practice page, review the activities completed throughout the week, and have students complete the weekly practice again before they complete the weekly test.

Week 3 — Patterns and Graphs
Practice

1 The Booster Club sells gourmet cookies for $1.50 each. Complete the input/output table, and graph the pattern.

Input (number of cookies)	Output (total cost)
1	$1.50
2	$3.00
3	$4.50
4	$6.00
5	$7.50
6	$9.00

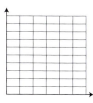

Students will title the graph Cookie Cost and place the numbers 1–6 across the horizontal axis labeled Number of Cookies. Students will place the dollar amounts $1.50–$9.00 by $1.50 increments on the vertical axis labeled Total Cost ($) and plot the following points: (1, 1.50), (2, 3.00), (3, 4.50), (4, 6.00), (5, 7.50), (6, 9.00). All points should be connected with a straight line.

2 Describe the pattern that is shown in the graph.
Each cookie costs $1.50. As the number of cookies purchased increases, the total cost increases. The graph shows a straight line when all points are connected.

3 How much would it cost to purchase 10 cookies?
$15

4 What remains the same in the graph?
the cost of each cookie

Week 4 — Variables and Equality
Practice

1 $x - 5 = 2$
$x = 7$

2 $10 - y = 6$
$y = 4$

3 $z + 5 = 15$
$z = 10$

4 $\square + \square + \square = 9$
What is 3?

5 $22 - \triangle = \triangle$
What is 11?

6 $\bigcirc + \bigcirc = 12$
What is 6?

7 $\triangle + \triangle = 18$
What is 9?

8 $m - 6 = 2$
$m = 8$

9 Explain how to find the values of the unknowns in the number sentence
$\square + \square + \square = 27$
The problem shows three of the same values that add to 27. Try 5 + 5 + 5. That equals 15, so try a greater number such as 7 + 7 + 7. That equals 21. Try 9 + 9 + 9. That equals 27. The unknown value in the number sentence is 9 for each square.

Unit 2
Number Patterns and Relationships Practice Pages

Weeks 3 and 4

1 Assign the Practice pages, **Workbook,** pages 44–45, at the end of Weeks 3 and 4. Students should complete these pages independently.

2 Check student answers using the annotated pages above.

3 If students have difficulty with a Practice page, review the activities completed throughout the week, and have students complete the weekly practice again before they complete the weekly test.

Addition and Subtraction Stories

Week at a Glance

This week, students begin *Number Worlds,* Level F, Addition and Subtraction. Students develop and extend their skills with addition and subtraction by solving word problems. Students decide which operation is needed, represent the problem with a number sentence, and solve the problem.

Math Background

Meaningful practice is necessary to develop fluency with basic number combinations and with strategies with multidigit numbers. Practice can be conducted in the context of other activities, including games that require computation . . . or focused activities that are a part of another mathematical investigation. Practice should be purposeful and should focus on developing thinking strategies and a knowledge of number relationships rather than drill isolated facts.

—*Principles and Standards for School Mathematics.* 2000. National Council of Teachers of Mathematics. Page 152.

How Students Learn

Four basic categories of word problems are covered this week: join problems, separate problems, part-part-whole problems, and compare problems. Within each of these categories are several subcategories based on which quantity is unknown. Although the week is structured around the different categories of problems, it is not intended that students learn to identify or label the categories.

Teaching for Understanding

Observe closely while evaluating the assigned tasks this week to see whether students can demonstrate the following understandings.

Benchmark after Lesson 2: Students can identify the operation (addition or subtraction) needed to solve the problem.

Benchmark after Lesson 3: Students can write number sentences that represent a word problem.

Benchmark after Lesson 4: Students can use addition and subtraction to solve a variety of word problems.

Skills Focus

- Interpret and solve word problems in which sets are joined or separated
- Interpret and solve word problems that involve the relationship between a set and its two subsets
- Interpret and solve word problems comparing two sets
- Use guiding questions to solve word problems

Math at Home

Give one copy of the Letter to Home, page A9, to each student. Encourage students to share and complete the activity with their caregivers.

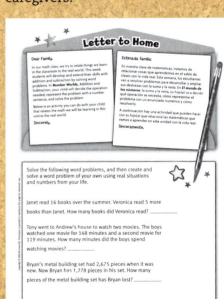

Letter to Home, Teacher Edition, p. A9

Week 1 Planner   Addition and Subtraction Stories

PACING	LESSON	LEARNING GOALS	NCTM	MATERIALS	TECHNOLOGY
DAY 1	pages 2–3	Students solve word problems in which sets are joined or separated.	• Number and Operations • Problem Solving • Communication	• Calculator • Base-Ten Blocks*	**Building Blocks** Off the Tree **MathTools** Function Machine
DAY 2	pages 4–5	Students solve problems in which a given set is related to its subsets.	• Number and Operations • Problem Solving • Communication • Connections	• Calculator • Number Cards (0–9)	**Building Blocks** Easy as Pie **MathTools** Function Machine
DAY 3	pages 6–7	Students solve problems comparing two sets.	• Number and Operations • Problem Solving • Communication • Connections	• Calculator • Index cards	**Building Blocks** Figure the Fact **MathTools** Function Machine
DAY 4	pages 8–9	Students use addition and subtraction to solve a variety of word problems.	• Number and Operations • Problem Solving • Communication	Calculator	**Building Blocks** Figure the Fact **MathTools** Function Machine
DAY 5	Review and Assess pages 10–11	Students review skills learned this week and complete the weekly assessment.	• Number and Operations • Problem Solving • Communication • Connections	Materials will be selected from Lessons 1–4.	**Building Blocks** **MathTools** Review previous activities

Math Vocabulary

sum The total when two quantities are added

difference The amount by which one quantity is greater or lesser than another quantity

unknown The answer you are looking for when you solve a word problem

solution An answer to a word problem and the way you find it

English Learners

SPANISH COGNATES

English	Spanish
problem	problema
separate	separar
solution	solución
verify	verificar

ALTERNATE VOCABULARY

sets Groups of objects or numbers

number sentence An equation; a complete mathematical expression using numerals and symbols

Lesson 1

Objective

Students solve word problems in which sets are joined or separated.

Materials

Additional Materials
- Calculator
- Base-Ten Blocks

Access Vocabulary

parade An organized march of celebration down a main street, usually including music, marching bands, horses, and floats

DVD A movie recorded on a small disc

Creating Context

Like all students, English Learners have many different life experiences. Part of developing academic language is learning more about typical experiences that may be included in school texts. In this lesson, it may be helpful to read Problems 8–10 with English Learners to make sure that they understand the math and the story elements. For example, Problem 8 mentions a slumber party. This may be a completely unfamiliar notion outside the United States and in some cultures, so a brief description of a slumber party expands cultural background for English Learners.

1 Warm Up 5

Concept Building COMPUTING

Provide students with Base-Ten Blocks. Instruct students to create two sets.
- **Name the value of each set.**
- **Now join the sets, trading as needed, and name the value of that set.**

Now instruct students to separate a group (any size) of blocks from the set just formed.
- **Name the value of the separated group and the new value of the remaining group.**

2 Engage 30

Skill Building

- **What happens to the size of a group if it is joined with another group?** It gets larger.
- **What happens to the size of a group if a part is separated?** It gets smaller.
- **Can you use this knowledge to predict the reasonableness of your answers to both types of problems? How?** Yes. Answers will vary.

Guiding Questions UNDERSTANDING

Explain to students that when solving problems they should follow a strategy of asking questions to arrive at a solution. Write the six **Guiding Questions** from *Workbook*, page 2, on the board.

For each problem below, guide students to a solution by asking the six questions.
- **Maggie selected 12 books to check out at the library. Maggie's mother selected 9 books to check out. How many books did they check out altogether?** 21 books
- **Travis brought home a 45-piece box of taffy from his vacation. He gave 14 pieces to his neighbors. How many pieces of taffy did he have left?** 31 pieces

Present additional problems to students, and encourage them to use the guiding questions to solve the problems on their own.

Monitoring Student Progress

If . . . students lack confidence in their computational skills,

Then . . . provide a calculator after they have completed the problems so that they can verify their answers.

Building Blocks For additional practice with joining and separating, students should complete *Building Blocks* Off the Tree.

MathTools Use Function Machine to demonstrate and explore addition and subtraction.

Using Student Pages

Have students complete *Workbook*, pages 2–3, on their own.

Lesson 1

Key Idea
For each problem ask yourself the following questions:
1. What is the problem asking?
2. Does the problem ask you to find a total or part of a total?
3. Can you record the problem in a number sentence?
4. Can you solve the problem?
5. Does your solution make sense?
6. How can you verify your solution?

Try This
Mrs. Jonas asked the students in her class to combine partial boxes of crayons. How many crayons are in each new group?

1 a box of 52 crayons and a box of 35
<u>87</u>

2 a box of 123 crayons and a box of 82
<u>205</u>

3 a box of 8 crayons and a box of 67
<u>75</u>

4 a box of 130 crayons and a box of 51
<u>181</u>

Find the size of the unknown part.

5 Janet had 479 beads. She gave away 372.   <u>107</u>

6 Bobby gave away 83 cars from his set of 399.   <u>316</u>

7 Linda shared 65 pieces of candy from a package of 345.   <u>280</u>

Practice
Answer the following questions for each problem.
a. What is the problem asking?
b. Are the sets *joined* or *separated*?
c. What is the solution?

8 Caitlin had 23 birthday party invitations to send out. She mailed 16 of them and delivered the rest by hand. How many did she deliver by hand?

a. <u>the number of invitations delivered by hand</u>

b. <u>separated</u>   c. <u>7 invitations</u>

9 Eduardo and Jimmy were comparing the number of DVDs each of them owns. Eduardo owns 56 DVDs. Jimmy owns 131 DVDs. How many DVDs do they own altogether?

a. <u>the number of DVDs Eduardo and Jimmy own combined</u>

b. <u>joined</u>   c. <u>187 DVDs</u>

10 While the Hernandez family traveled to the Grand Canyon, Jimmy did 56 word search puzzles. Andrea did 121 word search puzzles. How many puzzles did they do in all?

a. <u>the number of puzzles both children did altogether</u>

b. <u>joined</u>   c. <u>177 puzzles</u>

Reflect
In some of the above problems in which you used addition, what words helped you know that the problems were about joining? List other words that helped you know to use addition.

<u>Combined, altogether; other words will vary but may include *in all, more,*</u>
<u>and *total.*</u>

Reflect   10

Extended Response   REASONING
Review students' answers to the Reflect prompt at the bottom of student page 3.

■ **What words did you list that indicate addition?**

On the board, make a list of all of the words that students say. Instruct students to copy the list for themselves so that they can use it as a reference throughout this unit.

■ **What words would tell you to use subtraction?**

Real-World Application   APPLYING
■ **Name two situations at school that involve joining groups of people. An example is all the fourth-grade classes going to the auditorium to watch a play.**

■ **Name two situations outside the classroom in which one group is separated into two groups. An example is when neighborhood children gather to play ball, and two teams are formed.**

Assess

Informal Assessment
Use the Student Assessment Record, **Assessment,** page 100, to record informal observations.

UNDERSTANDING
Guiding Questions
Did the student
❏ make important observations?
❏ extend or generalize learning?
❏ provide insightful answers?
❏ pose insightful questions?

Addition and Subtraction Stories • Lesson 1   2–3

Lesson 2

Objective

Students solve word problems in which a given set is related to its subsets.

Materials

Additional Materials
- Calculator
- Number Cards (0–9)

Access Vocabulary

sports cards Small laminated cardboard cards with pictures of professional sports players on one side and their game statistics on the other

gaming system An electronic game machine with components and game cartridges

charity An organized effort to help others collect money for a cause

Creating Context

Emphasize to all students that the process described in the Key Ideas section is very important to solving problems systematically. English Learners may require some additional support to fully comprehend these instructions. Help students identify the clues in each problem that tell which operation is needed.

1 Warm Up 5

Concept Building COMPUTING

- **Choose one card from your Number Cards. Keep your card facedown until it is your turn.**

Organize students into groups of four or five. Each student shows his or her card on his or her turn. There should be no talking during the activity. As each student shows a card, all students will add that number (mentally) to the previous sum. When the last student has shown a card, compare sums. Students should then choose another card and play again.

2 Engage 30

Strategy Building

Write the six **Guiding Questions** from *Workbook,* page 4, on the board.

Guiding Questions UNDERSTANDING

For each problem, guide students to a solution by asking the six guiding questions.

- **Sandra knows that there are 35 images of a bird hidden in a picture. She has already found 18 birds. How many birds does Sandra still need to find?** 17 birds
- **Steve was building a castle that used 145 blocks. He has already placed 82 blocks in their positions. How many more blocks does Steve still need to place to finish the castle?** 63 blocks

Discuss with students that in each situation there are parts that add up to a whole. In the first problem, 35 birds is the whole, and 18 birds and 17 birds are the parts. Ask students to name the parts and the whole of the second problem. part: 82; part: 63; whole: 145

Tell students that they can use addition or subtraction to find the unknown part. If they use addition, they count on from one part until they reach the whole. The number they counted on is the size of the unknown part. Show students how to record this operation in the form of a missing addend number sentence:

18 birds + ☐ birds = 35 birds   ☐ = 17 birds

Monitoring Student Progress

If . . . students need more practice with mental math skills,

Then . . . do the Warm-Up activity again.

 Building Blocks For additional practice with sets and subsets, students should complete *Building Blocks* Easy as Pie.

 MathTools Use Function Machine to demonstrate and explore addition and subtraction.

Using Student Pages

Have students complete **Workbook,** pages 4–5, on their own.

Addition and Subtraction Stories

Lesson 2

Key Idea

For each problem ask yourself the following questions:
1. What is the problem asking?
2. Does the problem ask you to find a total or part of a total?
3. Can you record the problem in a number sentence?
4. Can you solve the problem?
5. Does your solution make sense?
6. How can you verify your solution?

Try This

Name the parts and the whole for each situation.

1. 38 baseball cards and 55 football cards in a collection of 93 sports cards

    part: __38 baseball cards__   part: __55 football cards__

    whole: __93 sports cards__

2. 825 girls and 762 boys in a school of 1,587 students

    part: __825 girls__   part: __762 boys__

    whole: __1,587 students__

3. 168 points scored when Yancy scored 82 points and Jessica scored 86 points

    part: __82 points by Yancy__   part: __86 points by Jessica__

    whole: __168 points__

Practice

Answer the following questions for each problem.
a. What is unknown?
b. Do you need to find the total or an unknown part?
c. What is the solution?

4. Quintin needs $232 to buy a new gaming system and two games. The two games cost a total of $103. What is the price of the gaming system?

    a. __the price of the gaming system__

    b. __unknown part__   c. __$129__

5. Zoey collected 837 aluminum-can tops between January and June. The rest of the year she collected 629 tops. How many aluminum-can tops was Zoey able to give to a charity altogether?

    a. __the total number of aluminum-can tops collected__

    b. __total__   c. __1,466 can tops__

6. Tony went to Andrew's house to watch two movies. The boys watched one movie for 148 minutes and a second movie for 119 minutes. How many minutes did the boys spend watching movies?

    a. __the total number of minutes in both movies__

    b. __total__   c. __267 minutes__

Reflect

Write one word problem that uses 514 miles and 262 miles to find a total distance. Show the solution to your problem.

_____

_____

Write a second word problem that uses 514 miles as part of a total distance of 776 miles. Show your solution.

Answers will vary, but the first problem should ask solvers to find a total of 776 miles. The second problem should ask solvers to find a part of the total that is the size of 262 miles.

---

3 Reflect   10

Extended Response   REASONING

Review students' answers to the Reflect prompt at the bottom of student page 5.

Invite students to display their word problems on the board. Discuss which problems have an unknown part and which problems have an unknown whole.

- **Was this activity easy or difficult?**
- **Is creating or solving a problem with an unknown part easier or more difficult than a problem with an unknown whole?**

Real-World Application   APPLYING

Many real-life situations involve both addition and subtraction.

- **A trip to the grocery store involves addition to determine the total for all the products purchased. It can involve subtraction to determine the amount of change when the customer pays an amount greater than the total.**
- **Name other situations in which both addition and subtraction are involved.** Possible answers: banking, shopping, diet and exercise

---

4 Assess

Informal Assessment

Use the Student Assessment Record, *Assessment,* page 100, to record informal observations.

UNDERSTANDING

Guiding Questions
Did the student
❏ make important observations?
❏ extend or generalize learning?
❏ provide insightful answers?
❏ pose insightful questions?

Lesson 3

Objective

Students solve problems that compare two disjoint sets.

Materials

Additional Materials
- Calculator
- Index cards

Access Vocabulary

Does your solution make sense? Is this solution possible, given what is known about the situation?

over the summer During the entire summer, may include time off from school

Creating Context

Academic language often assumes shared cultural experiences. When students have not shared in the same experiences as the author, information may be lost for the student. For example, in Problem 3, *Workbook,* page 6, a reader familiar with American sports may recognize this as a football reference. The Jets and the Giants are two professional football teams, and *Sunday's game* is a common reference to professional football, which is on television on Sundays throughout football season.

 Warm Up 5

Concept Building UNDERSTANDING

- **How many of you have pets at home?**
- **If you have three pets and I have two pets, how many more pets do you have than I have?**

Select two students who have pets, and compare the number of pets each has by modifying the previous statement.

- **Each person's pets are a disjoint or separate set from the other's pets.**

 **Engage** 30

Skill Building

- **When a problem asks us to figure out how many more one set has than another, what do we do to figure out the answer?** Possible answers: count up from the first set until you reach the second set; subtract the first set from the second set.

Discuss both strategies with students, and make sure they realize that both strategies are acceptable and useful solving this sort of problem.

Remind students that when solving problems, they should follow a strategy of asking questions to arrive at a solution. Write the six **Guiding Questions** from *Workbook,* page 6, on the board.

Guiding Questions UNDERSTANDING

For each problem, allow students time to work toward a solution using the guiding questions.

- **Mrs. Davee's class has 28 students. Mr. Vongard's class has 33 students. How many more students does Mr. Vongard's class have?** $33 - 28 = 5$ students or $28 + \square = 33$, $\square = 5$; Both solutions are acceptable.
- **Lisa slept for 9 hours yesterday. Karl slept 7 more hours than Lisa. How many hours did Karl sleep?** $9 + 7 = 16$ hours
- **Chloe biked 28 miles over the weekend. She biked 9 miles more than Jason. How many miles did Jason bike?** $28 - 9 = 19$ miles or $9 + \square = 28$, $\square = 19$

Allow time to discuss students' strategies and solutions to each problem.

Make sure students realize that two different number sentences can be written for the first problem and the third problem and both are acceptable.

Building Blocks For additional practice with disjoint sets, students should complete *Building Blocks* Figure the Fact.

eMathTools Use Function Machine to demonstrate and explore addition and subtraction.

Using Student Pages

Have students complete *Workbook,* pages 6–7, on their own.

Lesson 3

Key Idea
For each problem ask yourself the following questions:
1. What is the problem asking?
2. Does the problem ask you to find a total or part of a total?
3. Can you record the problem in a number sentence?
4. Can you solve the problem?
5. Does your solution make sense?
6. How can you verify your solution?

Try This
Write a number sentence for each situation.

1. Janet read 16 books. Veronica read 5 more books than Janet read. How many books did Veronica read?
   $16 + 5 = 21$

2. Dean runs 21 miles each week. Todd runs 26 miles each week. How many more miles does Todd run than Dean runs each week?
   $26 - 21 = 5$ or $21 + \square = 26; \square = 5$

3. The Jets scored 42 points in Sunday's game. The Jets scored 13 more points than the Giants scored. How many points did the Giants score?
   $42 - 13 = 29$ or $\square + 13 = 42; \square = 29$

Practice
Answer the following questions for each problem.
a. What is the problem asking?
b. Do you need to find a total or an unknown part?
c. What number sentence can you write for the problem? Include the solution.

4. Ethan won 43 tennis matches in one season. Brenda won 8 less matches than Ethan won. How many matches did Brenda win?
   a. the number of tennis matches Brenda won
   b. unknown part
   c. $43 - 8 = 35$, or $\square + 8 = 43; \square = 35$

5. Beads-A-Bunch comes with 1,225 beads. Be-Dazzle comes with 575 more beads than Beads-A-Bunch does. How many beads does Be-Dazzle have?
   a. the number of beads in Be-Dazzle
   b. total
   c. $1,225 + 575 = 1,800$

6. Tina has 129 postcards in her collection. David has 38 more postcards than Tina has. How many postcards does David have in his collection?
   a. the number of postcards David has
   b. total
   c. $129 + 38 = 167$

7. Cassandra bowled two games. She scored 174 in her first game. Her second game score was 133. How many more points did she score in her first game?
   a. how many more points Cassandra scored in her first game than her second game
   b. unknown part
   c. $174 - 133 = 41$ or $133 + \square = 174; \square = 41$

Reflect
Write a word problem that compares two sets and can be related to the number sentence below.

$136 - 77 = 59$

Answers will vary but will include two of the three numbers in the number sentence above.

---

3 Reflect 10 ▶

Extended Response REASONING
Review students' answers to the Reflect prompt at the bottom of student page 7.

■ **How did you decide how to use the numbers in the number sentence?**

Ask students to summarize what was discussed and to name the strategies they liked best.

Real-World Application APPLYING

■ **Name a situation in which you would need to compare two sets. Write a word problem that compares these sets. Have another student write a number sentence to solve the word problem.**

The students may want to discuss strategies used to solve each other's word problems.

4 Assess

Informal Assessment
Use the Student Assessment Record, **Assessment,** page 100, to record informal observations.

UNDERSTANDING

Guiding Questions
Did the student
❑ make important observations?
❑ extend or generalize learning?
❑ provide insightful answers?
❑ pose insightful questions?

Lesson 4

Objective

Students use addition and subtraction to solve a variety of problems.

Materials

Additional Materials
Calculator

Access Vocabulary

baseball season The months when official baseball games are played

Creating Context

Word problems often include sports references. Review with English Learners the scoring of various professional sports. For example, soccer scores one point per goal and basketball scores two points per basket, with exceptions for baskets made from far away or during penalty throws.

1 Warm Up 5

Concept Building UNDERSTANDING

Write the following word problems on the board. Ask students to write a number sentence to represent each problem and discuss the strategies they used and whether they used addition or subtraction.

- **Vincent ate 4 pieces of pizza. Rory ate 3 pieces of pizza. How many pieces did they eat altogether?**
  $4 + 3 = 7$
- **Mr. Black bought 24 tomato plants. He planted 15 before he stopped for lunch. How many must he plant after lunch?** $15 + \square = 24, \square = 9$; $24 - 15 = 9$
- **The math test was worth 45 points. Latifa scored 38 points. How many points did Latifa miss?**
  $45 - 38 = 7$; $38 + \square = 45, \square = 7$

2 Engage 30

Skill Building

Allow a student to act as the teacher and to guide students through the questions to the solution. Write the six **Guiding Questions** from **Workbook,** page 8, on the board.

Who Is the Teacher? UNDERSTANDING

For each of the problems, guide students to a solution by asking the guiding questions.

- **Aiden has a collection of action figures. For his birthday he received 12 more figures. Aiden now has 126 action figures in his collection. How many figures did Aiden have before his birthday?**
  114 action figures
- **Theresa rolled a score that was 15 pins more than her highest bowling score. Her previous highest score for a bowling game was 172. What was the score of the game she bowled?**
  187 points

Monitoring Student Progress

If . . . students want to model or draw a picture to accompany the problems,

Then . . . provide paper. Encourage all students to draw pictures to help them make sense of the problem.

 For additional practice with solving addition and subtraction problems, students should complete **Building Blocks** Figure the Fact.

 MathTools Use Function Machine to demonstrate and explore addition and subtraction.

Using Student Pages

Have students complete **Workbook,** pages 8–9, on their own.

Lesson 4

Key Idea

For each problem ask yourself the following questions:

1. What is the problem asking?
2. Does the problem ask you to find a total or part of a total?
3. Can you record the problem in a number sentence?
4. Can you solve the problem?
5. Does your solution make sense?
6. How can you verify your solution?

Try This

Tell what is unknown for each situation.

❶ Jude had 76 marbles. He gave some to Tate. Now Jude has 52 marbles.

the number of marbles Jude gave to Tate

❷ Max is saving money to buy a new video game. He has already saved $17. He still needs $8.

the price of the video game

❸ The temperature at midnight was 56 degrees. By 6:00 in the morning the temperature had risen to 70 degrees.

how many degrees the temperature had risen

Practice

Answer the following questions for each problem.

a. What is the problem asking?
b. Do you need to find a total or an unknown part?
c. What number sentence can you write for the problem? Include the solution.

---

❹ At the end of the 2004 baseball season, Barry Bonds had a career total of 703 home runs. During one season he hit 73 home runs. How many of his home runs were hit in other seasons?

a. the number of home runs Bonds hit in all seasons other than his highest season

b. unknown part

c. $703 - 73 = 630$ or ☐ $+ 73 = 703$; ☐ $= 630$

❺ Elyse has a rock collection. She picked up 12 more rocks on her hike. Now she has 147 rocks. How many rocks were in her collection before her hike?

a. the number of rocks in Elyse's collection before her hike

b. unknown part

c. $147 - 12 = 135$ or ☐ $+ 12 = 147$; ☐ $= 135$

❻ Ms. Ringer had a bag of coins that totaled $77. Ms. Ringer did not know that the bag had a hole in it. As she carried it to the bank, many coins fell out. When she counted the coins again, she had $65. How much money fell from the hole in the bag?

a. the amount of money lost

b. unknown part

c. $77 - 65 = 12$ or ☐ $+ 65 = 77$; ☐ $= 12$

Reflect

Choose one of Problems 4–6. Write a number sentence with pictures for the unknown quantity to show how you figured it out.

Answers will vary.

 Reflect 10

Extended Response REASONING

Review students' answers to the Reflect prompt at the bottom of student page 9.

- **How did you construct your number sentence?**
- **Will you be able to try anything new after hearing how other students solve word problems?**

Real-World Application APPLYING

Provide students with newspaper classified sections. Students need to locate and compare the prices of two cars.

- **Do you use addition or subtraction to compare the prices of the two cars?** Answers will vary depending on the strategy used.
- **Write a number sentence to show the difference in the prices of the two cars.** Answers will vary depending on the strategy used.
- **For what reason might you add the prices together?** You may want to buy both cars.

 Assess

Informal Assessment

Use the Student Assessment Record, *Assessment,* page 100, to record informal observations.

UNDERSTANDING

Who Is the Teacher?

Did the student
❏ make important observations?
❏ extend or generalize learning?
❏ provide insightful answers?
❏ pose insightful questions?

Addition and Subtraction Stories • Lesson 4   8–9

Lesson 5 Review

Objective

Students review skills learned this week and complete the weekly assessment.

Materials

Review materials will be selected from those used in previous activities.

Creating Context

English Learners may need additional support with written instructions. For example, one instruction states, "Name the parts and the whole for each situation." By itself, this direction may be quite puzzling, but when students put it together with the examples that follow, they may be able to better understand what is being asked. Remind students to ask for clarification if they do not understand the instructions.

2 Engage   20

Skill Building

Free-Choice Activity

For the last day of the week, allow students to choose an activity from the previous lessons. Some activities they may choose include the following:

- **Guiding Questions**
- **Who Is the Teacher?**

Make a note of the activities children select. Do they prefer easy or challenging activities? If you believe your students would benefit from extra practice on specific skills, choose an activity for them.

Monitoring Student Progress

If . . . students are not participating,

Then . . . continue the activity until all students have participated.

Using Student Pages

Have students complete **Workbook,** pages 10–11, on their own.

3 Reflect   10

Extended Response   APPLYING

Review students' answers to the Reflect prompts at the bottom of student pages 10–11.

Discuss these answers with the group to reinforce Week 1 concepts.

1 Warm Up   5

Concept Building   UNDERSTANDING

Select more than one student to tell a story for each of the bulleted situations below.

- Tell a story that has a missing number in it. Tell it so the missing number is the total of the two parts.
- Tell another story that has a missing number in it. Tell it so the missing number is the size of one of the parts.

Week 1 — Addition and Subtraction Stories

Review

This week you learned to solve word problems by choosing the operation needed and by representing the problem with a number sentence.

Lesson 1 Answer the following questions.
a. What is the problem asking?
b. Are the sets joined or separated?
c. What is the solution?

1. The concert before the play took 58 minutes. The play took 65 minutes. How many minutes were the concert and play combined?
   a. *the number of minutes of the concert and play together*
   b. *joined*
   c. *123 minutes*

Lesson 2 Name the parts and the whole.

2. The gardener planted 177 plants—98 petunias and 79 marigolds.
   part: *98 petunias*   part: *79 marigolds*
   whole: *177 plants*

Reflect
Write a word problem that uses 410 miles and 278 miles to find a total distance. Show the solution to your problem.
Answers will vary but should use addition to get a total of 688 miles.

10 Addition and Subtraction • Week 1

Lesson 3 Write a number sentence for the problem. Include the solution.

3. Kathy has 36 thin-tipped markers. Twelve of them are black or gray. How many markers are colors other than black or gray?
   36 − 12 = 24 or 12 + ☐ = 36; ☐ = 24

4. The tallest tree ever measured was in Australia. It was 435 feet. The tallest tree measured in the United States was 372 feet. How many feet taller was the Australian tree?
   435 − 372 = 63 or 372 + ☐ = 435; ☐ = 63

Lesson 4 Tell what is unknown for each situation.

5. Craig had some water balloons. He gave 14 balloons to Ken. Craig now has 18 balloons.
   the number of water balloons Craig had

6. It rained 22 inches during April. Between April 24 and April 30, it rained 3 inches.
   the number of inches of rain between April 1 and April 23

Reflect
Write a word problem that compares two sets and can be related to the following number sentence: 599 − 310 = 289
Answers will vary but should include two of the three numbers in the number sentence.

Addition and Subtraction Stories • Lesson 5 Review 11

---

4 Assess 10

A Gather Evidence

Formal Assessment

Have students complete the Weekly test on *Assessment,* pages 40–41. Record scores on the Student Assessment Record, *Assessment,* page 100.

B Summarize Findings

Determine whether students have Minimal, Basic, or Secure understanding of the concepts presented in Week 1.

C Differentiate Instruction

Based on your observations, use these teaching strategies next week to follow up.

Assessment, pp. 40–41

Minimal Understanding
- Repeat the Warm-Up and Engage activities to develop Week 1 concepts.
- Use *Building Blocks* Off the Tree and *eMathTools* Function Machine to develop and reinforce Week 1 concepts.

Basic Understanding
- Repeat Engage activities in subsequent weeks.
- Use *Building Blocks* Easy as Pie and *eMathTools* Function Machine to reinforce Week 1 concepts.

Secure Understanding
- Use *Building Blocks* Figure the Fact and *eMathTools* Function Machine to extend Week 1 concepts.

Addition and Subtraction Stories • Lesson 5 Review 10–11

Sharpening Computation Skills

Week at a Glance

This week, students continue **Number Worlds,** Level F, Addition and Subtraction. Students practice and more fully develop their computation skills. Students continue to use Base-Ten Blocks to model regrouping in addition and subtraction problems. Students will also explore which numbers in a trio have the greatest and least sum and the greatest and least difference.

Skills Focus

- Add two- and three-digit numbers, including those that require regrouping
- Subtract two- and three-digit numbers, including those that require regrouping
- Recognize the magnitude of each digit in a three-digit number

Math Background

As you teach mathematics, including various computational procedures and related concepts, you need to be concerned with more than just accurate learning. The experiences you plan for your students help form their present and continuing dispositions toward mathematics. Your use of models and manipulatives can contribute to a positive disposition if those experiences include explorations, problem solving, accurate modeling of the mathematics involved, and reflections by each student.

—Ashlock, Robert, 1998. *Error Patterns in Computation.* Merrill. Page 51.

How Students Learn

Students need to generalize a concept or procedure, not focus on the particulars of a single problem or model. Be on the lookout for indications that students do not understand the general concept. Interject often with probing questions that can help you assess developing skills and that can guide students to self-assess their own understanding.

Teaching for Understanding

Observe closely while evaluating the assigned tasks this week to see whether students can demonstrate the following understandings.

Benchmark after Lesson 2: Students can add and subtract two- and three-digit numbers.

Benchmark after Lesson 3: Students can determine which two numbers from a set of three numbers have the greatest or least sum.

Benchmark after Lesson 4: Students can determine which two numbers from a set of three numbers have the greatest or least difference.

Math at Home

Give one copy of the Letter to Home, page A10, to each student. Encourage students to share and complete the activity with their caregivers.

Letter to Home,
Teacher Edition, p. A10

Week 2 Planner | Sharpening Computation Skills

PACING	LESSON	LEARNING GOALS	NCTM	MATERIALS	TECHNOLOGY
DAY 1	pages 12–13	Students add two- and three-digit numbers.	• Number and Operations • Problem Solving • Communication	• Base-Ten Blocks* • Place Value Mat, p. B2 • Number Construction Mat, p. B1 • Index cards • Advertisements or sale papers	**B**uilding **B**locks Two-Digit Adding **e** MathTools Function Machine
DAY 2	pages 14–15	Students subtract two- and three-digit numbers.	• Number and Operations • Problem Solving • Communication	• Base-Ten Blocks* • Place Value Mat, p. B2 • Number Construction Mat, p. B1	**B**uilding **B**locks Word Problems **e** MathTools Function Machine
DAY 3	pages 16–17	Students identify which combination of numbers has a greater sum.	• Number and Operations • Problem Solving • Communication • Connections	• Number Cards (0–9) • 3-Digit Window, p. B4	**B**uilding **B**locks Eggcellent Addition **e** MathTools Function Machine
DAY 4	pages 18–19	Students identify which combination of numbers has a greater difference.	• Number and Operations • Problem Solving • Communication • Connections	• 3-Digit Window, p. B4 • Number Cards (0–9)	**B**uilding **B**locks Eggcellent Addition **e** MathTools Function Machine
DAY 5	**Review and Assess** pages 20–21	Students review skills learned this week and complete the weekly assessment.	• Operations • Problem Solving • Communication • Connections	Materials will be selected from Lessons 1–4.	**B**uilding **B**locks **e** MathTools Review previous activities

Math Vocabulary

unit block The Base-Ten Block that represents 1 in the base-ten number system

rod A single block that is equal in size and value to ten unit blocks

flat A single block that is equal in size and value to ten rods

cube A single block that is equal in size and value to ten flats

English Learners

SPANISH COGNATES

English	Spanish
addition	adición
sum	suma
standard form	forma estandar

ALTERNATE VOCABULARY

trades Exchanges; each side gives something and receives something of equal value

number sentence An equation; a complete mathematical expression using numerals and symbols

---

* Available from SRA

Lesson 1

Objective

Students add two- and three-digit numbers.

Materials

Program Materials
- Place Value Mat, p. B2
- Number Construction Mat, p. B1

Additional Materials
- Base-Ten Blocks
- Advertisements or sales papers

Access Vocabulary

sum The answer to an addition problem

trade To exchange ten blocks of a smaller size for one block of the next size block

Creating Context

Mathematics has many specialized terms that are necessary to understand the subject. English Learners may know the math concepts in their primary language but need to accelerate their acquisition of English academic language to continue to advance in mathematics. Help English Learners use the glossary to look up math terms and phrases that are unfamiliar.

1 Warm Up 5

Concept Building COMPUTING

- Place thirteen unit blocks on your Place Value Mat.
- Do you have enough to trade? yes What is the trade? 10 unit blocks for one rod
- How many unit blocks do you have left? 3
- What happened to the thirteen we just had there? We traded 10 away for a rod, leaving 3.
- This is called *regrouping*. We *regrouped* thirteen into 1 tens and 3 ones.

2 Engage 30

Skill Building

Trading at the Top UNDERSTANDING

Students should do each step at their seats as you model the problem to the class.

- We are going to use our mats to find the sum of 137 and 78. Do each step with me.
- On your Place Value Mat, model 137 with unit blocks, rods, and a flat.
- Without clearing any blocks, add the model for 78 to the mat. How many unit blocks of 8 can be placed before you have reached the top? 3
- When you place the block on the lightbulb, what does that signal you to do? Trade; remove the 10 unit blocks, and place 1 rod.
- Do you have to make another trade? Yes; only 6 of the 7 rods can be placed before the lightbulb is covered.
- How many unit blocks, rods, and flats do you have to transfer to the Number Construction Mat? Name the sum. 2 flats, 1 rod, 5 unit blocks; 215
- Write an addition problem of your own. Then exchange problems with a neighbor. Model and trade to find the sum. Check each other's answers.

Monitoring Student Progress

If . . . students want to name the sum without transferring blocks to the Number Construction Mat,	Then . . . make certain that students have correctly named at least two sums before you allow them to bypass that step.

Building Blocks For additional practice with adding two- and three-digit numbers, students should complete **Building Blocks** Two-Digit Adding.

@MathTools Use Function Machine to demonstrate and explore addition and subtraction.

Using Student Pages

Have students complete **Workbook,** pages 12–13, on their own.

Lesson 1

Key Idea
Using Base-Ten Blocks

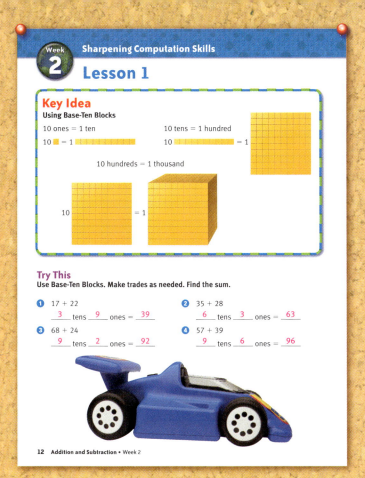

10 ones = 1 ten

10 ■ = 1 ▭▭▭▭▭▭▭▭▭▭

10 tens = 1 hundred

10 ▭▭▭▭▭▭▭▭▭▭ = 1

10 hundreds = 1 thousand

10 = 1

Try This
Use Base-Ten Blocks. Make trades as needed. Find the sum.

❶ 17 + 22
__3__ tens __9__ ones = __39__

❷ 35 + 28
__6__ tens __3__ ones = __63__

❸ 68 + 24
__9__ tens __2__ ones = __92__

❹ 57 + 39
__9__ tens __6__ ones = __96__

Solve each addition problem. Show your answer on the chart. Write the sum in standard form.

❺ 67 + 82   __149__

Hundreds	Tens	Ones
1 flat	4 rods	9 unit blocks

❻ 75 + 89   __164__

Hundreds	Tens	Ones
1 flat	6 rods	4 unit blocks

Practice
Find each sum. Write how many times you made trades.

❼ 123 + 49
__172__   __1__ trades

❽ 382 + 498
__880__   __2__ trades

❾ 867 + 645
__1,512__   __3__ trades

❿ 919 + 79
__998__   __1__ trades

⓫ 408 + 293
__701__   __2__ trades

⓬ 756 + 879
__1,635__   __3__ trades

Reflect
Write an addition problem in which ones are traded for a ten.
Answers will vary.

Write an addition problem in which ones are traded for a ten and tens are traded for a hundred.
Answers will vary.

Write an addition problem that has three trades in it.
Answers will vary.

---

③ Reflect   10

Extended Response   REASONING

Review students' answers to the Reflect prompt at the bottom of student page 13.

Select a student (a different student for each part of the question) to demonstrate that the problem he or she wrote meets the criteria. Instruct students to use models that show the trades so that the class can verify that the criteria were met.

Real-World Application   APPLYING

Provide students with an advertisement or sale paper for local retail stores.

■ Select two products with prices that can be added and that do not require regrouping of ones, tens, or hundreds. Find the sum of the products.

■ Select two products with prices that can be added and that do require regrouping. Find the sum of the products.

④ Assess

Informal Assessment

Use the Student Assessment Record, **Assessment,** page 100, to record informal observations.

UNDERSTANDING

Trading at the Top
Did the student
❏ make important observations?
❏ extend or generalize learning?
❏ provide insightful answers?
❏ pose insightful questions?

Lesson 2

Objective

Students subtract two- and three-digit numbers.

Materials

Program Materials
- Place Value Mat, p. B2
- Number Construction Mat, p.B1

Additional Materials
Base-Ten Blocks

Access Vocabulary

difference The answer to a subtraction problem
place value The value of a digit, based on its place or position

Creating Context

An excellent strategy for making content comprehensible for English Learners is to provide hands-on experience. Work with English Learners to model the trades by using Base-Ten Blocks. Provide Base-Ten Blocks for students to work through the subtraction problems in this lesson and to demonstrate their understanding of this concept. Remind students that in subtraction problems, the trades are made in the opposite direction of the trades made when adding.

1 Warm Up — 5

Concept Building COMPUTING

Provide students with a random number of units, rods, and flats. Instruct students to write the number modeled by all the blocks at their desks. Then instruct students to select a group of blocks to set aside. Students should now write a number sentence to show the subtraction they just performed.

2 Engage — 30

Skill Building

May I Regroup Blocks? UNDERSTANDING

Students should do each step at their seats as you model the problem for the class.

- We are going to use our mats to find the difference between 364 and 86. Do each step with me.
- On your Place Value Mat, model 364 with unit blocks, rods, and flats.
- To subtract 86, we need to remove blocks that represent 86 from the mat.
- Begin by removing 6 unit blocks. Notice that there are only 4 unit blocks.
- How can you make a trade so that you have at least 6 unit blocks? Remove 1 rod from the tens column, and place 10 unit blocks in the ones column.
- This is called regrouping. You regrouped blocks by exchanging 1 ten for 10 ones.
- Now remove 6 unit blocks. What do you need to remove next? Can you do this? 8 rods; no; there are only 5 rods on the mat.
- How can you make a trade so that you have at least 8 rods? Regroup 1 flat from the hundreds column, and exchange it for 10 tens. Place 10 rods in the tens column.
- What blocks are left on the mat after 86 have been removed? 2 flats, 7 rods, 8 unit blocks; 278

Monitoring Student Progress

If . . . students need more practice with subtraction problems,

Then . . . demonstrate and provide additional practice problems to work on in a small group.

Building Blocks For additional practice subtracting two- and three-digit numbers, students should complete **Building Blocks** Word Problems.

eMathTools Use Function Machine to demonstrate and explore addition and subtraction.

Using Student Pages

Have students complete **Workbook,** pages 14–15, on their own.

Lesson 2

Key Idea

In subtraction problems, the trades are made in the opposite direction of the trades you make when adding.

1 thousand = 10 hundreds   1 hundred = 10 tens

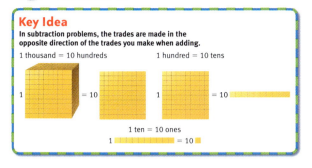

1 [cube] = 10 [flats]   1 [flat] = 10 [rods]

1 ten = 10 ones

1 [rod] = 10 [unit]

Try This

Use Base-Ten Blocks. Make trades as needed. Find the difference.

❶ 65 − 14
_5_ tens _1_ ones = _51_

❷ 76 − 48
_2_ tens _8_ ones = _28_

❸ 77 − 29
_4_ tens _8_ ones = _48_

❹ 83 − 67
_1_ tens _6_ ones = _16_

Hundreds	Tens	Ones

Solve each subtraction problem. Show your answers on the chart. Write the difference in standard form.

❺ 91 − 53 _38_

Hundreds	Tens	Ones
	3 rods	8 unit blocks

❻ 134 − 85 _49_

Hundreds	Tens	Ones
	4 rods	9 unit blocks

Practice

Find each difference. Write how many times you made trades.

❼ 246 − 68
_178_   _2_ trades

❽ 935 − 278
_657_   _2_ trades

❾ 281 − 76
_205_   _1_ trades

❿ 806 − 275
_531_   _1_ trades

⓫ 756 − 299
_457_   _2_ trades

⓬ 515 − 436
_79_   _2_ trades

Reflect

Find the difference. Explain your work for the problem below. Describe each trade and why it is necessary.

436 − 279

The difference is 157. Because you cannot subtract 9 from 6, trade 1 ten for 10 ones. Now subtract 9 from 16 to get 7 in the ones place. There are 2 in the tens place now, and 7 cannot be subtracted from 2. So trade 1 hundred for 10 tens. Now subtract 7 from 12 in the tens column. Place 5 in the tens place. There are 3 in the hundreds place now. 3 − 2 = 1. Place 1 in the hundreds place.

 Reflect 10

Extended Response REASONING

Review students' answers to the Reflect prompt at the bottom of student page 15. Compare the different ways that explanations were presented.

Real-World Application APPLYING

■ Suppose that you are in charge of supplies for a big birthday party. You buy a package of 300 paper plates. After the party, you have 36 plates. How can you find the number of plates used by the party goers? subtraction

■ How many plates were used? Write a number sentence. 300 − 36 = 264

For more practice, replace *plates* in the above scenario with each of the following:
cups: 500 − 163 = 337; forks: 325 − 89 = 236;
spoons: 275 − 117 = 158

 Assess

Informal Assessment

Use the Student Assessment Record, **Assessment,** page 100, to record informal observations.

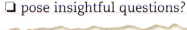

UNDERSTANDING

May I Regroup Blocks?
Did the student
❏ make important observations?
❏ extend or generalize learning?
❏ provide insightful answers?
❏ pose insightful questions?

Lesson 3

Objective

Students identify which combination of numbers has a greater sum.

Materials

Program Materials
3-Digit Window, p. B4

Additional Materials
Number Cards (0–9)

Access Vocabulary

greatest sum The largest sum
least sum The smallest sum

Creating Context

In order for English Learners to understand the instructions and the concept of least sum and lesser addends, they may need to learn or review the comparative and superlative forms in English. Review this chart with English Learners, and collect additional examples from lessons this week.

POSITIVE	COMPARATIVE	SUPERLATIVE
great	greater	greatest
less	lesser	least
small	smaller	smallest
big	bigger	biggest

1 Warm Up — 5

Concept Building COMPUTING

Give each student a set of Number Cards. Ask students to turn their cards facedown and randomly select three cards. Instruct students to use these cards to form the greatest number possible. Determine which student was able to form the greatest number in the class. Be aware that there could be more than one student with the greatest number.

2 Engage — 30

Skill Building

Each player begins with a blank 3-Digit Window. Shuffle the Number Cards, and turn them facedown.

Window Addition Game COMPUTING

- The goal of the game is to make the greatest sum.
- One number card at a time will be turned over. As soon as the card is shown, write the digit in any of the six windowpanes.

When the sixth card is shown, instruct students to calculate their sums.

- To make the greatest sum, which digits are best to place in the hundreds column? 8 or 9

Strategy Building

600 Addition Game REASONING

- The goal of the game this time to is to make a sum as close as possible to 600.
- We play the same way as in the previous game.

When the sixth card is shown, instruct students to calculate their sums.

- What were your strategies for reaching the target number 600? You should not choose to put any number greater than 5 in the left panes.

Monitoring Student Progress

If . . . students want to play another addition game,

Then . . . play **The Broken Calculator.** A key is broken on the calculator. Find a way to use the broken calculator to solve a problem.

Building Blocks For additional practice, students should complete **Building Blocks** Eggcellent Addition.

MathTools Use Function Machine to demonstrate and explore addition and subtraction.

Using Student Pages

Have students complete **Workbook,** pages 16–17, on their own.

Lesson 3

Key Idea

| 10 ones = 1 ten | 10 tens = 1 hundred | 10 hundreds = 1 thousand |

Try This
Find each sum.

1. 327 + 516 **843**
2. 812 + 658 **1,470**

516 + 327 **843**
658 + 812 **1,470**

3. Can you change the order of the numbers being added and get the same sum?
Yes

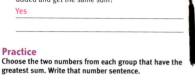

Practice
Choose the two numbers from each group that have the greatest sum. Write that number sentence.

4. 344 517 189
344 + 517 = 861

5. 954 451 658
954 + 658 = 1,612

6. 167 678 563
678 + 563 = 1,241

7. 972 872 927
972 + 927 = 1,899

Choose the two numbers from each group that have the smallest sum. Write that number sentence.

8. 692 453 634
453 + 634 = 1,087

9. 684 312 158
312 + 158 = 470

10. 527 338 491
338 + 491 = 829

Answer each question.

11. What is the greatest number you can add to 499 and have a sum less than 900? Write the number sentence.
400; 499 + 400 = 899

12. What is the least number you can add to 502 and have a sum greater than 600? Write the number sentence.
99; 502 + 99 = 601

Choose two numbers from each group that have the greatest sum. Write that number sentence.

13. 422 305 308
422 + 308 = 730

14. 224 243 202
243 + 224 = 467

Reflect
Explain how you decided which numbers to use to write the number sentence for Problem 8. Write equations to show that you made the correct choice of numbers.

The smallest sum comes from adding the smallest numbers together.
There are two possible ways to write the correct number sentence.
The other number sentences show that any combination of the other
numbers has a sum that is greater than 1,087.
453 + 634 = 1,087 or 634 + 453 = 1,087
453 + 692 = 1,145
634 + 692 = 1,326

Reflect 10

Extended Response REASONING

Review students' answers to the Reflect prompt at the bottom of student page 17.

Discuss the ways students can show that the correct choice of numbers was made. As a class, form a general rule about the relationship of the numbers. Then have students apply that rule to these numbers: 291, 335, 609. Order the numbers from greatest to least. Use the middle number minus the least number to get the least possible difference: 335 − 291 = 44.

Real-World Application APPLYING

Target numbers are commonly used in the real world. Discuss a situation, such as going to the grocery store to buy two items with $5. Use $5 as the target number, and talk about the need to estimate the prices of the two items to make sure that the sum of their prices is less than $5.

Assess

Informal Assessment

Use the Student Assessment Record, **Assessment,** page 100, to record informal observations.

COMPUTING	REASONING
Window Addition Game	**600 Addition Game**
Did the student	Did the student
❑ respond accurately?	❑ provide a clear explanation?
❑ respond quickly?	❑ communicate reasons and strategies?
❑ respond with confidence?	❑ choose appropriate strategies?
❑ self-correct?	❑ argue logically?

Sharpening Computation Skills • Lesson 3 16–17

Lesson 4

Objective

Students identify which combination of numbers has a greater difference.

Materials

Program Materials
3-Digit Window, p. B4

Additional Materials
Number Cards (0–9)

Access Vocabulary

greatest difference The largest difference
least difference The smallest difference

Creating Context

In the lessons this week, we see that the commutative property does not apply to subtraction. Work with English Learners to write and illustrate an explanation of this law in mathematics. A helpful exercise for English Learners is to provide sentence frames to complete. In this way students can show what they know while practicing the forms and they can gain the English fluency needed to advance in academic English.

1 Warm Up 5

Concept Building COMPUTING

Give each student a set of Number Cards. Ask students to turn their cards facedown and randomly select three cards. Instruct students to use these cards to build the smallest number possible.

Determine which student was able to form the smallest number in the class. Be aware that there could be more than one student with the smallest number.

2 Engage 30

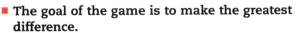
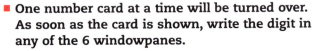

Skill Building

Window Subtraction Game COMPUTING

- The goal of the game is to make the greatest difference.
- One number card at a time will be turned over. As soon as the card is shown, write the digit in any of the 6 windowpanes.

When the sixth card is shown, instruct students to calculate their differences.

- **Which digits are best to place in the hundreds place to make the greatest difference?** one high number and one lower number

Strategy Building

Least Subtraction Game REASONING

- The goal of the game this time is to make the smallest difference.
- We play the same way as the previous game.

When the sixth card is shown, instruct students to calculate their differences.

- **What were your strategies for making the least difference?** You want to put numbers that are closest in value in the two left panes.

Monitoring Student Progress

| If . . . students want to vary the game or make it more challenging, | Then . . . encourage them to change the object to "make a difference of 200." |

Building Blocks For additional practice, students should complete **Building Blocks** Eggcellent Addition.

MathTools Use Function Machine to demonstrate and explore addition and subtraction.

Using Student Pages

Have students complete **Workbook,** pages 18–19, on their own.

Lesson 4

Key Idea

1 thousand = 10 hundreds   1 hundred = 10 tens   1 ten = 10 ones

Try This
Find each difference. Answer each question.

1. $961 - 735$ 226

Can you change the order of the numbers and get the same difference?
no

2. $372 - 186$ 186

Do you get the same answer if you write the problem as $186 - 372$?
no

3. What conclusion can you draw after finding the differences in Problems 1 and 2?
The order in which the numbers are written in the problem does affect the difference.

Practice
Choose the two numbers from each group that have the greatest difference. Write that number sentence.

4. 952   192   663
$952 - 192 = 760$

5. 825   764   736
$825 - 736 = 89$

6. 222   475   900
$900 - 222 = 678$

7. 820   360   690
$820 - 360 = 460$

Choose the two numbers from each group that have the smallest difference. Write that number sentence.

8. 315   682   794
$794 - 682 = 112$

9. 185   167   213
$185 - 167 = 18$

10. 956   188   117
$188 - 117 = 71$

Answer each question.

11. What is the greatest number you can subtract from 350 and have a difference of at least 200? Write the number sentence.
150; $350 - 150 = 200$

12. What is the smallest number you can subtract from 850 and have a difference no greater than 500? Write the number sentence.
350; $850 - 350 = 500$

Reflect
Briefly describe how you chose the numbers you used to write the number sentences in Problems 4–7. How did you choose the numbers in Problems 8–10?
Possible answers: I used the biggest number and the smallest number for 4–7, and I used the numbers that were closest together in 8–10.

---

3 Reflect   10

Extended Response   REASONING

Review students' answers to the Reflect prompt at the bottom of student page 19.

Invite several students to read their general statements aloud. As a class, work together to create a general statement that is accurate and understandable by all.

Real-World Application   APPLYING

Tell a story about a situation that involves subtraction. The main character in the story should be trying to achieve the greatest difference or the smallest difference. Include strategies that the character used.

---

4 Assess

Informal Assessment

Use the Student Assessment Record, *Assessment,* page 100, to record informal observations.

COMPUTING

Window Subtraction Game

Did the student
- ❏ respond accurately?
- ❏ respond quickly?
- ❏ respond with confidence?
- ❏ self-correct?

REASONING

Least Subtraction Game

Did the student
- ❏ provide a clear explanation?
- ❏ communicate reasons and strategies?
- ❏ choose appropriate strategies?
- ❏ argue logically?

Sharpening Computation Skills • Lesson 4   18–19

Lesson 5 Review

Objective

Students review skills learned this week and complete the weekly assessment.

Materials

Review materials will be selected from those used in previous activities.

Creating Context

Working in small groups is an excellent strategy for English Learners because it provides opportunities for verbal practice in a low-stress situation. For real-world application, have English Learners work with partners to create their story problems. English Learners may find it helpful to first discuss strategy with partners who speak the same primary language. For students of early English proficiency, it may be helpful to provide a story frame to be completed.

1 Warm Up 5

Concept Building COMPUTING

Provide students with a random number of units, rods, and flats. Instruct students to write the number modeled by all the blocks at their desks. Then instruct students to form two groups with the blocks and to write an addition sentence that shows the sum. Then instruct students to form two groups with the blocks and to write a subtraction sentence that shows the difference.

2 Engage 20

Skill Building

Free-Choice Activity

For the last day of the week, allow students to choose an activity from the previous lessons. Some activities they may choose include the following:

- **Trading at the Top**
- **May I Regroup Blocks?**
- **Window Addition Game**
- **600 Addition Game**
- **Window Subtraction Game**
- **Least Subtraction Game**

Make a note of the activities students select. Do they prefer easy or challenging activities? If you believe your students would benefit from extra practice on specific skills, choose an activity for them.

Monitoring Student Progress

If . . . students are not participating, **Then . . .** continue the activity until all students have participated.

Using Student Pages

Have students complete **Workbook,** pages 20–21, on their own.

3 Reflect 10

Extended Response APPLYING

Review students' answers to the Reflect prompts at the bottom of student pages 20–21.

Discuss these answers with the group to reinforce Week 3 concepts.

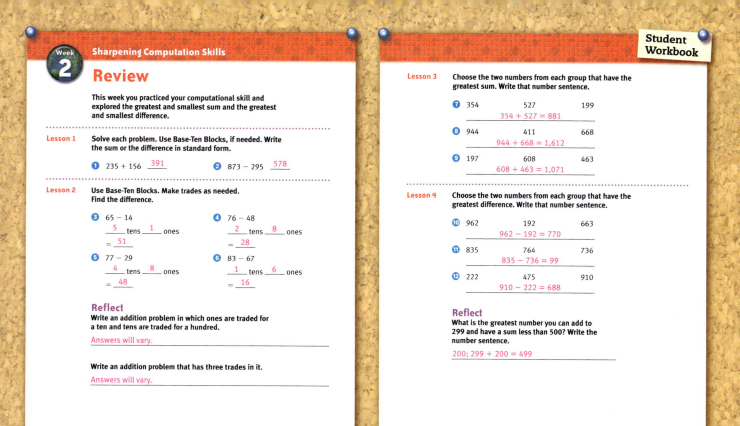

Week 2 — Sharpening Computation Skills

Review

This week you practiced your computational skill and explored the greatest and smallest sum and the greatest and smallest difference.

Lesson 1
Solve each problem. Use Base-Ten Blocks, if needed. Write the sum or the difference in standard form.

❶ 235 + 156 = __391__   ❷ 873 − 295 = __578__

Lesson 2
Use Base-Ten Blocks. Make trades as needed. Find the difference.

❸ 65 − 14
__5__ tens __1__ ones
= __51__

❹ 76 − 48
__2__ tens __8__ ones
= __28__

❺ 77 − 29
__4__ tens __8__ ones
= __48__

❻ 83 − 67
__1__ tens __6__ ones
= __16__

Reflect
Write an addition problem in which ones are traded for a ten and tens are traded for a hundred.
Answers will vary.

Write an addition problem that has three trades in it.
Answers will vary.

Lesson 3
Choose the two numbers from each group that have the greatest sum. Write that number sentence.

❼ 354   527   199
354 + 527 = 881

❽ 944   411   668
944 + 668 = 1,612

❾ 197   608   463
608 + 463 = 1,071

Lesson 4
Choose the two numbers from each group that have the greatest difference. Write that number sentence.

❿ 962   192   663
962 − 192 = 770

⓫ 835   764   736
835 − 736 = 99

⓬ 222   475   910
910 − 222 = 688

Reflect
What is the greatest number you can add to 299 and have a sum less than 500? Write the number sentence.
200; 299 + 200 = 499

---

4 Assess   10

A Gather Evidence

Formal Assessment

Have students complete the Weekly test on **Assessment,** pages 42–43. Record scores on the Student Assessment Record, **Assessment,** page 100.

B Summarize Findings

Determine whether students have Minimal, Basic, or Secure understanding of the concepts presented in Week 2.

C Differentiate Instruction

Based on your observations, use these teaching strategies next week to follow up.

Minimal Understanding
- Repeat the Warm-Up and Engage activities to develop Week 2 concepts.
- Use **Building Blocks** Two-Digit Adding and **eMathTools** Function Machine to develop and reinforce Week 2 concepts.

Basic Understanding
- Repeat Engage activities in subsequent weeks.
- Use **Building Blocks** Eggcellent Addition and **eMathTools** Function Machine to reinforce Week 2 concepts.

Secure Understanding
Use Building Blocks Eggcellent Addition and **eMathTools** Function Machine to extend Week 2 concepts.

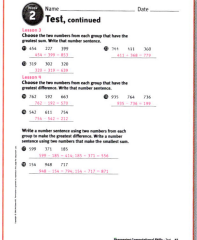

Assessment, pp. 42–43

Week 3

Computational Estimation

Week at a Glance

This week, students continue with *Number Worlds,* Level F, Addition and Subtraction. Students will explore when and why to estimate sums and differences as well as several methods for how to estimate.

Math Background

Estimation serves as an important companion to computation. It provides a tool for judging the reasonableness of calculator, mental, and paper-and-pencil computations. However, being able to compute exact answers does not automatically lead to an ability to estimate or judge the reasonableness of answers.

—*Principles and Standards for School Mathematics.* 2000. National Council of Teachers of Mathematics. Page 155.

How Students Learn

Estimating helps students develop confidence in their abilities to reason with numbers. Students should value estimation as an important part of computation, not as an "add-on" or "extra" activity. For some problems an estimated answer is sufficient, but many problems require an exact answer. Whenever an exact answer is needed, estimation is still useful. Students can estimate before or after computing to check for reasonableness. Students should make estimating a habit when solving problems.

Teaching for Understanding

Observe closely while evaluating the assigned tasks this week to see whether students can demonstrate the following understandings.

Benchmark after Lesson 2: Students can identify an estimate and an exact answer, as well as when each type of answer is appropriate.

Benchmark after Lesson 3: Students can use multiple strategies for estimating.

Benchmark after Lesson 4: Students should be able to problem solve by selecting a strategy that is best for each problem and their own abilities.

Skills Focus

- Recognize estimation as a valuable way of checking a computation.
- Determine a result as reasonable or not.
- Distinguish between situations in which exact answers are needed and situations in which estimated answers are sufficient.
- Learn different strategies of how to estimate.

Math at Home

Give one copy of the Letter to Home, page A11, to each student. Encourage students to share and complete the activity with their caregivers.

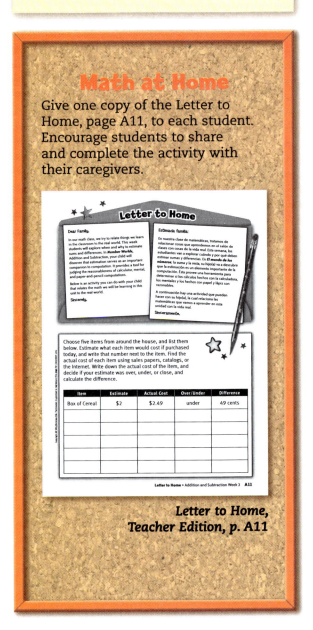

Letter to Home,
Teacher Edition, p. A11

PACING	LESSON	LEARNING GOALS	NCTM	MATERIALS	TECHNOLOGY
DAY 1	pages 22–23	Students recognize that a "nice number" can be used to estimate an answer.	• Number and Operations • Problem Solving • Communication	• Calculator • Base-Ten Blocks*	**Building Blocks** Number Snapshots
DAY 2	pages 24–25	Students determine when an estimated answer is sufficient.	• Number and Operations • Problem Solving • Communication	Paddle Signs, p. B10	**Building Blocks** Rocket Blast
DAY 3	pages 26–27	Students use multiple strategies for estimating.	• Number and Operations • Problem Solving • Communication	Calculator	**Building Blocks** Number Snapshots
DAY 4	pages 28–29	Students use rounding for estimating.	• Number and Operations • Problem Solving • Communication	Calculator	**Building Blocks** Rocket Blast
DAY 5	Review and Assess pages 30–31	Students review skills learned this week and complete the weekly assessment.	• Number and Operations • Problem Solving • Communication	Materials will be selected from Lessons 1–4.	**Building Blocks** Review previous activities

Math Vocabulary

overestimate An estimate that is greater than the exact answer

underestimate An estimate that is less than the exact answer

front-end estimation Using the leftmost digits to estimate and then making adjustments based on the rest of both numbers

rounding Estimating a number to a given place value

English Learners

SPANISH COGNATES

English	Spanish
estimation	estimación
exact	exacto
compare	comparar
reasonable	razonable

ALTERNATE VOCABULARY

the most reasonable estimate The estimate that seems to fit best, given what you know

strategy A plan of action based on knowledge and experience with the problem

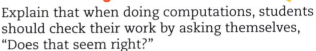
Lesson 1

Objective

Students recognize that "nice numbers" close to the numbers in a problem can be used to estimate the answer.

Materials

Additional Materials
- Calculator
- Base-Ten Blocks

Access Vocabulary

overestimate An estimate that is more than the exact answer

underestimate An estimate that is less than the exact answer

show your work Write down all the calculations you made to find the answer

Creating Context

Prepositions in English are difficult for students of beginning English proficiency but are especially important in mathematics. In this lesson we learn about *over*estimates and *under*estimates. A helpful strategy for English Learners is to post each key preposition next to a picture so that students can refer back to check its meaning. English Learners should keep a word book with key words and phrases they learn to reference at any time.

1 Warm Up   5

Concept Building UNDERSTANDING

- **What is estimation?**
- **How is estimation different from guessing?**
- **Can you think of situations in everyday life when you might want to estimate?**
- **Can you think of situations when using an estimation problem would not be best?** Possible answer: when paying money to buy something; when counting the number of people in a class to make sure that you have enough books

2 Engage   30

Skill Building

Explain that when doing computations, students should check their work by asking themselves, "Does that seem right?"

- **What are some words we use when we estimate?** Possible answers: *about, around, almost, about right, close to*

Mistake or Not? UNDERSTANDING

Display the following problem on the board.

$$301 - 48 = 153$$

- **Without solving the problem, does the answer seem about right, or reasonable?**
- **Use estimation to make your judgment.**

Give students time to consider the problem, and then call on several students to report their answers and reasoning.

If the discussion does not yield the following explanation, offer it as a strategy for estimation. Write each step on the board next to the original problem.

- **301 is about 300. 48 is about 50. 300 − 50 = 250. The answer is not reasonable. The correct solution should be about 250.**

Let students see that you are not using the actual numbers, but *easier* numbers that are *close* to the actual numbers.

- **Can anyone explain the strategy in his or her own words?**

Allow one or more students to repeat the strategy. If no student is able to do this, repeat your explanation until students are able to offer it on their own.

Monitoring Student Progress

If . . . students do not understand the concept of "reasonableness,"

Then . . . help students see that in one sense *reasonable* means "close to."

 Building Blocks For additional practice with estimation, students should complete **Building Blocks** Number Snapshots.

Using Student Pages

Have students complete **Workbook,** pages 22–23, on their own.

Lesson 1

Key Idea
You can use "nice numbers" that are close to the numbers in a problem to estimate the answer.

Try This
Circle the estimation that is most reasonable. Do not calculate.

1 39 + 28
   50   (70)

2 237 − 140
   (100)   200

3 96 + 244
   (350)   250

4 658 − 429
   100   (200)

Write a "nice number" that is close to the given number and would be useful in an estimation. There may be more than one right answer.

5 497 _____ 500
6 552 _____ 550
7 9,938 _____ 10,000; 9,950
8 103 _____ 100
9 412 _____ 400; 410
10 137 _____ 140; 150

Practice
Circle the phrase that best describes an estimate for each total. Do not calculate.

11 56 + 53
   under 100
   (over 100)

12 874 − 228
   (under 1,000)
   over 1,000

13 98 + 72
   (under 200)
   over 200

14 121 − 42
   (under 100)
   over 100

15 682 + 45
   under 700
   (over 700)

16 57 − 49
   (under 10)
   over 10

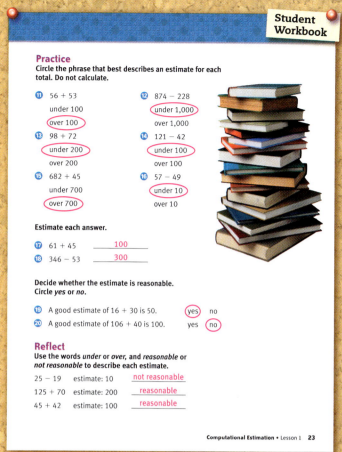

Estimate each answer.

17 61 + 45 _____ 100
18 346 − 53 _____ 300

Decide whether the estimate is reasonable. Circle *yes* or *no*.

19 A good estimate of 16 + 30 is 50.   (yes)   no
20 A good estimate of 106 + 40 is 100.   yes   (no)

Reflect
Use the words *under* or *over*, and *reasonable* or *not reasonable* to describe each estimate.

25 − 19   estimate: 10 _____ not reasonable
125 + 70   estimate: 200 _____ reasonable
45 + 42   estimate: 100 _____ reasonable

3 Reflect   10

Extended Response   REASONING
Review students' answers to the Reflect prompt at the bottom of student page 23.

- **How do you decide whether an estimate is reasonable?**

Discuss different ways to check for the reasonableness of the result.

Real-World Application   APPLYING
- **Your school has 517 students enrolled, and 59 are absent today. Use "nice numbers" to estimate how many students have come to school today.** 500 − 50 = 450; about 450.
- **What is the actual number?** 517 − 59 = 458
- **Was your estimate reasonable?** yes

4 Assess

Informal Assessment
Use the Student Assessment Record, **Assessment,** page 100, to record informal observations.

UNDERSTANDING

Mistake or Not?
Did the student
- ❏ make important observations?
- ❏ extend or generalize learning?
- ❏ provide insightful answers?
- ❏ provide insightful questions?

Lesson 2

Objective

Students determine when an estimated answer is sufficient.

Materials

Program Materials
Paddle signs (one *Quick Estimate,* one *Exact Answer*), p. B10

Access Vocabulary

daily life Common routines we experience every day

store clerk A store employee who helps customers

White House The place where the President of the United States works and lives in Washington, D.C.

Tip Gratuity, a percentage of the food bill in a restaurant given to the server for good service

Creating Context

One difficulty that English Learners encounter in the classroom is the use of instructions and directions for completing assignments. Help students know that if they do not understand the instructions, they should ask someone before they begin, because they may not be alone in wondering what they are supposed to do. Review the written directions in this lesson by asking students to describe what the directions mean.

1 Warm Up 5

Concept Building REASONING

For each situation listed below, ask students to describe an instance in which you would need to know an exact amount and an instance in which you would be fine with an estimated amount.

- shopping at the grocery store
- allowance
- movie theater

2 Engage 30

Skill Building

Provide each student with two paddle signs (similar to those used in auctions). One sign reads *Quick Estimate,* and the other reads *Exact Answer.*

Estimate or Exact? UNDERSTANDING

Present each of the real-world scenarios listed below one at a time. When you name the scenario, students are to hold up one of their signs to indicate whether a quick estimate is sufficient for the situation or an exact answer is needed. If most of the class agrees, move on to the next scenario. If the class is divided in their responses, then discuss the situation and guide students to agree.

- **the amount you pay a cashier at a bookstore** exact
- **the length of time the previews and the movie plays** estimate
- **the number of people at the football game** estimate
- **the distance the shuttle will travel in space as reported on the news** estimate
- **the amount of money paid for a job completed** exact
- **the amount of money paid for the check at a restaurant** exact
- **the length of time spent shopping at a mall** estimate
- **the payment for an automatic car wash** exact

Monitoring Student Progress

If . . . one or two students answer differently from the rest of the class,

Then . . . talk with each student about his or her responses to help the student better understand the scenario.

Building Blocks For additional practice with estimation, students should complete **Building Blocks** Rocket Blast.

Using Student Pages

Have students complete **Workbook,** pages 24–25, on their own.

Lesson 2

Key Idea
Many computational problems arise in daily life. Often it is not necessary to find the exact answer. An estimate may be enough.

Try This
Write whether the story tells about an estimated amount or an exact amount.

1. The grocery store clerk said that Karen owed $18.34. — exact amount
2. Marcos needed about $10 to get refreshments. — estimated amount
3. Janine's phone bill in May was $56.05. — exact amount
4. Wade wants his car payment to be less than $250 a month. — estimated amount

Practice
Underline the item that would be most likely to be described by using an exact amount.

5. a. the number of people in the sports arena   b. the admission price to the zoo
6. a. the distance from your home to the White House   b. the cost to mail a package
7. a. the score on Geoff's spelling quiz   b. the length of time a baseball game lasts
8. a. the prices on the menu at Jay's Steakhouse   b. the amount of tip to leave

Describe something that matches each statement.

9. A time today when you will need an exact amount — Answers will vary.
10. A time yesterday when an estimate was fine — Answers will vary.

Choose the best type of estimate for each situation.
A. overestimate   B. underestimate   C. no estimate, exact amount
Explain your choice.

11. Heidi's mother gives her money to go to the store to get milk, eggs, and bread. — A
Her mother wants Heidi to have enough money to buy the items.

12. Mr. Young has a meeting at 1:00. He checks the distance from his office to the meeting site to estimate when he needs to leave. — A
He needs to be on time for the meeting, so he should allow extra time.

Reflect
Write a story about something that happened in your life when you estimated an amount. Explain whether estimating was the best choice or whether you should have used an exact amount.

Stories and explanations will vary.

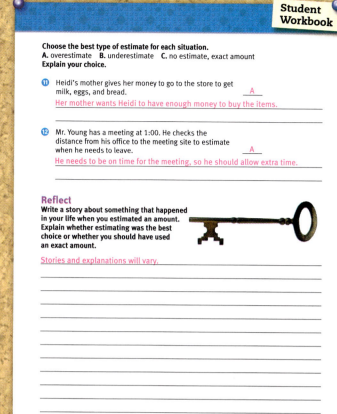

Reflect   10

Extended Response   REASONING
Before students answer the Reflect prompt, lead the class in a brainstorming session. This will get students thinking about different situations that have occurred in their lives. Then review students' answers to the Reflect prompt at the bottom of student page 25.

- **Do you think that people estimate more often than they calculate exact answers? Explain your answer.**

Real-World Application   APPLYING
- **Do bus schedules require exact times or estimates? Explain your answer.** Answers will vary, and disagreement is acceptable. Some students may feel that a bus driver cannot arrive at every bus stop at exactly the same time every day. Others may express that the bus arrives and departs each stop at close enough to the correct time to be considered exact.
- **Can you think of a situation where a reasonable estimate is not good enough?** Answers will vary.

Assess

Informal Assessment
Use the Student Assessment Record, **Assessment,** page 100, to record informal observations.

UNDERSTANDING

Estimate or Exact?
Did the student
❏ make important observations?
❏ extend or generalize learning?
❏ provide insightful answers?
❏ pose insightful questions?

Computational Estimation • Lesson 2   24–25

Lesson 3

Objective

Students use different strategies for estimating.

Materials

Additional Materials
Calculator

Access Vocabulary

Front-End Estimation Using the leftmost digits to estimate and then making adjustments based on the rest of both numbers

Reference-Point Estimation Using "nice numbers" to estimate

Creating Context

This lesson uses Front-End Estimation. Think of a car as having two ends, a *front end* and a *back end*. Multidigit numbers have a front end which is the digit with the greatest place value and a back end which is the digit in the ones place.

1 Warm Up   5

Concept Building   COMPUTING

- **Why is estimation important?** Answers will vary.
- Sometimes it is more important to arrive at a reasonable answer very quickly rather than to arrive at an exact answer slowly.
- Today you will learn two different estimation strategies, Front-End Estimation and Reference-Point Estimation.

2 Engage   30

Strategy Building

Estimation Strategies   COMPUTING

- **I will first demonstrate how to Front-End Estimate this problem: 785 + 219.**
- **Add the two highest place values of each number.**
- **Change all the other place values to 0.**
- **Add 780 + 210 to get 990. The Front-End Estimate is 990.**
- **Now try this problem on your own: 449 + 308.** Ask volunteers to write their estimates on the board and to explain the process they used to find the estimated sum. 740
- **Can anyone explain why we call this Front-End Estimating?** Allow several students to answer.

Make sure students understand that they are adding only the largest place values in each number and forgetting about the smaller place values to obtain an estimate of the sum.

- **Another method of estimation is Reference-Point Estimation.** Explain that reference points are an informal way to estimate. Reference points are "nice numbers" that are easy to work with. Common reference points are 5, 10, 50, and 100.
- **Estimate 116 to the nearest 50. Is it closer to 100 or 150?** 100
- **Estimate 527 to the nearest 100. Is it closer to 500 or 600?** 500
- **Which estimate is closer to the actual number?** estimating to the nearest 50
- **Which is easier, estimating to the nearest 50 or estimating to the nearest 100?** Probable answer: estimating to the nearest 100
- **Can anyone tell me why we might estimate to the nearest 50 even if it is harder to do?** It gives you a more precise estimate, and sometimes you might need to be as close to the actual number as possible.

 Building Blocks For additional practice estimating, students should complete *Building Blocks* Number Snapshots.

Using Student Pages

Have students complete **Workbook,** pages 26–27, on their own.

Lesson 3

Key Idea

Two strategies for good estimating include the following:

- Reference Point Estimation
- Front-End Estimation

Try This

Estimate by using 50 and multiples of 100. Show your work.

1. 461 + 911
   450 + 900 = 1,350

2. 844 − 187
   850 − 200 = 650

3. 2,631 + 1,487
   2,500 + 1,500 = 4,000

4. 921 − 448
   900 − 450 = 450

Estimate each sum by using Front-End Estimation. Explain your work.

5. 3,089 + 5,748
   9,000; 3,000 + 5,000 = 8,000, but 89 + 748 is greater than 500, so adjust to 9,000.

6. 623 + 311
   900; 600 + 300 = 900, and 23 + 11 is less than 50, so no adjustment is needed.

Practice

Estimate by using any strategy. Find the exact answer. Then use a calculator to check your work.

7. 428 + 796   estimate: 1,200   exact answer: 1,224   check: ✔
8. 9,236 − 7,058   estimate: 2,000   exact answer: 2,178   check: ✔
9. 276 − 79   estimate: 200   exact answer: 197   check: ✔
10. 1,926 − 208   estimate: 1,800   exact answer: 1,718   check: ✔
11. 2,568 + 3,537   estimate: 6,500   exact answer: 6,105   check: ✔

Reflect

Estimate the following problem by using both methods. Find the exact answer. Compare each estimation to the exact answer and to the other estimations.

5,617 + 2,933

Estimate by using 500:   5,500 + 3,000 = 8,500

Estimate by using Front-End Estimation: _____
5,000 + 2,000 = 7,000, but 617 is greater than 500, and 933 is very close to 1,000, so adjust by increasing 1,500; 8,500

Exact answer: 8,550

Comparisons: _____
Both estimates are very close to the exact answer. The first estimate was slightly easier to do than the second estimate.

 Reflect   10

Extended Response   REASONING

Review students' answers to the Reflect prompt at the bottom of student page 27.

Discuss the different strategies used in this lesson.

- **Which strategy is your favorite?**
- **Did you use the same strategy for each problem? Explain.**

Helping students understand that the best strategy to use may be different for different problems.

Real-World Application   APPLYING

- **Before you make a purchase, you should estimate the total cost. This way when the sales clerk tells you the amount you owe, you should know whether an error was made computing the sale.**

---

4 Assess

Informal Assessment

Use the Student Assessment Record, **Assessment,** page 100, to record informal observations.

COMPUTING

Estimation Strategies

Did the student
- ❏ respond accurately?
- ❏ respond quickly?
- ❏ respond with confidence?
- ❏ self-correct?

Lesson 4

Objective

Students use rounding for estimating.

Materials

Additional Materials
Calculator

Access Vocabulary

rounding Estimating a number to a given place value, approximating the nearest 10 or 100

Creating Context

An excellent strategy for English Learners is to illustrate the concept by making a graphic display of information or a drawing of the situation. Throughout this lesson, there are examples of situations for estimation. Examine closely the drawings of English Learners to assess their comprehension of the estimation.

1 Warm Up   5

Concept Building   COMPUTING

■ A third estimation strategy is *rounding*.

Use the numbers below to show how to round.

43 40   768 800   221 220
1,462 1,500   6,712 7,000

■ If the digit to the right of the underlined digit is less than 5, change all digits to the right of it to 0.

■ If the digit is 5 or greater than 5, increase the underlined digit by 1, and change all digits to the right of it to 0.

2 Engage   30

Strategy Building

■ If you have a two-digit number, which digit should be rounded? the digit in the tens column

■ If you have a three-digit number, which digit should be rounded? either the digit in the tens column or the digit in the hundreds column

Comparing Strategies   COMPUTING

Arrange students into groups of three so each can use a different estimation strategy.

■ **Estimate the sum: $7.24 + $10.53.** Reference-Point = $17.75; Rounding = $18; Front-End = $17

■ **Compare estimates. Who has the same amount as their estimates? Are all estimates reasonable?**

■ **Switch strategies so each student is using a different strategy than last time.**

■ **Estimate the difference: 3,265 − 1,897.** Rounding = 1,000; Front-End = 1,400; Reference-Point = 1,300

Instruct students to make comparisons again and to switch strategies again.

■ **Estimate the sum: 482 + 250.** Reference-Point = 750; Rounding = 800; Front-End = 730

Instruct students to make comparisons again.

Monitoring Student Progress

If . . . students' estimates differ,

Then . . . encourage them to use calculators to find the exact answers before switching strategies and moving to the next problem.

 For additional practice estimating, students should complete **Building Blocks** Rocket Blast.

Using Student Pages

Have students complete **Workbook,** pages 28–29, on their own.

Lesson 4

Key Idea
Rounding is an additional strategy that can be used for estimating.

Try This
Estimate by rounding. Show your work.

1. Carrie has three piggy banks in which she saves coins. One bank has 59¢, another has 38¢, and the third one has 86¢. About how much money does Carrie have altogether?

$2.00;
60¢ + 40¢ + 90¢

2. Daniel played in four basketball games. He scored 8 points in the first and fourth games. He scored 11 points in the second game and 4 points in the third game. About how many points did Daniel score in all four games combined?

35;
10 + 10 + 10 + 5

3. Kanisha watched five videos in one week. Two of the videos were 98 minutes each. Two of the videos were 138 minutes each. One video was 79 minutes. How many minutes of videos did Kanisha watch that week altogether?

600;
100 + 100 + 140 + 140 + 80

Practice
Estimate the total amount. Use any strategy.

4.   45 67 84 _____200_____ jelly beans

Estimate. Show your work, and explain the strategy you used.

5. Heiko has $4.00. Which two items can he buy and have more than 10 cents but less than 50 cents left?

$2.49 $3.99 $1.51 $1.22

ball and truck; The lunch box uses all but one cent of the $4. The ball and airplane are exactly $4. The airplane and truck total less than $3. Strategies will vary.

Reflect
Write a word problem that fits the picture. Estimate the answer.

761   1,089   675   94

Word problems will vary.
Estimate: 800 + 1,100 + 700 + 100 = 2,700.

3 Reflect   10

Extended Response   REASONING
Review students' answers to the Reflect prompt at the bottom of student page 29.

- **Is rounding easier or more difficult than the other methods? Explain.** Answers and explanations will vary.

Real-World Application   APPLYING
Many times in real-life situations an exact number may be known, but estimation is used instead.

- **A survey asks for the age of the person using ranges. For example, choices are 10–15, 16–25, and over 25.**
- **Name other situations in which the exact answer is known but estimation is used.** Possible answers: distance, money, calories

4 Assess

Informal Assessment
Use the Student Assessment Record, **Assessment**, page 100, to record informal observations.

COMPUTING

Comparing Strategies
Did the student
- ❏ respond accurately?
- ❏ respond quickly?
- ❏ respond with confidence?
- ❏ self-correct?

Lesson 5 Review

Objective

Students review skills learned this week and complete the weekly assessment.

Materials

Review materials will be selected from those used in previous activities.

Creating Context

The prepositions of English are often part of idiomatic phrases that can be puzzling to English Learners. These idioms add meaning to a situation but can exclude listeners who are unfamiliar with their use. For example, the word *over* has a number of different meanings, such as *above*, as in *He has a roof over his head*, and *finished*, as in *The movie is over*. English Learners may collect these phrases in a word book and illustrate the literal meanings along with the common meanings.

1 Warn Up   5

Concept Building UNDERSTANDING

■ Yancy went to the store to buy fruit. He spent $5.89. Is this an estimated amount or an exact amount? exact Name a reasonable estimate for the amount Yancy spent. about $6

■ Constance gave her children $4.75 each week as an allowance. Is this an estimated amount or an exact amount? exact Name a reasonable estimate for the amount Constance gave. about $5.00

2 Engage   20

Skill Building

Free-Choice Activity

For the last day of the week, allow students to choose an activity from the previous lessons. Some activities they may choose include the following:

- **Mistake or Not?**
- **Estimate or Exact?**
- **Estimation Strategies**
- **Comparing Strategies**

Make a note of the activities students select. Do they prefer easy or challenging activities? If you believe your students would benefit from extra practice on specific skills, choose an activity for them.

> **Monitoring Student Progress**
>
> **If . . .** students are not participating,
>
> **Then . . .** continue the activity until all students have participated.

Using Student Pages

Have students complete **Workbook,** pages 30–31, on their own.

3 Reflect   10

Extended Response APPLYING

Review students' answers to the Reflect prompts at the bottom of student pages 30–31.

Discuss these answers with the group to reinforce Week 3 concepts.

■ **Why does drawing pictures help you solve some problems?**

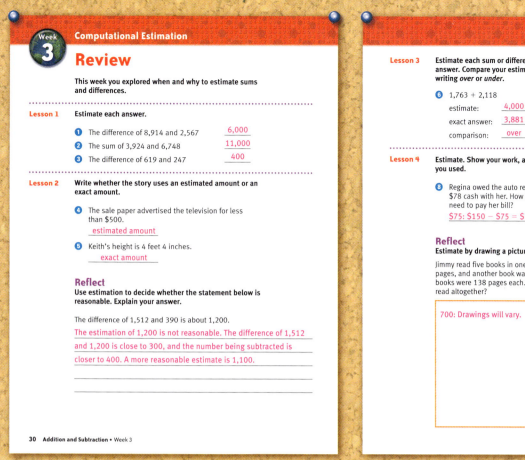

Review

This week you explored when and why to estimate sums and differences.

Lesson 1 Estimate each answer.

1. The difference of 8,914 and 2,567 — 6,000
2. The sum of 3,924 and 6,748 — 11,000
3. The difference of 619 and 247 — 400

Lesson 2 Write whether the story uses an estimated amount or an exact amount.

4. The sale paper advertised the television for less than $500.
   estimated amount
5. Keith's height is 4 feet 4 inches.
   exact amount

Reflect

Use estimation to decide whether the statement below is reasonable. Explain your answer.

The difference of 1,512 and 390 is about 1,200.

The estimation of 1,200 is not reasonable. The difference of 1,512 and 1,200 is close to 300, and the number being subtracted is closer to 400. A more reasonable estimate is 1,100.

Lesson 3 Estimate each sum or difference. Then calculate the exact answer. Compare your estimation to the exact answer by writing *over* or *under*.

6. 1,763 + 2,118
   estimate: 4,000
   exact answer: 3,881
   comparison: over

7. 4,882 − 1,764
   estimate: 3,000
   exact answer: 3,118
   comparison: under

Lesson 4 Estimate. Show your work, and explain the strategy you used.

8. Regina owed the auto repair shop $143. She has only $78 cash with her. How much more money does she need to pay her bill?
   $75: $150 − $75 = $75. Strategies will vary.

Reflect

Estimate by drawing a picture.

Jimmy read five books in one month. One book was 145 pages, and another book was 98 pages. The other three books were 138 pages each. How many pages did Jimmy read altogether?

700: Drawings will vary.

4 Assess 10

A Gather Evidence

Formal Assessment

Have students complete the Weekly test on **Assessment,** pages 44–45. Record scores on the Student Assessment Record, **Assessment,** page 100.

B Summarize Findings

Determine whether students have Minimal, Basic, or Secure understanding of the concepts presented in Week 3.

C Differentiate Instruction

Based on your observations, use these teaching strategies next week to follow up.

Assessment, pp. 44–45

Minimal Understanding

- Repeat the Warm-Up and Engage activities to develop Week 3 concepts.
- Use **Building Blocks** Number Snapshots to develop and reinforce Week 3 concepts.

Basic Understanding

- Repeat Engage activities in subsequent weeks.
- Use **Building Blocks** Rocket Blast to reinforce Week 3 concepts.

Secure Understanding

Use **Building Blocks** Rocket Blast to extend Week 3 concepts.

Week 4 Computing Units of Measurement

Week at a Glance

This week, students conclude **Number Worlds,** Level F, Addition and Subtraction. Students experience addition and subtraction in the context of real-world situations. Although many of the contexts presented deal with units of measure, the focus is not on teaching measurement but rather on using measurement in context for practicing addition and subtraction. Problems have been carefully selected to avoid the need to convert to smaller or larger units.

Math Background

Computational fluency refers to having efficient and accurate methods for computing. Students exhibit computational fluency when they demonstrate flexibility in the computational methods they choose, understand and can explain these methods, and produce accurate answers efficiently.

—*Principles and Standards for School Mathematics.* 2000. National Council of Teachers of Mathematics. Page 152.

How Students Learn

In the beginning, the teacher usually decides whether students need additional instruction. But a student's self-assessment is also an important aspect of the assessment process. To facilitate self-assessment, a checklist should be provided and then revised and expanded as students reflect and review their written work.

Teaching for Understanding

Observe closely while evaluating the assigned tasks this week to see whether students can demonstrate the following understandings.

Benchmark after Lesson 2: Students can estimate reasonable answers and then compare exact answers to addition and subtraction problems as a form of self-assessment.

Benchmark after Lesson 3: Students can compute using different units of measure, with the focus being on addition and subtraction, not measurement.

Benchmark after Lesson 4: Students can informally compute with commonly used fractions in the context of the kitchen.

Skills Focus

- Add and subtract two- and three-digit numbers in measurement contexts
- Use reasoning to solve problems that involve common fractions

Math at Home

Give one copy of the Letter to Home, page A12, to each student. Encourage students to share and complete the activity with their caregivers.

Letter to Home, Teacher Edition, p. A12

Week 4 Planner — Computing Units of Measurement

PACING	LESSON	LEARNING GOALS	NCTM	MATERIALS	TECHNOLOGY
DAY 1	pages 32–33	Students estimate and compute exact answers to currency problems.	• Number and Operations • Problem Solving • Communication • Representation	• Newspaper advertisements • Classroom items	**Building Blocks** Word Problems **e MathTools** Coins and Money
DAY 2	pages 34–35	Students estimate and compute exact answers to problems related to time.	• Number and Operations • Problem Solving • Communication • Connections • Representation	No additional materials needed	**Building Blocks** Word Problems **e MathTools** Stopwatch
DAY 3	pages 36–37	Students estimate and compute exact answers to problems related to weight and distance.	• Number and Operations • Problem Solving • Communication • Connections • Representation	No additional materials needed	**Building Blocks** Word Problems **e MathTools** Metric/Customary Conversion
DAY 4	pages 38–39	Students use reasoning to solve problems that involve commonly used fractions.	• Number and Operations • Problem Solving • Communication • Connections	• Measuring cups • Measuring spoons • Sand or rice	**Building Blocks** Word Problems **e MathTools** Estimating Proportion
DAY 5	**Review and Assess** pages 40–41	Students review skills learned this week and complete the weekly assessment.	• Number and Operations • Problem Solving • Communication • Connections • Representation	Materials will be selected from Lessons 1–4.	**Building Blocks** **e MathTools** Review previous activities

Math Vocabulary

sum The answer to an addition problem

difference The answer to a subtraction problem

English Learners

SPANISH COGNATES

English	Spanish
bills	billetes
estimate	estimar
possible	posible

ALTERNATE VOCABULARY

change Money you get back when you overpay for a purchase

purchase Something you buy

Lesson 1

Objective

Students estimate and compute exact answers to currency problems.

Materials

Additional Materials
- Newspaper advertisements
- Classroom items for **Engage** section

Access Vocabulary

cashier A person at a store who totals the bill and collects the money to pay for it

Creating Context

Have students select from a catalog or advertisement four or five items they would like to buy. Ask them to make quick estimates of how much money they would need to buy these things. Use a calculator to add up the exact amount and compare totals.

1 Warm Up 5

Concept Building COMPUTING

Write the following amounts on the board:

$82 $1,670 $619 $328

Tell students that these are the exact prices for four items.

- **If a friend wanted to know about how much each item cost in easy numbers, what answer would you give?**

Discuss student answers for each item, and determine what would happen if the friend wanted to buy the items using the estimates as a guide. Would he or she have enough money to do so?

2 Engage 30

Skill Building

Select three to five items that are around the classroom or from a catalog or newspaper ad. Ask students to think about how much each of these items might cost.

Shopping Day UNDERSTANDING

- **Make two columns on your papers. Label one *Exact Price*. Label the other *Estimated Price*.**
Hold up one item.
- **Write an estimated price.**
Hold up each of the other items so that students can write on their papers the estimated prices. When students have finished estimating, ask them to add their estimates.
- **If you bought all of these items, about how much would you spend?** Estimates will vary.

Tell the exact price of each item, and have students write on their papers the exact prices. When students have finished writing, ask them to add the exact prices.

- **If you bought all of these items, exactly how much would you spend?** Students should all have the same answer.
- **If you had a ten-dollar bill to purchase these items, estimate the amount of change you would receive from your purchase.**

Monitoring Student Progress

If . . . students need more practice,

Then . . . vary **Shopping Day** by giving students an amount they can spend and having them select the items they can afford.

Building Blocks For additional practice, students should complete **Building Blocks** Word Problems.

MathTools Use Coins and Money to demonstrate and explore addition and subtraction of currency.

Using Student Pages

Have students complete **Workbook,** pages 32–33, on their own.

Lesson 1

Key Idea

Having money models can be useful when solving problems related to money (currency).

When you solve currency problems, follow a plan similar to the following:

1. Visualize the amount in bills and coins. Then estimate a reasonable answer.
2. Decide whether the problem is an addition or a subtraction problem.
3. Solve the problem.
4. Compare your answer with your estimate. Rework if necessary.

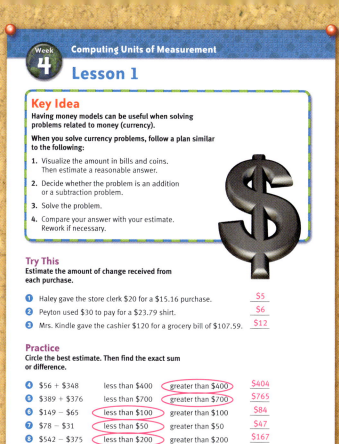

Try This

Estimate the amount of change received from each purchase.

1. Haley gave the store clerk $20 for a $15.16 purchase. — $5
2. Peyton used $30 to pay for a $23.79 shirt. — $6
3. Mrs. Kindle gave the cashier $120 for a grocery bill of $107.59. — $12

Practice

Circle the best estimate. Then find the exact sum or difference.

4. $56 + $348   less than $400   (greater than $400) — $404
5. $389 + $376   less than $700   (greater than $700) — $765
6. $149 − $65   (less than $100)   greater than $100 — $84
7. $78 − $31   (less than $50)   greater than $50 — $47
8. $542 − $375   (less than $200)   greater than $200 — $167

Select one or two items that can be purchased for each amount. Spend as much of the money as possible.

$12.95 $26.95 $1.75 $7.95

9. $10 — videotape and glider
10. $28 — baseball glove
11. $22 — videotape and globe
12. $13 — globe
13. $20 — globe and glider
14. $30 — baseball glove and glider

Reflect

About how much money would it take to purchase all four items above? Most likely, how much would you give the store clerk to pay for the purchase? Describe the amount of change you would expect to receive using *less than* or *greater than*.

$50; $50; less than $1

---

3 Reflect   10

Extended Response   REASONING

Review students' answers to the Reflect prompt at the bottom of student page 33.

- **How did you arrive at your estimate?**
- **How can you check your answer?**

Real-World Application   APPLYING

Have students draw pictures of items they would like to buy, including prices. Post the pictures around the classroom. Tell students to suppose that they have $25 to spend and to make a list of the items they would select from among the pictured items. Students should have a total for their purchases and the amount of change they would receive. Pairs of students can exchange lists and check each other's work.

4 Assess

Informal Assessment

Use the Student Assessment Record, **Assessment,** page 100, to record informal observations.

UNDERSTANDING

Shopping Day

Did the student
- ❏ make important observations?
- ❏ extend or generalize learning?
- ❏ provide insightful answers?
- ❏ pose insightful questions?

Lesson 2

Objective

Students estimate and compute exact answers to problems related to time.

Materials

No additional materials needed

Access Vocabulary

units for measuring time Day, hour, minute, and second

analog clock A clock that has hands, not a digital display

Creating Context

How many times a day do you think you look at a clock? How many times a day do you rely on a clock to do something on time or to get somewhere on time?

1 Warm Up 5

Concept Building COMPUTING

■ Name two things you can do in fifteen minutes. Tell how long it takes you to do each. Possible answer: get dressed (ten minutes) and make the bed (five minutes)

■ Name two things you can do in three hours. Tell how long it takes you to do each. Possible answer: play a board game (one hour) and watch a movie (two hours)

2 Engage 30

Skill Building

Explain that for this activity all times will be listed in minutes.

Good Morning Alarm APPLYING

■ Write at the top of your paper the amount of time it takes you to get ready for school.

■ Make a list of everything you do after you wake up to get ready for school.

Invite students to share items from their lists. This may help other students think of things they missed when making their lists.

■ Next to each activity on your lists, write the number of minutes it takes you to complete the task.

■ Total your minutes.

■ Compare your total with the time you wrote at the top of the page.

■ The times are probably not exactly the same. Why might they be different? The time written at the top of the page was an estimate written without doing any computations.

■ Are the times close? The times are probably reasonably close.

Monitoring Student Progress

If . . . students' estimates and totals differ greatly,

Then . . . have students review their lists again to make sure that the times listed are reasonable.

Building Blocks For further practice with solving addition and subtraction problems related to time, students should complete **Building Blocks** Word Problems with Tools.

MathTools Use Real Time Stopwatch to demonstrate and explore addition and subtraction of time.

Using Student Pages

Have students complete **Workbook**, pages 34–35, on their own.

Lesson 2

Key Idea

When you solve time problems, follow a plan similar to the following:

1. Decide the unit of time in the problem. If there is a combination of units, add or subtract only like units.
2. Decide whether the problem is an addition or a subtraction problem.
3. Estimate a reasonable range of time for the answer.
4. Solve the problem.
5. Compare your answer with your estimate. Rework if necessary.

Try This

Find each sum or difference. Include the unit of time in your answer.

1 63 days + 32 days
95 days

2 68 minutes − 27 minutes
41 minutes

3 365 days − 263 days
102 days

4 55 hours − 38 hours
17 hours

5 165 minutes − 84 minutes
81 minutes

6 45 seconds + 11 seconds
56 seconds

Practice

Solve each problem.

7 George ran a mile in 11 minutes 23 seconds. Kelly ran a mile in 9 minutes 10 seconds. By how many minutes and seconds was Kelly faster than George?
__2__ minutes __13__ seconds

---

8 The world's record for the longest read-aloud marathon team is 81 hours 15 minutes. Craig, Tina, Paula, and Steve tried to break the record, but they read for only 63 hours 8 minutes. By how much time did they miss breaking the record?
__18__ hours __7__ minutes

Solve each problem. Explain the order in which you added the times.

9 Kareem has been running across the country for four weeks to raise money for cancer research. The first week he ran 55 hours. During the second week it stormed on several days, so he ran only 28 hours. The third week he ran 75 hours, and the fourth week he ran 62 hours. How many hours did he run during the four weeks?
__220__ hours   Add 55 and 75. Add 28 and 62. Then add both sums: 130 + 90.

10 Corinna rented three movies to watch this afternoon. She plans to watch one right after the other. How many minutes will she watch movies if one movie is 112 minutes, another is 98 minutes, and the third is 105 minutes?
__315__ minutes   Add 112 and 98. Then add that sum and 105: 210 + 105.

Reflect

Choose Problem 9 or 10. Explain why you added the numbers in the order you did. Try adding the numbers in a different order. Which order was easier? Explain why.

Answers and explanations will vary.

Reflect   10 ▶

Extended Response   REASONING

Review students' answers to the Reflect prompt at the bottom of student page 35. Invite students to share their explanations with the class.

■ **When a problem about time is an addition problem, does the order in which the numbers are added matter? Why or why not?** No; explanations will vary.

Real-World Application   APPLYING

Pose the following situation, and allow students time to find a solution.

■ **You and a friend want to see two movies. The movie theater is 15 minutes from your house, and you can only be out between 12:00 noon and 5:00 P.M. Which two movies could you see and still be home on time?**

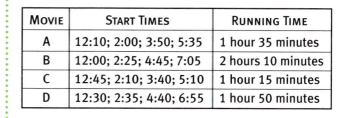

Movie	Start Times	Running Time
A	12:10; 2:00; 3:50; 5:35	1 hour 35 minutes
B	12:00; 2:25; 4:45; 7:05	2 hours 10 minutes
C	12:45; 2:10; 3:40; 5:10	1 hour 15 minutes
D	12:30; 2:35; 4:40; 6:55	1 hour 50 minutes

Assess

Informal Assessment

Use the Student Assessment Record, *Assessment*, page 100, to record informal observations.

APPLYING

Good Morning Alarm
Did the student
❏ apply learning to new situations?
❏ contribute concepts?
❏ contribute answers?
❏ connect mathematics to the real world?

Lesson 3

Objective

Students estimate and compute exact answers to problems related to weight and distance.

Materials

No additional materials needed

Access Vocabulary

units for measuring distance Miles, yards, feet, inches, kilometers, meters

units for measuring weight Tons, pounds, ounces, kilograms, grams

Creating Context

Looking at the root word and any affixes of a word is one reading strategy that can help us figure out new words. Word parts that are added to the beginning of a root word are called *prefixes* because *pre-* means "before." Word parts that are added to the end of a root word are called *suffixes.* The prefix *re-* is found in many words in this lesson. Help English Learners scan the lesson for words such as *rework, record, reflect,* and *relocate,* and discuss what they mean. Make sure to explain that although the word *reasonable* begins with *re-,* it is part of the root word reason.

1 Warm Up   5

Concept Building   COMPUTING

Students should make a list of the units given in Access Vocabulary. Ask students to write a real-life situation in which those units of measure might be used. Share students' ideas with the class.

2 Engage   30

Strategy Building

Organize students into groups of three. Groups are to write a brief story using measures of distance or measures of weight. Instruct students to include two- and three-digit measures in their stories. Give students an example of what a story might be like.

Example story: Bob works at the zoo feeding the elephants. The largest elephant weighs 7 tons and eats about 375 pounds of food every day. The mother elephant weighs 5 tons and eats about 250 pounds of food every day. The baby elephant weighs only 1 ton. He eats only about 70 pounds of food per day, but he is growing fast.

Example questions: How much do the elephants weigh altogether? The elephants weigh about 13 tons altogether.

How much food do the elephants eat altogether? about 700 pounds of elephant food every day

Show Me   COMPUTING

Take turns having groups read their stories. Then invite students to ask the groups questions about their stories so that each group must solve an addition or subtraction problem to answer the question. Each group should demonstrate how they solved the problem posed by the class and justify that their answer is reasonable by using estimation.

Building Blocks For further practice with solving addition and subtraction problems related to weight and distance, students should complete **Building Blocks** Word Problems with Tools.

MathTools Use Metric/Customary Conversion to demonstrate and explore addition and subtraction of weight and distance.

Using Student Pages

Have students complete **Workbook,** pages 36–37, on their own.

Lesson 3

Key Idea

When you solve weight and distance problems, follow a plan similar to the following:

1. Determine the unit of weight or distance in the problem. If there is a combination of units, add or subtract only like units.
2. Decide whether the problem is an addition or a subtraction problem.
3. Estimate a reasonable answer.
4. Solve the problem.
5. Compare your answer with your estimate. Rework if necessary.

Try This
Find the sum or difference.

1. 158 miles + 465 miles
   623 miles
2. 45 feet − 29 feet
   16 feet
3. 225 meters − 24 meters
   201 meters
4. 872 miles − 571 miles
   301 miles
5. 100 yards − 37 yards
   63 yards
6. 318 kilometers + 674 kilometers
   992 kilometers

Find the sum or difference.

7. 741 pounds + 874 pounds
   1,615 pounds
8. 126 ounces − 45 ounces
   81 ounces
9. 838 grams − 627 grams
   211 grams
10. 473 tons − 67 tons
    406 tons

Solve each problem.

11. The UpThere Balloon Ride Company has a policy that the passengers can weigh no more than 365 pounds altogether. Charlie weighs 197 pounds. What is the most the other passengers can weigh combined?
    168 pounds

12. Molly recorded the distances her family traveled on the interstate each day of their vacation. Her log is shown below. How many miles did the family travel during the four days?

Monday	Tuesday	Wednesday	Thursday
286 miles	247 miles	203 miles	214 miles

950 miles

13. In the Olympic triathalon event, athletes swim 1 mile, bike 25 miles, and run 6 miles. What is the total distance each triathalon athlete travels during the competition?
    32 miles

Reflect

In one national park in Alaska, two families of brown bears needed to be relocated to a mountain range that had a greater food supply available. The rangers' trailer could hold up to 5,000 kilograms. Among the bears that needed to be transported were two males that weighed 778 and 689 kilograms and two females that weighed 645 and 503 kilograms. One female had two cubs that weighed 314 and 281 kilograms. The other female had one cub that weighed 389 kilograms.

Will the rangers' trailer carry all of the bears and cubs at one time? What is the total weight of the bears?

Yes; the adult bears weighed 2,615 kilograms. The cubs weighed 984 kilograms. Altogether the weight is 3,599 kilograms.

---

3 Reflect   10

Extended Response   REASONING

Review students' answers to the Reflect prompt at the bottom of student page 37.

- **Could you graphically organize the information in a table or diagram?**
- **Could you break the problem into smaller parts and solve each part before finding the final answer?**

Real-World Application   APPLYING

Present students with the following list of produce items and their prices per pound. Have students make a grocery list for the produce section that includes several of the items, a total weight, and a total price for their purchase.

- **What is in the produce section of a grocery store?** fruits and vegetables

Item	Apples	Bananas	Oranges	Carrots
Price per Pound	$ 0.50	$ 0.40	$ 0.90	$1.00
Item	Peaches	Tomatoes	Potatoes	Grapes
Price per Pound	$1.00	$ 0.80	$ 0.40	$ 0.80

4 Assess

Informal Assessment

Use the Student Assessment Record, **Assessment,** page 100, to record informal observations.

COMPUTING

Show Me

Did the student
- ❏ respond accurately?
- ❏ respond quickly?
- ❏ respond with confidence?
- ❏ self-correct?

Lesson 4

Objective

Students use reasoning to solve problems that involve commonly used fractions.

Materials

Additional Materials
- Measuring cups
- Sand or uncooked rice

Access Vocabulary

measuring cups Cups used for measuring; usually provided in units of $\frac{1}{4}$, $\frac{1}{3}$, $\frac{1}{2}$, $\frac{2}{3}$, $\frac{3}{4}$, and 1 cup

measuring spoons Spoons used for measuring, usually provided in units of teaspoons and tablespoons

recipe A list of ingredients and how to combine them to prepare foods

Creating Context

Using real objects to make concepts more concrete is an excellent strategy for English Learners because it helps students visualize the concept and learn the new labels. For English Learners who may be more familiar with the metric system, this is especially important when working with measurements in the customary system.

1 Warm Up   5

Concept Building   UNDERSTANDING

Instruct students to draw a drinking glass on their papers. Instruct students to color the picture so the glass looks half full. As a class, discuss what the glass looks like when it is one-quarter full and three-quarters full.

2 Engage   30

Skill Building

Explain that for this activity, actual computation is not necessary. Students should model the problem with the measuring tools and then use reasoning to find the solution. If available, students can use sand or rice in place of empty measuring cups.

Culinary Math   UNDERSTANDING

A recipe that you are following includes the following ingredients:

$2\frac{3}{4}$ cups flour

$\frac{2}{3}$ cup packed light brown sugar

- If you have 5 cups of flour in your bin, how much will you have left after you make this recipe? $2\frac{1}{4}$ cups
- Do you have enough flour to double the recipe? no
- If you had only a $\frac{1}{4}$-cup measuring cup to use, how many times would you have to fill it to measure the flour? 11 times
- If you had only a $\frac{1}{3}$-cup measuring cup to use, how many times would you have to fill it to measure the brown sugar? 2 times

Monitoring Student Progress

If . . . students do not have measuring cups and spoons to use,

Then . . . provide pictures of each size cup and spoon. Encourage students to look at the pictures and to draw pictures to match the word problems.

 Building Blocks For additional practice with fractions, students should complete **Building Blocks** Word Problems.

MathTools Use Estimating Proportions to demonstrate and explore fractions.

Using Student Pages

Have students complete **Workbook,** pages 38–39, on their own.

Week 4 — Computing Units of Measurement

Lesson 4

Key Idea

Many times in the real world, such as when following a recipe, you need to compute with commonly used fractions. Having measuring cups and measuring spoons when you are solving problems related to a recipe can be useful.

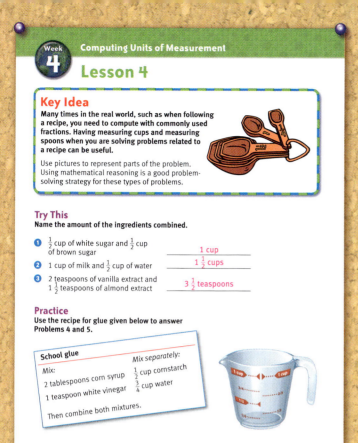

Use pictures to represent parts of the problem. Using mathematical reasoning is a good problem-solving strategy for these types of problems.

Try This

Name the amount of the ingredients combined.

1. $\frac{1}{2}$ cup of white sugar and $\frac{1}{2}$ cup of brown sugar   **1 cup**

2. 1 cup of milk and $\frac{1}{2}$ cup of water   **1 $\frac{1}{2}$ cups**

3. 2 teaspoons of vanilla extract and 1 $\frac{1}{2}$ teaspoons of almond extract   **3 $\frac{1}{2}$ teaspoons**

Practice

Use the recipe for glue given below to answer Problems 4 and 5.

School glue	
Mix:	Mix separately:
2 tablespoons corn syrup	$\frac{1}{2}$ cup cornstarch
1 teaspoon white vinegar	$\frac{3}{4}$ cup water
Then combine both mixtures.	

4. Olivia is making glue. She has a tablespoon and a teaspoon for measuring ingredients, but she can find only a $\frac{1}{4}$-cup measuring cup. How many times does she need to fill the measuring cup for each ingredient?

    3 times for water   **2** times for cornstarch

5. Mr. Longfellow made four batches of glue for his art class. List the amount of each ingredient he used.

    8 tablespoons corn syrup   **4** teaspoons white vinegar

    2 cups cornstarch   **3** cups water

Reflect

Most complete sets of measuring cups come with the following sizes:

$\frac{1}{4}$ cup $\frac{1}{3}$ cup $\frac{1}{2}$ cup $\frac{2}{3}$ cup $\frac{3}{4}$ cup   1 cup

If you had misplaced all but two sizes of measuring cups, which two sizes would you prefer to still have? Explain your answer.

$\frac{1}{4}$ cup and $\frac{1}{3}$ cup; with these two cups, all other measuring cup sizes can be made by filling the cup more than one time. You can use the $\frac{1}{4}$ cup to make $\frac{1}{2}$ cup by filling it two times; you can make $\frac{3}{4}$ cup by filling it three times; and you can make 1 cup by filling it four times. You can use the $\frac{1}{3}$ cup to make $\frac{2}{3}$ cup by filling it two times, and you can make 1 cup by filling it three times.

---

3 — Reflect   10

Extended Response   REASONING

Review students' answers to the Reflect prompt at the bottom of student page 39.

Using measuring cups, students can experiment before they decide which two cups they prefer to have.

■ **Explain your choice.** Explanations will vary.

Real-World Application   APPLYING

Have students bring their favorite recipes to class to share. Tell students to write one question about their recipe that is similar to the questions in their exercises. Students can share recipes and questions as a class or in small groups.

4 — Assess

Informal Assessment

Use the Student Assessment Record, **Assessment,** page 100, to record informal observations.

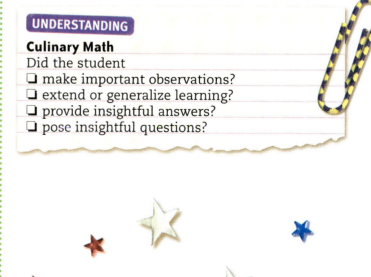

UNDERSTANDING

Culinary Math

Did the student
❏ make important observations?
❏ extend or generalize learning?
❏ provide insightful answers?
❏ pose insightful questions?

Week 4

Lesson 5 Review

Objective

Students review skills learned this week and complete the weekly assessment.

Materials

Review materials will be selected from those used in previous activities.

Creating Context

Name some everyday situations in which weight, distance, time, and other measurements are important.

Unit Assessment

Students should complete the Addition and Subtraction Test found on *Assessment,* pages 76–77. Using the key on *Assessment,* page 98, identify incorrect responses. Reteach and review the Warm-Up and Engage activities to reinforce concept understanding.

1 Warm Up   5

Concept Building UNDERSTANDING

Students should make a list of the units of measurement used this week. Ask students to write a real-life situation in which these units of measure might be used. Share students' ideas with the class.

2 Engage   20

Skill Building

Free-Choice Activity

For the last day of the week, allow students to choose an activity from the previous lessons. Some activities they may choose include the following:

- **Shopping Day**
- **Good Morning Alarm**
- **Show Me**
- **Culinary Math**

Make a note of the activities children select. Do they prefer easy or challenging activities? If you believe your students would benefit from extra practice on specific skills, choose an activity for them.

Monitoring Student Progress

If . . . students are not participating,

Then . . . continue the activity until all students have participated.

Using Student Pages

Have students complete **Workbook,** pages 40–41, on their own.

3 Reflect   10

Extended Response APPLYING

Review students' answers to the Reflect prompts at the bottom of student pages 40–41.

Discuss these answers with the group to reinforce Week 4 concepts.

Review

This week you experienced addition and subtraction in the context of real-world situations dealing with units of measure.

Lesson 1 · Estimate the amount of change received from each purchase.

1. Corie gave the cashier $30 for a $24.64 purchase. **$5**

2. Lynn gave the counter clerk $10 to pay for a $3.69 sandwich. **$6**

3. Mr. Kim used $100 to pay a $78.03 dinner bill. **$20**

Lesson 2 · Find the exact sum or difference.

4. 365 days + 267 days **632 days**

5. 165 minutes − 86 minutes **79 minutes**

6. Samantha read for 2 hours 16 minutes in the morning and then again for 1 hour 42 minutes in the evening. How much time did Samantha spend reading in all?

   3 hours **58** minutes

Reflect
Explain how you figured out the answer to Problem 5. Could you have solved this problem differently? Explain.

Answers and explanations will vary.

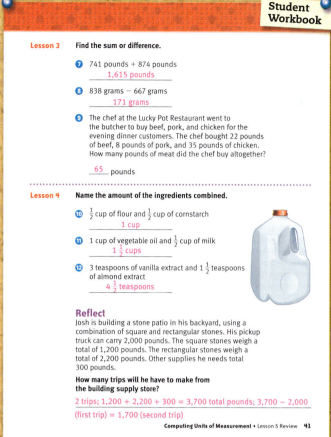

Lesson 3 · Find the sum or difference.

7. 741 pounds + 874 pounds **1,615 pounds**

8. 838 grams − 667 grams **171 grams**

9. The chef at the Lucky Pot Restaurant went to the butcher to buy beef, pork, and chicken for the evening dinner customers. The chef bought 22 pounds of beef, 8 pounds of pork, and 35 pounds of chicken. How many pounds of meat did the chef buy altogether?

   65 pounds

Lesson 4 · Name the amount of the ingredients combined.

10. $\frac{1}{2}$ cup of flour and $\frac{1}{2}$ cup of cornstarch **1 cup**

11. 1 cup of vegetable oil and $\frac{1}{2}$ cup of milk **1 $\frac{1}{2}$ cups**

12. 3 teaspoons of vanilla extract and $1\frac{1}{2}$ teaspoons of almond extract **4 $\frac{1}{2}$ teaspoons**

Reflect
Josh is building a stone patio in his backyard, using a combination of square and rectangular stones. His pickup truck can carry 2,000 pounds. The square stones weigh a total of 1,200 pounds. The rectangular stones weigh a total of 2,200 pounds. Other supplies he needs total 300 pounds.

How many trips will he have to make from the building supply store?

2 trips; 1,200 + 2,200 + 300 = 3,700 total pounds; 3,700 − 2,000 (first trip) = 1,700 (second trip)

4 Assess 10

A Gather Evidence

Formal Assessment
Have students complete the weekly test on *Assessment,* pages 46–47. Record scores on the Student Assessment Record, *Assessment* page 100.

B Summarize Findings
Determine whether students have Minimal, Basic, or Secure understanding of the concepts presented in Week 4.

C Differentiate Instruction
Based on your observations, use these teaching strategies next week to follow up.

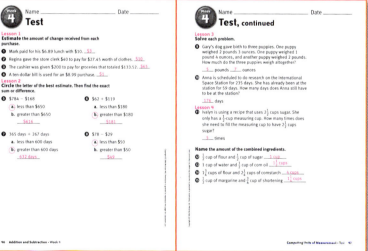

Assessment, pp. 46–47

Minimal Understanding
- Repeat the Warm-Up and Engage activities to develop Week 4 concepts.
- Use *Building Blocks* Word Problems and *eMathTools* Coins and Money to develop and reinforce Week 4 concepts.

Basic Understanding
- Repeat Engage activities in subsequent weeks.
- Use *Building Blocks* Word Problems with Tools and *eMathTools* Stopwatch to reinforce Week 4 concepts.

Secure Understanding
Use *Building Blocks* Word Problems and *eMathTools* Metric/Customary Conversion to extend Week 4 concepts.

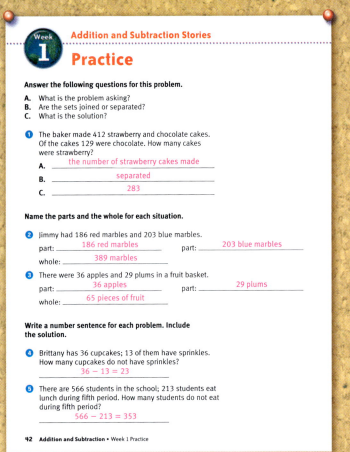

Addition and Subtraction Stories

Practice

Answer the following questions for this problem.

A. What is the problem asking?
B. Are the sets joined or separated?
C. What is the solution?

1. The baker made 412 strawberry and chocolate cakes. Of the cakes 129 were chocolate. How many cakes were strawberry?

   A. _the number of strawberry cakes made_

   B. _separated_

   C. _283_

Name the parts and the whole for each situation.

2. Jimmy had 186 red marbles and 203 blue marbles.

   part: _186 red marbles_   part: _203 blue marbles_

   whole: _389 marbles_

3. There were 36 apples and 29 plums in a fruit basket.

   part: _36 apples_   part: _29 plums_

   whole: _65 pieces of fruit_

Write a number sentence for each problem. Include the solution.

4. Brittany has 36 cupcakes; 13 of them have sprinkles. How many cupcakes do not have sprinkles?

   _36 − 13 = 23_

5. There are 566 students in the school; 213 students eat lunch during fifth period. How many students do not eat during fifth period?

   _566 − 213 = 353_

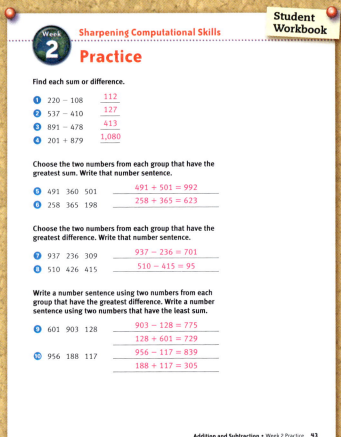

Sharpening Computational Skills

Practice

Find each sum or difference.

1. $220 − 108$ _112_
2. $537 − 410$ _127_
3. $891 − 478$ _413_
4. $201 + 879$ _1,080_

Choose the two numbers from each group that have the greatest sum. Write that number sentence.

5. 491 360 501   _491 + 501 = 992_
6. 258 365 198   _258 + 365 = 623_

Choose the two numbers from each group that have the greatest difference. Write that number sentence.

7. 937 236 309   _937 − 236 = 701_
8. 510 426 415   _510 − 415 = 95_

Write a number sentence using two numbers from each group that have the greatest difference. Write a number sentence using two numbers that have the least sum.

9. 601 903 128   _903 − 128 = 775_
   _128 + 601 = 729_

10. 956 188 117   _956 − 117 = 839_
    _188 + 117 = 305_

Unit 3
Addition and Subtraction
Practice Pages

Weeks 1 and 2

1 Assign the Practice pages, *Workbook,* pages 42–43, at the end of Weeks 1 and 2. Students should complete these pages independently.

2 Check student answers using the annotated pages above.

3 If students have difficulty with a Practice page, review the activities completed throughout the week, and have students complete the weekly practice again before they complete the weekly test.

42–43   Addition and Subtraction • Weeks 1 and 2

Week 3 | Computational Estimation
Practice

Describe each estimate as over or under the exact answer.

1. The difference of 8,717 and 2,190 is about 7,000. _over_
2. The sum of 3,717 and 5,534 is about 10,000. _over_
3. The difference of 937 and 614 is about 300. _under_

Estimate each sum or difference. Then calculate the exact answer. Compare your estimate to the exact answer using *over* or *under*. Answers will vary; sample answers:

4. 1,776 + 2,582

   estimate: _5,000_   exact answer: _4,358_   comparison: _over_

Estimate using any strategy. Find the exact answer. Then use a calculator to check your work.

5. 1,980 + 2,005   estimate: _4,000_   exact answer: _3,985_   check: ✔
6. 8,461 − 4,260   estimate: _4,000_   exact answer: _4,201_   check: ✔

Tell whether the story uses an estimated amount or an exact amount.

7. The number of students at school is 128. _exact amount_
8. New swing sets cost more then $500. _estimated amount_

Estimate. Show your work, and explain the strategy you used.

9. Jaden needs $434 to buy a new swimming pool. He has $240 right now. How much more money does he need to buy his swimming pool?
   Strategies will vary; around 200.

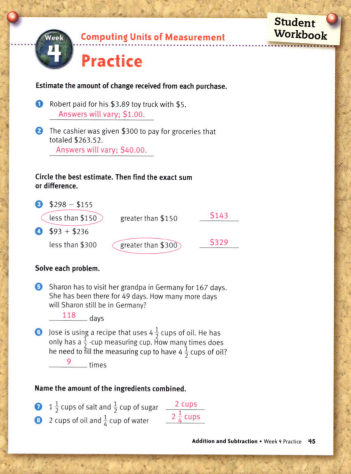

Week 4 | Computing Units of Measurement
Practice

Estimate the amount of change received from each purchase.

1. Robert paid for his $3.89 toy truck with $5.
   Answers will vary; $1.00.
2. The cashier was given $300 to pay for groceries that totaled $263.52.
   Answers will vary; $40.00.

Circle the best estimate. Then find the exact sum or difference.

3. $298 − $155
   (less than $150)   greater than $150   _$143_
4. $93 + $236
   less than $300   (greater than $300)   _$329_

Solve each problem.

5. Sharon has to visit her grandpa in Germany for 167 days. She has been there for 49 days. How many more days will Sharon still be in Germany?
   _118_ days

6. Jose is using a recipe that uses $4\frac{1}{2}$ cups of oil. He has only has a $\frac{1}{2}$-cup measuring cup. How many times does he need to fill the measuring cup to have $4\frac{1}{2}$ cups of oil?
   _9_ times

Name the amount of the ingredients combined.

7. $1\frac{1}{2}$ cups of salt and $\frac{1}{2}$ cup of sugar   _2 cups_
8. 2 cups of oil and $\frac{1}{4}$ cup of water   _$2\frac{1}{4}$ cups_

Unit 3
Addition and Subtraction
Practice Pages
Weeks 3 and 4

1. Assign the Practice pages, *Workbook,* pages 44–45, at the end of Weeks 3 and 4. Students should complete these pages independently.

2. Check student answers using the annotated pages above.

3. If students have difficulty with a Practice page, review the activities completed throughout the week, and have students complete the weekly practice again before they complete the weekly test.

Week 1 Models for Multiplication

Week at a Glance

This week, students begin *Number Worlds,* Level F, Multiplication and Beginning Division. Students should explore grouping and repeated addition as models for multiplication.

Math Background

Multiplication and division have always been important topics in elementary school mathematics. However, in the past, instruction has focused primarily on helping students develop procedural competency with basic facts and paper-and-pencil algorithms. This procedural competency is an important goal, but recent research makes it clear that students also need to develop deep conceptual knowledge of multiplication and division.

—Chapin, Suzanne, and Art Johnson. 2000. *Math Matters: Grades K–6, Understanding the Math You Teach.* Math Solutions Publications. Page 33.

How Students Learn

Being able to perform calculations accurately does not guarantee that students actually understand the operations they are performing. Conceptual knowledge is based on understanding the relationships represented by the numbers in the computations. Students can use pictures, charts, manipulatives, symbols, and words to interpret and define these relationships.

Teaching for Understanding

Observe closely while evaluating the assigned tasks this week to see whether students can demonstrate the following understandings.

Benchmark after Lesson 2: Students can determine grouping and repeated addition to find a product.

Benchmark after Lesson 3: Students can record a multiplication problem using the × symbol.

Benchmark after Lesson 4: Students can write multiplication sentences from information presented in pictures and charts.

Skills Focus

- Use objects and pictures to create equal groups and to identify a product
- Relate repeated addition to pictures and groups to identify a product
- Develop a conceptual knowledge of multiplication through the use of models

Math at Home

Give one copy of the Letter to Home, page A13, to each student. Encourage students to share and complete the activity with their caregivers.

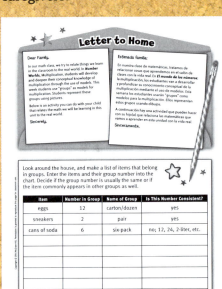

Letter to Home, Teacher Edition, p. A13

PACING	LESSON	LEARNING GOALS	NCTM	MATERIALS	TECHNOLOGY
DAY 1	pages 2–3	Students use pictures of equal groups to create models for multiplication.	• Number and Operations • Problem Solving • Communication • Representation	• Counters* • Drawing paper	**B**uilding **B**locks Comic Book Shop ⓔ MathTools 100 Table
DAY 2	pages 4–5	Students use repeated addition to create models for multiplication.	• Number and Operations • Problem Solving • Communication • Connections	• Counters* • Picture cards, pp. B11–B12	**B**uilding **B**locks Comic Book Shop ⓔ MathTools 100 Table
DAY 3	pages 6–7	Students use the × symbol to write multiplication problems.	• Number and Operations • Problem Solving • Communication • Connections	• Counters* • Pictures created in Lesson 1 • Recording Chart, p. B13	**B**uilding **B**locks Comic Book Shop ⓔ MathTools 100 Table
DAY 4	pages 8–9	Students describe groups that come in sets.	• Number and Operations • Problem Solving • Communication • Connections • Representation	• Counters* • Recording Chart, p. B13 • Number 1–6 Cube	**B**uilding **B**locks Comic Book Shop ⓔ MathTools 100 Table
DAY 5	**Review and Assess** pages 10–11	Students review skills learned this week and complete the weekly assessment.	• Number and Operations • Problem Solving • Communication • Connections • Representation	Materials will be selected from Lessons 1–4.	**B**uilding **B**locks Review previous activities

Math Vocabulary

repeated addition An addition problem that has the same addend several times

× symbol The operation sign that means multiplication, or groups of

English Learners

SPANISH COGNATES

English	Spanish
doubles	dobles
sum	suma
product	produto
equal	igual

ALTERNATE VOCABULARY

doubles An addition problem in which both addends are the same number

group A collection of objects or pictures

Lesson 1

Objective

Students use pictures of equal groups to create models for multiplication.

Materials

Additional Materials
- Drawing paper
- Counters

Access Vocabulary

match A card that is the same as the one you are holding

fund-raiser An organized event or effort to collect money for a cause

Creating Context

Help English Learners practice the names of things that come in groups. Create a large chart with counting numbers 1 to 20. Have students write next to each number something that comes in this grouping. Some spaces may remain blank.

1 Warm Up 5

Concept Building ENGAGING

Brainstorm with the entire class to name items that normally come in groups. Write every item named on the board. Do not allow students to judge or comment on others' contributions to the brainstorming list.

2 Engage 30

Skill Building

Pick items from the list of brainstormed ideas. Provide drawing paper for each student.

Circle the Groups UNDERSTANDING

Choose one item that comes in groups of two for the following dialogue. Replace the italicized words as needed. Draw items on the board as students draw on their papers.

- If I buy *shoes* for myself, how many *shoes* do I buy? 2
- Show this in a picture.
- If I buy *shoes* for myself and *shoes* for one of you, how many groups of *shoes* do I buy? 2
- How many *shoes* are in each group? How many *shoes* are there altogether? 2; 4
- Show the two groups of two *shoes* in a picture.
- Add another group of *shoes*. 2 *shoes* + 2 *shoes* + 2 *shoes* equals 6 *shoes*.
- Draw a circle around each group of *shoes*.
- How many circles or groups of *shoes* are there? How many *shoes* are in each group? How many *shoes* are there altogether in the picture? 3; 2; 6
- Next to your drawing, write "three groups of shoes equals six shoes" and 3 × 2 = 6.

Repeat the series of questions and drawings for groups of three and groups of four.

Collect pictures for use later in the week.

Monitoring Student Progress

If . . . students would rather use objects to make groups,

Then . . . provide students with Counters that they can physically move into groups.

Building Blocks For additional practice identifying groups, students should complete **Building Blocks** Comic Book Shop.

MathTools Use the 100 Table to demonstrate and explore multiplication.

Using Student Pages

Have students complete **Workbook,** pages 2–3, on their own.

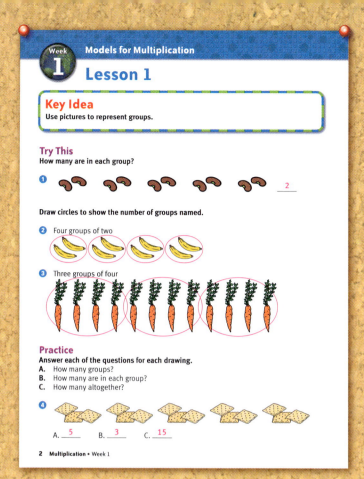

Week 1 — Models for Multiplication

Lesson 1

Key Idea
Use pictures to represent groups.

Try This
How many are in each group?

1  2

Draw circles to show the number of groups named.

2 Four groups of two

3 Three groups of four

Practice
Answer each of the questions for each drawing.
A. How many groups?
B. How many are in each group?
C. How many altogether?

4

A. 5 B. 3 C. 15

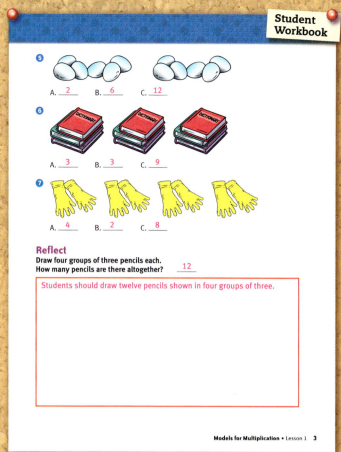

5
A. 2 B. 6 C. 12

6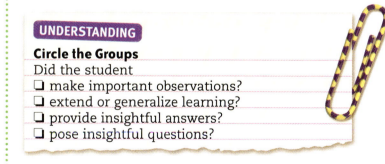
A. 3 B. 3 C. 9

7
A. 4 B. 2 C. 8

Reflect
Draw four groups of three pencils each.
How many pencils are there altogether?   12

Students should draw twelve pencils shown in four groups of three.

3 Reflect   10

Extended Response REASONING
Review students' answers to the Reflect prompt at the bottom of student page 3.

- **Can twelve pencils be divided into a different number of equal groups?** yes; 2 groups of 6; 3 groups of 4; 6 groups of 2

Real-World Application APPLYING
- **Name an item that is bought in groups of 12.** Possible answers: eggs or flowers.
- **Name an item that is bought in groups of more than 20. Describe where and how the item is sold.** Possible answer: Trash bags come in boxes of 30 and are sold in grocery stores.

4 Assess

Informal Assessment
Use the Student Assessment Record, *Assessment,* page 100, to record informal observations.

UNDERSTANDING
Circle the Groups
Did the student
❑ make important observations?
❑ extend or generalize learning?
❑ provide insightful answers?
❑ pose insightful questions?

Lesson 2

Objective

Students use repeated addition to create models for multiplication.

Materials

Program Materials
Picture cards, pp. B11–B12

Additional Materials
Counters

Access Vocabulary

bumper stickers Adhesive strips that have a short slogan or phrase, placed on the rear bumper of a car

check your work Look at the list of answers for the problems you completed to see whether they are done correctly

Creating Context

Ask students what bumper stickers are. Bumper sticker slogans often rely on double meanings for humor. Discuss appropriate bumper sticker slogans that you or students have seen. Work with English Learners to understand their meanings.

1 Warm Up 5

Concept Building COMPUTING

An addition strategy that students usually learn and practice is adding doubles. Write the following doubles addition problems on the board. Students should find each sum.

7 + 7 = 14   6 + 6 = 12   9 + 9 = 18
5 + 5 = 10

2 Engage 30

Strategy Building

Mix up the picture cards, and then hand one card to each student. Not all cards have to be distributed.

Picture Match-Up ENGAGING

Students should move about and find other students who have a picture card that matches the card they are holding.

- **How many people have cards that match your card?**
- **Line up the cards on a sheet of paper. Below each card write the number of items shown on the card. Place a plus sign between each number.**
- **Look at the expression you have written. Does anyone know what type of addition problem this is?**
- **It is called repeated addition because you are adding the same quantity more than two times.**
- **Rewrite your addition statement by describing the number of groups and how many there are in each group. For example, 3 + 3 + 3 + 3 means four groups of three.**

Groups should share their picture cards, repeated addition expression, and meaning with the class.

Monitoring Student Progress

If . . . only a few students are participating in **Picture Match-Up,**	**Then . . .** distribute more than one card to each student so that students will be part of more than one group.

Building Blocks For additional practice modeling multiplication using repeated addition, students should complete **Building Blocks** Comic Book Shop.

e MathTools Use the 100 Table to demonstrate and explore multiplication.

Using Student Pages

Have students complete **Workbook,** pages 4–5, on their own.

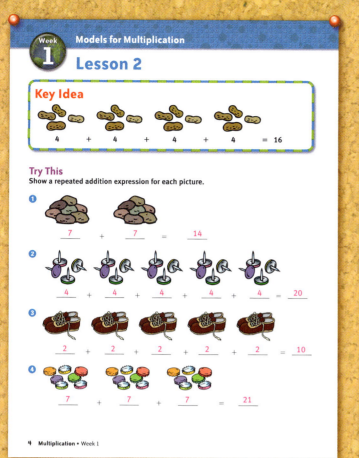

Week 1

Models for Multiplication

Lesson 2

Key Idea

4 + 4 + 4 + 4 = 16

Try This

Show a repeated addition expression for each picture.

1. 7 + 7 = 14

2. 4 + 4 + 4 + 4 + 4 = 20

3. 2 + 2 + 2 + 2 + 2 = 10

4. 7 + 7 + 7 = 21

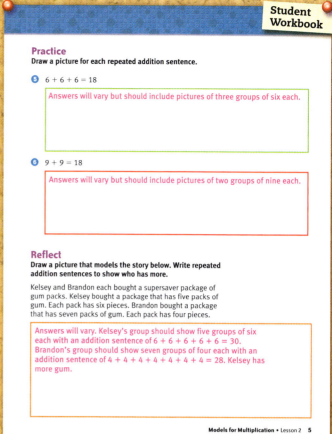

Practice

Draw a picture for each repeated addition sentence.

5. 6 + 6 + 6 = 18

Answers will vary but should include pictures of three groups of six each.

6. 9 + 9 = 18

Answers will vary but should include pictures of two groups of nine each.

Reflect

Draw a picture that models the story below. Write repeated addition sentences to show who has more.

Kelsey and Brandon each bought a supersaver package of gum packs. Kelsey bought a package that has five packs of gum. Each pack has six pieces. Brandon bought a package that has seven packs of gum. Each pack has four pieces.

Answers will vary. Kelsey's group should show five groups of six each with an addition sentence of 6 + 6 + 6 + 6 + 6 = 30. Brandon's group should show seven groups of four each with an addition sentence of 4 + 4 + 4 + 4 + 4 + 4 + 4 = 28. Kelsey has more gum.

3 Reflect   10

Extended Response REASONING

Review students' answers to the Reflect prompt at the bottom of student page 5.

Arrange students in pairs. Students should write and solve a problem similar to the one about Kelsey and Brandon. If time permits, invite students to share their story and work.

- **How would you explain this to someone who has never done this before?**
- **Can you think of other times outside school when you would use this skill?**

Real-World Application APPLYING

Suppose that your school had a fund-raiser by selling bumper stickers for $3 each.

- **Write a repeated addition sentence that shows how much money you made if you sold seven stickers in one day.** 3 + 3 + 3 + 3 + 3 + 3 + 3 = 21
- **Write an addition sentence that shows how much money you made if you sold seven stickers every day for one week.** 21 + 21 + 21 + 21 + 21

4 Assess

Informal Assessment

Use the Student Assessment Record, *Assessment,* page 100, to record informal observations.

ENGAGING

Picture Match-Up

Did the student

❏ pay attention to the contributions of others?
❏ contribute information and ideas?
❏ improve on a strategy?
❏ reflect on and check accuracy of work?

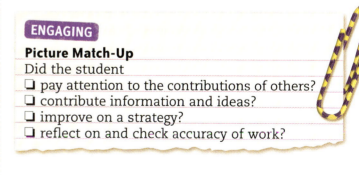

Models for Multiplication • Lesson 2   4–5

Lesson 3

Objective

Students use the × symbol to write multiplication problems.

Materials

Program Materials
Recording Chart, p. B13

Additional Materials
- Pictures created in Lesson 1
- Counters

Access Vocabulary

tricycle A three-wheeled bike for young children
advertisements Announcements of things for sale that often include pictures of the items, prices, and dates of availability

Creating Context

An excellent strategy to use with English Learners is to chart information visually. This gives English Learners one more reference point for comprehension and assists in critical thinking. In the skill-building section of this lesson, expand the chart and repeat any areas that need additional explanation for English Learners.

1 Warm Up 5

Concept Building UNDERSTANDING

Write a multiplication symbol on the board.

- **Does anyone know what it means when this symbol appears between two numbers?** Allow students to answer.

Explain that this symbol is the times sign, or multiplication symbol, and it tells us to multiply. Continue by saying that it may be helpful to think of that symbol as meaning "groups of" for now. Show several examples on the board, using "groups of" and the times sign interchangeably.

2 Engage 30

Skill Building

Instruct students to retrieve the pictures they drew in Lesson 1. Distribute a Recording Chart to each student.

Draw It, Chart It APPLYING

- **Complete a line in the Recording Chart for each picture created in Lesson 1.**

For each line in the chart, students should answer the following questions:

- **How many groups are shown?**
- **How many things are in each group?**
- **What does the number sentence look like that uses this information and a × symbol?**

For each line completed on the chart, students should read the information that follows aloud, changing the numbers and names of the items as needed.

- **I have two groups of tricycles. There are three wheels on each tricycle. I have two groups of three. Two times 3 is 6. There are six wheels in all.**

Monitoring Student Progress

| **If . . .** students are working too rapidly or finishing early, | **Then . . .** instruct students to create two more pictures and to fill in the Recording Chart for each picture. |

Building Blocks For additional practice writing multiplication sentences, students should complete **Building Blocks** Comic Book Shop.

eMathTools Use the 100 Table to demonstrate and explore multiplication.

Using Student Pages

Have students complete **Workbook,** pages 6–7, on their own.

Lesson 3

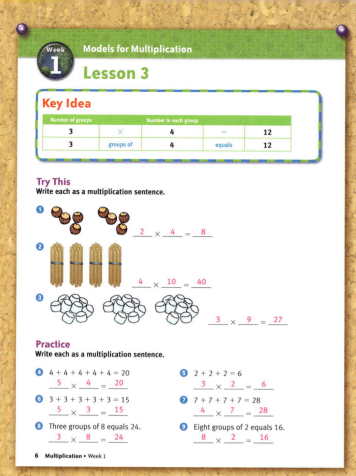

Key Idea

Number of groups		Number in each group		
3	×	4	=	12
3	groups of	4	equals	12

Try This
Write each as a multiplication sentence.

1. $\underline{2} \times \underline{4} = \underline{8}$

2. $\underline{4} \times \underline{10} = \underline{40}$

3. $\underline{3} \times \underline{9} = \underline{27}$

Practice
Write each as a multiplication sentence.

4. $4 + 4 + 4 + 4 + 4 = 20$
$\underline{5} \times \underline{4} = \underline{20}$

5. $2 + 2 + 2 = 6$
$\underline{3} \times \underline{2} = \underline{6}$

6. $3 + 3 + 3 + 3 + 3 = 15$
$\underline{5} \times \underline{3} = \underline{15}$

7. $7 + 7 + 7 + 7 = 28$
$\underline{4} \times \underline{7} = \underline{28}$

8. Three groups of 8 equals 24.
$\underline{3} \times \underline{8} = \underline{24}$

9. Eight groups of 2 equals 16.
$\underline{8} \times \underline{2} = \underline{16}$

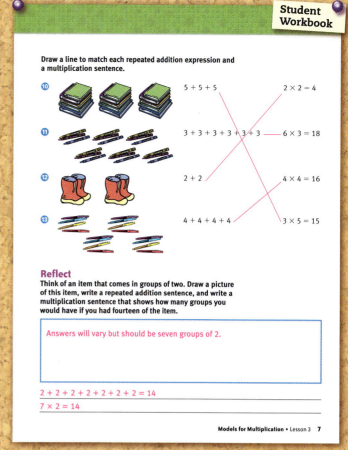

Draw a line to match each repeated addition expression and a multiplication sentence.

10. $5 + 5 + 5$ — $3 \times 5 = 15$

11. $3 + 3 + 3 + 3 + 3 + 3$ — $6 \times 3 = 18$

12. $2 + 2$ — $2 \times 2 = 4$

13. $4 + 4 + 4 + 4$ — $4 \times 4 = 16$

Reflect
Think of an item that comes in groups of two. Draw a picture of this item, write a repeated addition sentence, and write a multiplication sentence that shows how many groups you would have if you had fourteen of the item.

Answers will vary but should be seven groups of 2.

$2 + 2 + 2 + 2 + 2 + 2 + 2 = 14$
$7 \times 2 = 14$

 ## Reflect   10

Extended Response   REASONING

Review students' answers to the Reflect prompt at the bottom of student page 7.

Invite students to share with the class their pictures and number sentences at the board.

- **Was it easy or difficult to think of an item that comes in a group of two?**
- **What is the lowest number you can think of that is not used to group items together?**

Real-World Application   APPLYING

Advertisements often include pictures to show items that can be bought in groups. Have students create an advertisement that shows how many dinner rolls a customer gets when he or she buys three packages of rolls with six rolls per package.

- **How many groups of rolls are purchased?** 3
- **How many rolls are in each group?** 6
- **Write a repeated addition expression for the rolls purchased.** $6 + 6 + 6$
- **Write a multiplication sentence using the × symbol.** $3 \times 6 = 18$

 ## Assess

Informal Assessment

Use the Student Assessment Record, *Assessment*, page 100, to record informal observations.

APPLYING

Draw It, Chart It
Did the student
❏ apply learning to new situations?
❏ contribute concepts?
❏ contribute answers?
❏ connect mathematics to the real world?

Models for Multiplication • Lesson 3   6–7

Lesson 4

Objective

Students describe groups that come in sets.

Materials

Program Materials
- Recording Chart, p. B13
- Number 1–6 Cube

Additional Materials
Counters

Access Vocabulary

skip counting Counting by skipping an interval such as every other number or every three numbers

antennae The feelers that an insect has on its head to help it sense predators or prey

Creating Context

Some students may be unfamiliar with concession stands at sporting events. Review with English Learners the types of food available at concession stands.

---

1 Warm Up 5

Concept Building UNDERSTANDING

- If you are counting items that come in pairs, by what do you skip count? 2
- If you are counting items that come in half a dozen, by what do you skip count? 6
- If you are counting 3, 6, 9, 12, and so on, what number are you using to skip count? 3

---

2 Engage 30

Skill Building

Each student needs a Recording Chart and a Number 1–6 Cube.

A Group by Another Name UNDERSTANDING

- How many gloves are in a pair of gloves? 2
- What word in the previous question tells you that gloves come in groups of two? pair
- Roll the Number Cube. Use that number as the number of groups of gloves.
- Complete a line on your Recording Chart for the pairs of gloves.
- How many sides does a triangle have? 3
- What word in the previous question tells you that there are three sides? triangle
- Roll the Number Cube, and complete a line in the Recording Chart.

Accept suggestions from students of other words or phrases that describe a group of items. Students should roll the Number Cube and complete a line on the Recording Chart for each suggestion.

Monitoring Student Progress

If . . . students do not know the meaning of the specialty words used to describe things that come in sets,

Then . . . have them find the meaning in a dictionary. Active learning fosters deeper understanding.

Building Blocks For additional practice recording groups as a multiplication sentence, students should complete **Building Blocks** Cube 100.

MathTools Use the 100 Table to demonstrate and explore multiplication.

Using Student Pages

Have students complete **Workbook,** pages 8–9, on their own.

Week 1

Models for Multiplication

Lesson 4

Key Idea

Groups can be described in specific ways. Below are some examples.

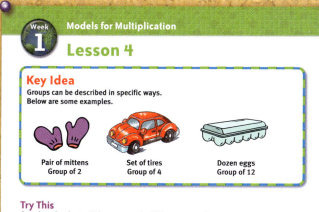

Pair of mittens
Group of 2

Set of tires
Group of 4

Dozen eggs
Group of 12

Try This

Complete the charts. Write a repeated addition sentence for each problem if you need a model to figure out the answer.

1

Number of Students	Number of Mittens per Student	Multiplication Sentence
1	2	$1 \times 2 = 2$
2	2	$2 \times 2 = 4$
3	2	$3 \times 2 = 6$
4	2	$4 \times 2 = 8$
5	2	$5 \times 2 = 10$

2

Number of Cars	Number of Tires on a Car	Multiplication Sentence
1	4	$1 \times 4 = 4$
2	4	$2 \times 4 = 8$
3	4	$3 \times 4 = 12$
4	4	$4 \times 4 = 16$
5	4	$5 \times 4 = 20$

Practice

Write a multiplication sentence for each situation. Write a repeated addition sentence if you need a model to figure out the answer.

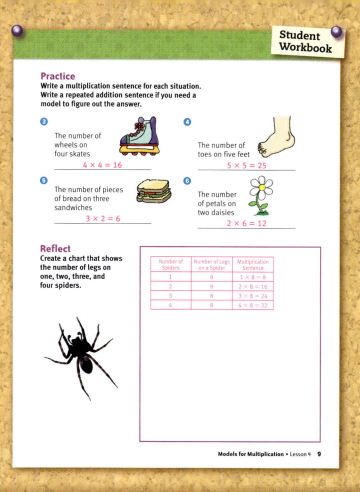

3 The number of wheels on four skates

$4 \times 4 = 16$

4 The number of toes on five feet

$5 \times 5 = 25$

5 The number of pieces of bread on three sandwiches

$3 \times 2 = 6$

6 The number of petals on two daisies

$2 \times 6 = 12$

Reflect

Create a chart that shows the number of legs on one, two, three, and four spiders.

Number of Spiders	Number of Legs on a Spider	Multiplication Sentence
1	8	$1 \times 8 = 8$
2	8	$2 \times 8 = 16$
3	8	$3 \times 8 = 24$
4	8	$4 \times 8 = 32$

3 Reflect   10

Extended Response   REASONING

Review students' answers to the Reflect prompt at the bottom of student page 9.

Discuss whether this chart fits other types of bugs. If any student has extensive knowledge about a particular type of insect, invite him or her to play the role of teacher and guide the class to create a chart showing the number of legs, eyes, or antennae.

- **What was easy about this activity?**
- **What was difficult about this activity?**

Real-World Application   APPLYING

People who work at concession stands often have charts similar to those used in this lesson to help them quote prices quickly.

- **Make a chart for selling one to ten hot dogs. Each hot dog costs $2.**
- **Often, people order the same number of drinks and hot dogs. Make a chart for selling one drink and one hot dog to ten drinks and ten hot dogs. Each hot dog and drink together cost $3.**

4 Assess

Informal Assessment

Use the Student Assessment Record, *Assessment,* page 100, to record informal observations.

UNDERSTANDING

A Group by Another Name

Did the student
- ❏ make important observations?
- ❏ extend or generalize learning?
- ❏ provide insightful answers?
- ❏ pose insightful questions?

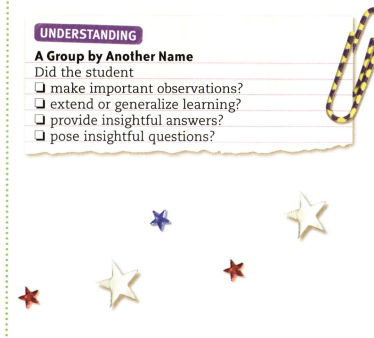

Lesson 5 Review

Objective

Students review skills learned this week and complete the weekly assessment.

Materials

Review materials will be selected from those used in previous activities.

Creating Context

English Learners may benefit from clarification of some common phrases and words that proficient English speakers probably know. Occasionally words have more than one meaning or are used in potentially puzzling idiomatic expressions. Before or during the lesson, be sure to clarify the words and phrases that may be confusing.

1 Warm Up — 5

Concept Building UNDERSTANDING

- Name the ways we looked at multiplication this week. grouping, repeated addition, using the × symbol
- Show how many there are altogether in four groups of three apples each.
  $3 + 3 + 3 + 3 = 12; 4 \times 3 = 12$

2 Engage — 20

Skill Building

Free-Choice Activity ENGAGING

For the last day of the week, allow students to choose an activity from the previous lessons. Some activities they may choose are:

- Circle the Groups
- Picture Match-Up
- Draw It, Chart It
- A Group by Another Name

Make a note of the activities students select. Do they prefer easy or challenging activities? If you believe that students would benefit from extra practice on specific skills, choose an activity for them.

Monitoring Student Progress

If . . . students are still relying heavily on drawing pictures to model the multiplication,

Then . . . encourage those students to choose **Draw It, Chart It.**

Using Student Pages

Have students complete **Workbook,** pages 10–11, on their own.

3 Reflect — 10

Extended Response REASONING

Review students' answers to the Reflect prompt at the bottom of student page 11.

Discuss the answer with the group to reinforce Week 1 concepts.

Week 1 — Models for Multiplication

Lesson 5 Review

This week you looked at multiplication sentences. You learned that multiplication is repeated addition. You also learned that the × symbol means "times" or "groups of."

Lesson 1 Draw circles to show the number of groups named.

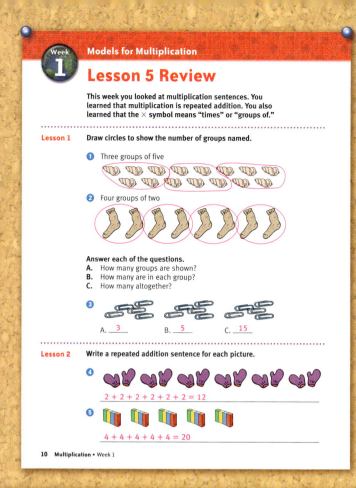

❶ Three groups of five

❷ Four groups of two

Answer each of the questions.
A. How many groups are shown?
B. How many are in each group?
C. How many altogether?

❸

A. __3__   B. __5__   C. __15__

Lesson 2 Write a repeated addition sentence for each picture.

❹

2 + 2 + 2 + 2 + 2 + 2 = 12

❺

4 + 4 + 4 + 4 + 4 = 20

Lesson 3 Write each as a multiplication sentence.

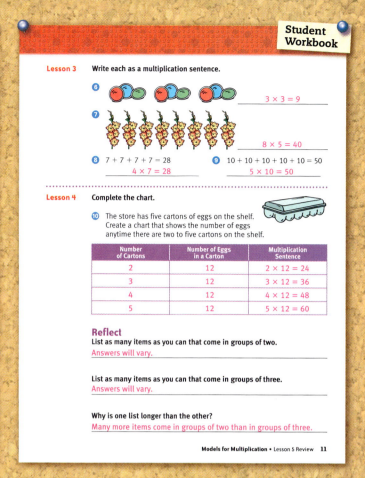

❻   3 × 3 = 9

❼   8 × 5 = 40

❽ 7 + 7 + 7 + 7 = 28   ❾ 10 + 10 + 10 + 10 + 10 = 50
   4 × 7 = 28   5 × 10 = 50

Lesson 4 Complete the chart.

❿ The store has five cartons of eggs on the shelf. Create a chart that shows the number of eggs anytime there are two to five cartons on the shelf.

Number of Cartons	Number of Eggs in a Carton	Multiplication Sentence
2	12	2 × 12 = 24
3	12	3 × 12 = 36
4	12	4 × 12 = 48
5	12	5 × 12 = 60

Reflect
List as many items as you can that come in groups of two.
Answers will vary.

List as many items as you can that come in groups of three.
Answers will vary.

Why is one list longer than the other?
Many more items come in groups of two than in groups of three.

---

4 Assess   10

A Gather Evidence

Formal Assessment

Have students complete the weekly test on *Assessment,* pages 48–49. Record progress on the Student Assessment Record, *Assessment,* page 100.

B Summarize Findings

Determine whether students have Minimal, Basic, or Secure understanding of the concepts presented in Week 1.

C Differentiate Instruction

Based on your observations, use these teaching strategies next week to follow up.

Minimal Understanding
- Repeat the Warm-Up and Engage activities to develop Week 1 concepts.
- Use *Building Blocks* Comic Book Shop and *eMathTools* 100 Table to develop and reinforce this week's concepts.

Basic Understanding
- Repeat Engage activities to reinforce Week 1 concepts.
- Use *Building Blocks* Comic Book Shop and *eMathTools* 100 Table to reinforce Week 1 concepts.

Secure Understanding
- Use *Building Blocks* Comic Book Shop and *eMathTools* 100 Table to reinforce Week 1 concepts.
- Use variations of the weekly Warm-Up and Engage activities, using higher numbers and multiple steps.

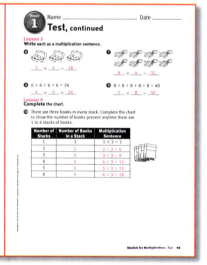

Assessment, pp. 48–49

Week 2 Number Lines and Arrays

Week at a Glance

This week, students continue *Number Worlds,* Level F, Multiplication. Students should use number lines and rectangular arrays to create models for multiplication.

Math Background

It is important that students become familiar with using models other than the groups and repeated addition for multiplication. Number lines and rectangular arrays are appropriate "next" models because repeated addition can be linked to rectangular arrays, and the array models open the door for investigating commutative and distributive properties as well as illustrating multiplication of decimals and fractions when the time comes.

How Students Learn

To understand multiplication, students need to see the data as units, not simply numbers. In multiplication the number 5 might represent five discrete items, or it might represent five groups of items. Students need to understand the concept that composite groups (groups of more than one item) can be counted multiple times and also be treated as a single number.

Teaching for Understanding

Observe closely while evaluating the assigned tasks this week to see whether students can demonstrate the following understandings.

Benchmark after Lesson 2: Students can visualize multiplication as a dot pattern and write multiplication sentences to describe these patterns.

Benchmark after Lesson 3: Students can create an array to show a multiplication sentence.

Benchmark after Lesson 4: Students can write a multiplication sentence when shown an array.

Skills Focus

- Use a number line or chart to skip count to find a product
- Visualize multiplication patterns
- Model multiplication facts, using rectangular arrays
- Write multiplication sentences by looking at a rectangular array

Math at Home

Give one copy of the Letter to Home, page A14, to each student. Encourage students to share and complete the activity with their caregivers.

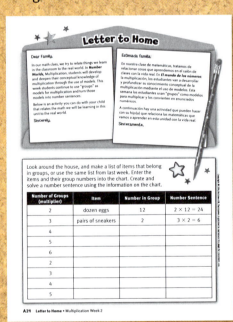

Letter to Home, Teacher Edition, p. A14

Number Lines and Arrays

PACING	LESSON	LEARNING GOALS	NCTM	MATERIALS	TECHNOLOGY
DAY 1	pages 12–13	Students skip count and use a number line to create a model for multiplication.	• Number and Operations • Problem Solving • Communication • Connections	• Number Line • Counters*	**Building Blocks** Comic Book Shop **e MathTools** Number Line
DAY 2	pages 14–15	Students visualize multiplication as a dot pattern and write multiplication sentences to describe these patterns.	• Number and Operations • Problem Solving • Communication • Representation	• Number Chart 0–99, p. B14 • Multiplication Bingo, pp. B15–16 • Dot 1–6 Cubes, 2 • Number line • Counters*	**Building Blocks** Comic Book Shop **e MathTools** 100 Table
DAY 3	pages 16–17	Students write multiplication sentences to describe information presented in a rectangular array.	• Number and Operations • Problem Solving • Communication • Representation	• Scissors • Graph paper • Crayons	**Building Blocks** Arrays in Area **e MathTools** Array Tool
DAY 4	pages 18–19	Students investigate the Commutative Property of multiplication, using rectangular arrays.	• Number and Operations • Problem Solving • Communication • Connections • Representation	• Graph paper • Scissors • Crayons	**Building Blocks** Arrays in Area **e MathTools** Array Tool
DAY 5	Review and Assess pages 20–21	Students review skills learned this week and complete the weekly assessment.	• Number and Operations • Problem Solving • Communication • Connections • Representation	Materials will be selected from Lessons 1–4.	**Building Blocks** **e MathTools** Review previous activities

Math Vocabulary

rectangular array An arrangement of tiles or objects in rows and columns

multiplication sentence A number sentence that contains factors separated by × and the product after an = sign

English Learners

SPANISH COGNATES

English	Spanish
product	producto
multiplication	multiplicación

ALTERNATE VOCABULARY

array A set number of items organized and arranged in rows and columns

Lesson 1

Objective

Students skip count and use a number line to create a model for multiplication.

Materials

Additional Materials
- Number Line
- Counters

Access Vocabulary

number line A line that is marked with numbers, used in counting

Creating Context

An excellent strategy for improving comprehension for English Learners is to have students draw a picture of the situation described in the math problem. This helps the students envision the problem and helps the teacher determine the students' understanding of the problem.

1 Warm Up · 5

Concept Building COMPUTING

- Draw a picture to show four groups of six flowers each. Write a multiplication sentence that describes your picture. $4 \times 6 = 24$
- Draw a picture of a window with four panes. Now draw two more of the same window. Write a multiplication sentence that describes your picture. $3 \times 4 = 12$

2 Engage · 30

Skill Building

Sarah prepared eight treat bags with four pieces of candy in each bag. Altogether, how many pieces of candy did she use to make the treat bags?

Provide students with drawing paper and number lines, and instruct them to re-create on their papers what you draw on the board.

Jump the Line UNDERSTANDING

- How many bags of candy does Sarah need? 8
- Draw all eight bags.
- How many pieces of candy go into each bag? 4
- Draw four candies in each bag.
- Use skip counting on a number line to show how many pieces of candy Sarah needs.

Draw a number line on the board and draw the arches to show eight jumps of four, counting each jump as you do so.

- What do the jumps show? the number of groups
- How many jumps did I make? 8
- How many spaces did I jump each time? 4
- What does the four-number leap represent? the number of candies in each group
- Write a multiplication sentence to describe the number of pieces of candy you drew altogether. $8 \times 4 = 32$

Repeat the series of questions and drawings, using an item that comes in groups of two and groups of three.

Collect the drawings for use later in the week.

Building Blocks For additional practice skip counting, students should complete **Building Blocks** Comic Book Shop.

MathTools Use the Number Line to demonstrate and explore multiplication by skip counting.

Using Student Pages

Have students complete **Workbook,** pages 12–13, on their own.

Lesson 1

Key Idea
You can use skip counting as a model for multiplication. A number line can help you skip count.

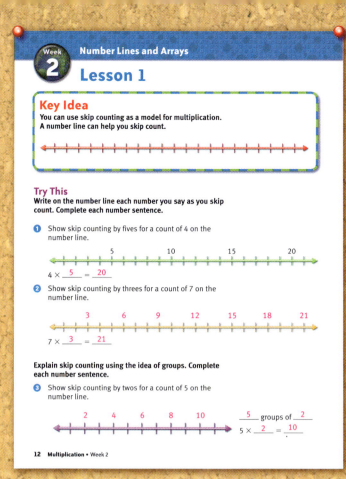

Try This
Write on the number line each number you say as you skip count. Complete each number sentence.

1. Show skip counting by fives for a count of 4 on the number line.

5   10   15   20

$4 \times \underline{5} = \underline{20}$

2. Show skip counting by threes for a count of 7 on the number line.

3   6   9   12   15   18   21

$7 \times \underline{3} = \underline{21}$

Explain skip counting using the idea of groups. Complete each number sentence.

3. Show skip counting by twos for a count of 5 on the number line.

2   4   6   8   10

$\underline{5}$ groups of $\underline{2}$

$5 \times \underline{2} = \underline{10}$.

---

4. Show skip counting by threes for a count of 3 on the number line.

3   6   9

$\underline{3}$ groups of $\underline{3}$

$3 \times \underline{3} = \underline{9}$

Practice
Use the number line on page 12 to skip count each problem. Write by which number you are skip counting and how many times you count. Complete each number sentence.

5. $2 \times 9 = \underline{18}$ Skip count by $\underline{9}$ for a count of $\underline{2}$.

6. $6 \times 4 = \underline{24}$ Skip count by $\underline{4}$ for a count of $\underline{6}$.

7. $3 \times 8 = \underline{24}$ Skip count by $\underline{8}$ for a count of $\underline{3}$.

8. $9 \times 3 = \underline{27}$ Skip count by $\underline{3}$ for a count of $\underline{9}$.

Skip count to complete each number sentence.

9.

$3 \times \underline{3} = \underline{9}$

10.

$\underline{2} \times 12 = \underline{24}$

Reflect
Draw three bicycles. Skip count to find the number of wheels altogether. Describe the wheels as groups. Write a multiplication sentence to show the number of wheels altogether.

Students should draw 3 bicycles.

2 , 4 , 6

$\underline{3}$ groups of $\underline{2}$ wheels each

$\underline{3} \times \underline{2} = \underline{6}$

 Reflect   10

Extended Response   REASONING

Review students' answers to the Reflect prompt at the bottom of student page 13.

Invite students to the board to draw pictures, skip count, show repeated addition, and write the multiplication sentences for items in groups of four, five, and six.

- **Did this activity help you do anything you could not do before?**
- **What was easy about this activity?**
- **What was difficult about this activity?**

Real-World Application   APPLYING

Skip counting people in a line is one way to organize individuals into several groups. Have students line up at the front of the classroom. Walk in front of the line, and tap every other person in line. Have the 'tapped' students step forward.

- **By what number did I skip count to select these people?** 2

 Assess

Informal Assessment

Use the Student Assessment Record, **Assessment,** page 100, to record informal observations.

UNDERSTANDING

Jump the Line

Did the student
- ❏ make important observations?
- ❏ extend or generalize learning?
- ❏ provide insightful answers?
- ❏ pose insightful questions?

Lesson 2

Objective

Students visualize multiplication as a dot pattern and write multiplication sentences to describe these patterns.

Materials

Program Materials
- Number Chart 0–99, p. B14
- Multiplication Bingo, pp. B15–B16
- Dot Cubes 1–6, 2

Additional Materials
- Number line
- Counters

Access Vocabulary

quarters Coins worth twenty-five cents
baking sheets Flat metal pans used for baking cookies in the oven

Creating Context

An excellent strategy for building vocabulary and concepts with English Learners is to use real objects. Bring in items such as an ice cube tray, an egg carton, or a fast food drink holder. Have students identify each item and the groups they represent.

1 Warm Up 5

Concept Building COMPUTING

On a number line on the board, draw seven arched arrows showing leaps of three spaces. Skip count aloud as you draw each arrow.

- **How many groups are shown in this skip counting?** 7
- **How many spaces are in each group?** 3
- **At what number does the skip counting end?** 21
- **What number sentence can we write to show what we just did?** 7 × 3 = 21

2 Engage 30

Skill Building

Multiplication Bingo UNDERSTANDING

- Distribute a bingo card and Counters to each student.
- Carefully explain the rules. (Multiplication Bingo Activity Sheet, p. B15)
- Play one or two practice rounds to make sure that everyone understands the rules.

Strategy Building

Organize students into groups of three. Give each group a sheet of paper to write the results of each step.

Skip to the Answer COMPUTING

- One student in the group randomly chooses a number between 2 and 9 to represent the number of groups in the problem.
- Another student in the group randomly chooses a number between 2 and 5 to represent how many items are in each group.
- The third student states the problem and skip counts on the number line to find the answer.
- Each group should do the activity three times so that each student can play each role.

After all groups have finished, demonstrate how to use Number Chart 0–99 in place of a number line.

Monitoring Student Progress

If . . . students finish the activity early, **Then . . .** have them repeat the activity.

Building Blocks For additional practice using a number line and chart to model multiplication, students should complete **Building Blocks** Comic Book Shop.

MathTools Use the 100 Table to demonstrate and explore multiplication.

Using Student Pages

Have students complete **Workbook,** pages 14–15, on their own.

Week 2 — Number Lines and Arrays

Lesson 2

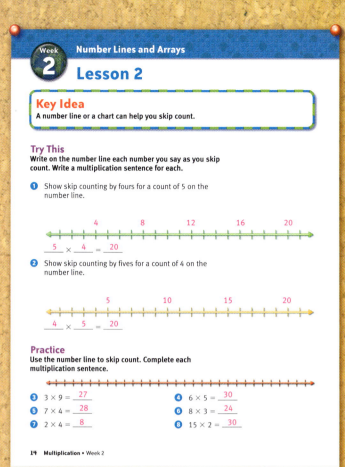

Key Idea
A number line or a chart can help you skip count.

Try This
Write on the number line each number you say as you skip count. Write a multiplication sentence for each.

1 Show skip counting by fours for a count of 5 on the number line.

4   8   12   16   20

$\underline{5} \times \underline{4} = \underline{20}$

2 Show skip counting by fives for a count of 4 on the number line.

5   10   15   20

$\underline{4} \times \underline{5} = \underline{20}$

Practice
Use the number line to skip count. Complete each multiplication sentence.

3 $3 \times 9 = \underline{27}$

4 $6 \times 5 = \underline{30}$

5 $7 \times 4 = \underline{28}$

6 $8 \times 3 = \underline{24}$

7 $2 \times 4 = \underline{8}$

8 $15 \times 2 = \underline{30}$

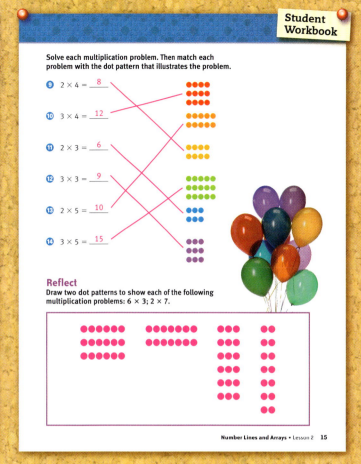

Solve each multiplication problem. Then match each problem with the dot pattern that illustrates the problem.

9 $2 \times 4 = \underline{8}$

10 $3 \times 4 = \underline{12}$

11 $2 \times 3 = \underline{6}$

12 $3 \times 3 = \underline{9}$

13 $2 \times 5 = \underline{10}$

14 $3 \times 5 = \underline{15}$

Reflect
Draw two dot patterns to show each of the following multiplication problems: 6×3; 2×7.

---

3 Reflect   10

Extended Response   REASONING

Review students' answers to the Reflect prompt at the bottom of student page 15.

Discuss why a dot pattern is a good way to visualize multiplication.

- **Why can each multiplication sentence be shown two ways?** There are two numbers: the number of groups and the number in each group. We can make rows and columns with either number because the answer includes both.

Real-World Application   APPLYING

Counting money is a common use of skip counting.

- **Suppose that I skip count a group of nickels. I say 5, 10, 15, 20. Write a multiplication sentence that shows how many nickels I have.** $4 \times 5 = 20$
- **Suppose that I skip count a group of quarters. I say 25, 50, 75, 100, 125, 150. Write a multiplication sentence that shows how many quarters I have.** $6 \times 25 = 150$

4 Assess

Informal Assessment

Use the Student Assessment Record, **Assessment,** page 100, to record informal observations.

UNDERSTANDING	COMPUTING
Multiplication Bingo	**Skip to the Answer**
Did the student	Did the student
❏ make important observations?	❏ respond accurately?
❏ extend or generalize learning?	❏ respond quickly?
❏ provide insightful answers?	❏ respond with confidence?
❏ pose insightful questions?	❏ self-correct?

Lesson 3

Objective

Students write multiplication sentences to describe information presented in a rectangular array.

Materials

Additional Materials
- Graph paper
- Scissors
- Crayons

Access Vocabulary

rows Objects arranged in horizontal straight lines

muffins Small, individual, sweet cakelike baked goods

Creating Context

In this lesson, students learn the Commutative Property of multiplication. Remind students that the commutative property means that you can multiply the numbers in any order and still get the same answer.

1 Warm Up 5

Concept Building COMPUTING

Give students graph paper and scissors, and instruct them to cut out twenty squares to use in forming arrays.

- **Line up four squares in a column.**
- **This is one group of four.**
- **Place three more columns of four squares next to the first column.**
- **How many squares are in the array altogether?** 16

2 Engage 30

Skill Building

Students need squares and paper on which to write. Explain and repeat each step several times as the students are working, and write the equations on the board.

Build an Array COMPUTING

- **Use two squares to make a vertical rectangle. This is a model of one group of two. Write $1 \times 2 = 2$.**
- **Build a second vertical rectangle, and place it beside the first rectangle. This models two groups of two. Write the multiplication sentence modeled.** $2 \times 2 = 4$
- **Build a third vertical rectangle, and place it beside the other rectangles. This models three groups of two. Write the multiplication sentence modeled.** $3 \times 2 = 6$
- **Build a fourth vertical rectangle, and place it beside the other rectangles. How many groups of two? What number sentence is modeled?** $4; 4 \times 2 = 8$
- **Build another vertical rectangle, and write the number sentence modeled.** $5 \times 2 = 10$

Allow time for students to build arrays of groups of three and then groups of four. Students should write the multiplication sentence modeled each time.

Monitoring Student Progress

If . . . students are not catching on to the idea of the Commutative Property,

Then . . . have them color several arrays on graph paper. Draw a 2×3 array next to a 3×2 array and a 3×4 array next to a 4×3 array.

 Building Blocks For additional practice with arrays, students should complete **Building Blocks** Arrays in Area.

eMathTools Use Array Tool to demonstrate and explore arrays.

Using Student Pages

Have students complete **Workbook,** pages 16–17, on their own.

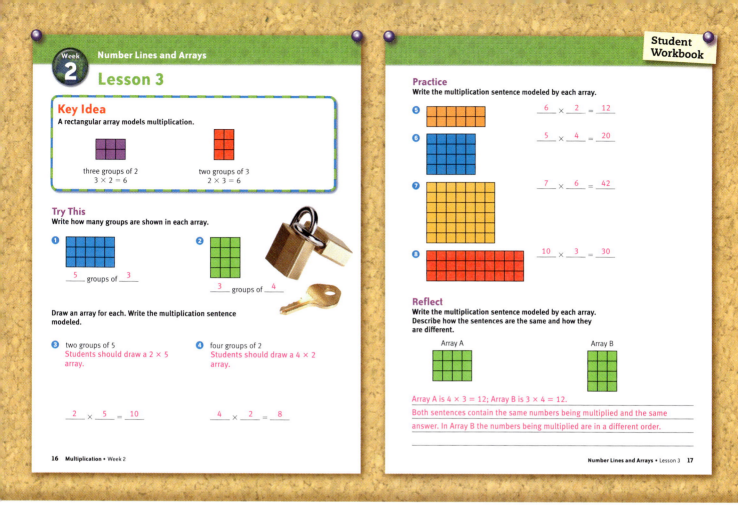

Lesson 3

Key Idea
A rectangular array models multiplication.

three groups of 2
$3 \times 2 = 6$

two groups of 3
$2 \times 3 = 6$

Try This
Write how many groups are shown in each array.

① <u> 5 </u> groups of <u> 3 </u>

② <u> 3 </u> groups of <u> 4 </u>

Draw an array for each. Write the multiplication sentence modeled.

③ two groups of 5
Students should draw a 2 × 5 array.

<u> 2 </u> × <u> 5 </u> = <u> 10 </u>

④ four groups of 2
Students should draw a 4 × 2 array.

<u> 4 </u> × <u> 2 </u> = <u> 8 </u>

Practice
Write the multiplication sentence modeled by each array.

⑤ <u> 6 </u> × <u> 2 </u> = <u> 12 </u>

⑥ <u> 5 </u> × <u> 4 </u> = <u> 20 </u>

⑦ <u> 7 </u> × <u> 6 </u> = <u> 42 </u>

⑧ <u> 10 </u> × <u> 3 </u> = <u> 30 </u>

Reflect
Write the multiplication sentence modeled by each array. Describe how the sentences are the same and how they are different.

Array A   Array B

Array A is 4 × 3 = 12; Array B is 3 × 4 = 12.
Both sentences contain the same numbers being multiplied and the same answer. In Array B the numbers being multiplied are in a different order.

3 Reflect   10

Extended Response   REASONING

Review students' answers to the Reflect prompt at the bottom of student page 17.

Discuss the Commutative Property of multiplication.

- Choose one array, and hold it in any direction. Record the multiplication sentence.
- Have students rotate the array 90 degrees and record the multiplication sentence.
- Emphasize that for each multiplication sentence, it was the same array and did not change sizes. Guide students to draw the conclusion that the order of the factors does not affect the product.

Real-World Application   APPLYING

■ Describe how baking cookies or muffins uses the concept of arrays. The baked goods are usually arranged in rows and columns.

4 Assess

Informal Assessment

Use the Student Assessment Record, **Assessment,** page 100, to record informal observations.

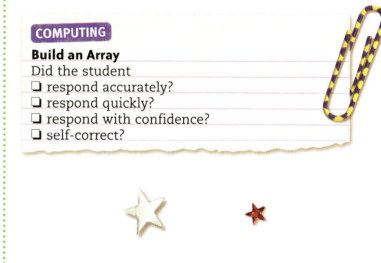

COMPUTING

Build an Array
Did the student
❏ respond accurately?
❏ respond quickly?
❏ respond with confidence?
❏ self-correct?

Lesson 4

Objective

Students investigate the Commutative Property of multiplication, using rectangular arrays.

Materials

Additional Materials
- Graph paper
- Scissors
- Crayons

Access Vocabulary

patterns Predictable designs made of repeated shapes, numbers, or colors

Creating Context

Have students work together to look through a newspaper advertisement for food items that are packaged and sold in arrays. Have them cut out and paste examples on a display and label them to show the name of each item and the array.

1 Warm Up 5

Concept Building UNDERSTANDING

Draw a carton of eggs on the board.

- **Do you see an array? What multiplication sentence is modeled?** $6 \times 2 = 12$ or $2 \times 6 = 12$
- **Think of a box of crayons. Do you see an array? What multiplication sentence is modeled?** Answers will vary, depending on the number and arrangement of crayons in the box.

2 Engage 30

Skill Building

Students need graph paper and scissors.

Array to the Product REASONING

Write the following multiplication problems on the board:

$$5 \times 7 \quad 2 \times 10 \quad 8 \times 6 \quad 15 \times 3$$

- Use your scissors to cut out four arrays that match the four expressions on the board.
- Who cut their array by counting five squares over and then seven squares down? Select one student to come forward to show the array.
- Who cut their array by counting seven squares over and then five squares down? Select one student to come forward to show the array.
- **Are these arrays the same?** yes

Repeat questions for the remaining problems.

Strategy Building

Share and Match ENGAGING

- Cut out three arrays in any sizes you like.
- On a different sheet of paper, write the multiplication problems that describe your arrays. Do not list them in any particular order.
- Trade your problems and arrays with another student's problems and arrays. Match each array to each problem. When you have finished matching, check each other's matches.

Monitoring Student Progress

| If . . . students do not have good motor skills that allow them to cut with scissors, | Then . . . provide crayons so that they can color an array on the graph paper instead of cutting it out. |

Building Blocks For additional practice with arrays, students should complete **Building Blocks** Arrays in Area.

e MathTools Use Array Tool to demonstrate and explore arrays.

Using Student Pages

Have students complete **Workbook,** pages 18–19, on their own.

Lesson 4

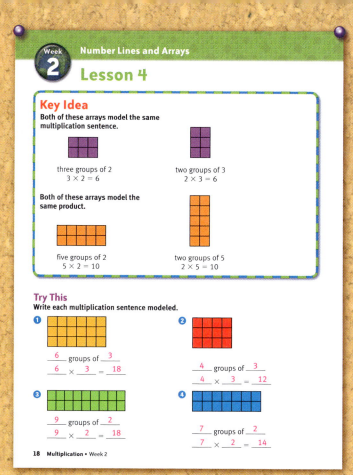

Key Idea

Both of these arrays model the same multiplication sentence.

three groups of 2
$3 \times 2 = 6$

two groups of 3
$2 \times 3 = 6$

Both of these arrays model the same product.

five groups of 2
$5 \times 2 = 10$

two groups of 5
$2 \times 5 = 10$

Try This
Write each multiplication sentence modeled.

1
<u>6</u> groups of <u>3</u>
<u>6</u> × <u>3</u> = <u>18</u>

2
<u>4</u> groups of <u>3</u>
<u>4</u> × <u>3</u> = <u>12</u>

3
<u>9</u> groups of <u>2</u>
<u>9</u> × <u>2</u> = <u>18</u>

4
<u>7</u> groups of <u>2</u>
<u>7</u> × <u>2</u> = <u>14</u>

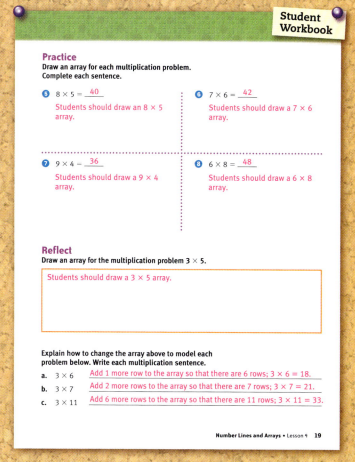

Practice
Draw an array for each multiplication problem. Complete each sentence.

5 $8 \times 5 = $ <u>40</u>
Students should draw an 8 × 5 array.

6 $7 \times 6 = $ <u>42</u>
Students should draw a 7 × 6 array.

7 $9 \times 4 = $ <u>36</u>
Students should draw a 9 × 4 array.

8 $6 \times 8 = $ <u>48</u>
Students should draw a 6 × 8 array.

Reflect
Draw an array for the multiplication problem 3×5.

Students should draw a 3 × 5 array.

Explain how to change the array above to model each problem below. Write each multiplication sentence.

a. 3×6 Add 1 more row to the array so that there are 6 rows; $3 \times 6 = 18$.

b. 3×7 Add 2 more rows to the array so that there are 7 rows; $3 \times 7 = 21$.

c. 3×11 Add 6 more rows to the array so that there are 11 rows; $3 \times 11 = 33$.

---

3 Reflect   10

Extended Response   REASONING

Review students' answers to the Reflect prompt at the bottom of student page 19.

Have students describe their answers by discussing patterns that are visible when building arrays. Relate the columns to the number of groups, and relate the rows to the number of items in each group.

- **Did this activity help you do anything you couldn't do before?**
- **Are there other strategies that could be used to solve this problem?**

Real-World Application   APPLYING

- **Name a product that you or someone in your family buys that is packaged in an array.**
Answers will vary, but may include cookies, golf balls, box of chocolates, and batteries.

- **If you buy a package of batteries that is six batteries across and two batteries deep, how many batteries are in the package?** 12

---

4 Assess

Informal Assessment

Use the Student Assessment Record, **Assessment**, page 100, to record informal observations.

REASONING	ENGAGING
Array to the Product	**Share and Match**
Did the student	Did the student
❑ provide a clear explanation?	❑ pay attention to the contributions of others?
❑ communicate reasons and strategies?	❑ contribute information and ideas?
❑ choose appropriate strategies?	❑ improve on a strategy?
❑ argue logically?	❑ reflect on and check accuracy of work?

Lesson 5 Review

Objective

Students review skills learned this week and complete the weekly assessment.

Materials

Review materials will be selected from those used in previous activities.

Creating Context

An excellent way to check for understanding with English Learners who are at early levels of vocabulary proficiency is to have them draw a picture, draw symbols, or sketch a representation of their response.

1 Warm Up   5

Concept Building   APPLYING

- Which visual aid did you prefer to use this week: number line, Number Chart 0–99, dot patterns, or graph-paper arrays? Answers will vary.
- Name a place you may go that uses the idea of an array for a seating arrangement. Possible answers: theater, classroom

2 Engage   20

Strategy Building

Free-Choice Activity   ENGAGING

For the last day of the week, allow students to choose an activity from the previous lessons. Some activities they may choose include the following:

- **Jump the Line**
- **Multiplication Bingo**
- **Skip to the Answer**
- **Build an Array**
- **Array to the Product**
- **Share and Match**

Make a note of the activities children select. Do they prefer easy or challenging activities? If you believe your students would benefit from extra practice on specific skills, choose an activity for them.

Monitoring Student Progress

If . . . students favor visual learning opportunities,

Then . . . encourage those students to choose **Multiplication Bingo.**

Using Student Pages

Have students complete **Workbook,** pages 20–21, on their own.

3 Reflect   10

Extended Response   REASONING

Review students' answers to the Reflect prompts at the bottom of student pages 20–21.

Discuss the answers with the group to reinforce Week 2 concepts.

Lesson 5 Review

This week you looked at more ways to represent multiplication sentences. You modeled by using skip counting on a number line and by using Number Chart 0–99. You also modeled multiplication by using rectangular arrays.

Lesson 1
Write the numbers you say as you skip count. Complete the number sentence.

❶ Show skip counting by threes for a count of 6 on the number line.

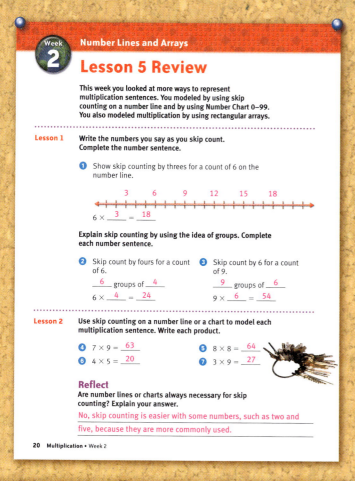

3   6   9   12   15   18

$6 \times \underline{3} = \underline{18}$

Explain skip counting by using the idea of groups. Complete each number sentence.

❷ Skip count by fours for a count of 6.
$\underline{6}$ groups of $\underline{4}$
$6 \times \underline{4} = \underline{24}$

❸ Skip count by 6 for a count of 9.
$\underline{9}$ groups of $\underline{6}$
$9 \times \underline{6} = \underline{54}$

Lesson 2
Use skip counting on a number line or a chart to model each multiplication sentence. Write each product.

❹ $7 \times 9 = \underline{63}$

❺ $8 \times 8 = \underline{64}$

❻ $4 \times 5 = \underline{20}$

❼ $3 \times 9 = \underline{27}$

Reflect
Are number lines or charts always necessary for skip counting? Explain your answer.
No, skip counting is easier with some numbers, such as two and five, because they are more commonly used.

20 Multiplication • Week 2

Lesson 3
Write how many groups are shown in each array.

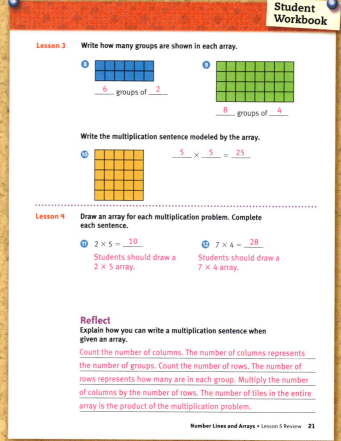

❽ $\underline{6}$ groups of $\underline{2}$

❾ $\underline{8}$ groups of $\underline{4}$

Write the multiplication sentence modeled by the array.

❿ $\underline{5} \times \underline{5} = \underline{25}$

Lesson 4
Draw an array for each multiplication problem. Complete each sentence.

⓫ $2 \times 5 = \underline{10}$
Students should draw a 2×5 array.

⓬ $7 \times 4 = \underline{28}$
Students should draw a 7×4 array.

Reflect
Explain how you can write a multiplication sentence when given an array.
Count the number of columns. The number of columns represents the number of groups. Count the number of rows. The number of rows represents how many are in each group. Multiply the number of columns by the number of rows. The number of tiles in the entire array is the product of the multiplication problem.

Number Lines and Arrays • Lesson 5 Review 21

---

4 Assess   10

A Gather Evidence

Formal Assessment
Have students complete the weekly test on **Assessment,** pages 50–51. Record progress on the Student Assessment Record, **Assessment,** page 100.

B Summarize Findings
Determine whether students have Minimal, Basic, or Secure understanding of the concepts presented in Week 2.

C Differentiate Instruction
Based on your observations, use these teaching strategies next week to follow up.

Minimal Understanding
- Repeat the Warm-Up and Engage activities to develop Week 2 concepts.
- Use **Building Blocks** Comic Book Shop and **eMathTools** Number Line to develop and reinforce this week's concepts.

Basic Understanding
- Repeat Engage activities to reinforce Week 2 concepts.
- Use **Building Blocks** Arrays in Area and **eMathTools** Array Tool to develop and reinforce this week's concepts.

Secure Understanding
- Use **Building Blocks** Arrays in Area and **eMathTools** Array Tool to reinforce this week's concepts.
- Use variations of the weekly activities, using higher numbers and multiple steps.

Assessment, pp. 50–51

Number Lines and Arrays • Lesson 5 Review **20–21**

Building Multiplication Facts

Week at a Glance

This week, students continue *Number Worlds,* Level F, Multiplication. Students should learn the basic multiplication facts and build flash cards to use for practice with a goal to memorize all basic facts with 2 through 9 as factors.

Math Background

Students develop competent skills in computation by exposure to a balance between conceptual understanding and computational proficiency. Computation methods that are practiced without regard to conceptual understanding are often forgotten or remembered incorrectly. However, conceptual understanding without memorization of basic facts can hinder the problem-solving process.

How Students Learn

Not only do students need time to practice basic multiplication facts and the tools with which to practice, but they also need opportunities other than drill and practice. Students benefit from instructional environments in which they feel secure to recall a fact, even if they might make a mistake. Games can provide such an instructional environment. Students are often willing to try even when they are not absolutely sure of an answer because it is acceptable to take chances in games.

Teaching for Understanding

Observe closely while evaluating the assigned tasks this week to see whether students can demonstrate the following understandings.

Benchmark after Lesson 2: Students know the multiplication facts for twos, threes, fours, and fives.

Benchmark after Lesson 3: Students know the multiplication facts for sixes and sevens.

Benchmark after Lesson 4: Students know the multiplication facts for eights and nines.

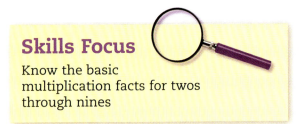

Skills Focus

Know the basic multiplication facts for twos through nines

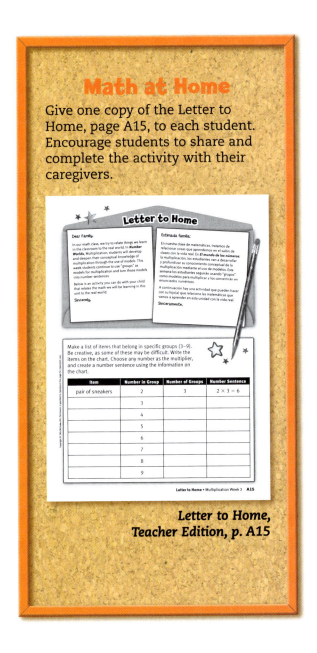

Math at Home

Give one copy of the Letter to Home, page A15, to each student. Encourage students to share and complete the activity with their caregivers.

Letter to Home,
Teacher Edition, p. A15

Week 3 Planner | Building Multiplication Facts

PACING	LESSON	LEARNING GOALS	NCTM	MATERIALS	TECHNOLOGY
DAY 1	pages 22–23	Students know multiplication facts for twos and threes.	• Number and Operations • Problem Solving • Communication	• Building Facts Charts, p. B17 • Index cards • Multiplication Table, p. B18	**Building Blocks** Function Machine 2 **MathTools** Multiplication Table
DAY 2	pages 24–25	Students know multiplication facts for fours and fives.	• Number and Operations • Problem Solving • Communication	• Building Facts Charts, p. B17 • Index cards • Multiplication Table, p. B18 • Multiplication Memory Game Cards, pp. B19–B20	**Building Blocks** Function Machine 2 **MathTools** Multiplication Table
DAY 3	pages 26–27	Students know multiplication facts for sixes and sevens.	• Number and Operations • Problem Solving • Communication	• Building Facts Charts, p. B17 • Index cards • Multiplication Table, p. B18	**Building Blocks** Function Machine 2 **MathTools** Multiplication Table
DAY 4	pages 28–29	Students know multiplication facts for eights and nines.	• Number and Operations • Problem Solving • Communication	• Building Facts Charts, p. B17 • Index cards • Multiplication Table, p. B18 • Product Bingo Cards, pp. B21–B22	**Building Blocks** Function Machine 2 **MathTools** Multiplication Table
DAY 5	**Review and Assess** pages 30–31	Students review skills learned this week and complete the weekly assessment.	• Number and Operations • Problem Solving • Communication	Materials will be selected from Lessons 1–4.	**Building Blocks** Review previous activities

Math Vocabulary

basic multiplication fact A basic number sentence that has two one-digit factors, a × symbol, and an answer

English Learners

SPANISH COGNATES

English	Spanish
multiplication	multiplicación
factor	factor
product	producto
activity	actividad

ALTERNATE VOCABULARY

product The answer to a multiplication problem

Lesson 1

Objective

Students know multiplication facts for twos and threes.

Materials

Program Materials
- Building Facts Charts, p. B17
- Multiplication Table, p. B18

Additional Materials

Index cards

Access Vocabulary

flash cards Small cards with a math fact written on each one

Creating Context

Some English Learners who have attended school outside the United States until recently may continue to count and skip count in their primary language because it is faster for them. Allow students to do so, but encourage students to practice counting numbers in English as well.

1 Warm Up 5

Concept Building COMPUTING

Practice repeated addition problems that use 2 and 3 as addends. Begin with the following:

2+2+2 6   2+2+2+2 8   2+2+2+2+2 10

3+3+3 9   3+3+3+3 12   3+3+3+3+3 15

2 Engage 30

Skill Building

Students need all the materials listed for the Skill-Building Activity.

2 and 3 Flash Cards UNDERSTANDING

- Draw a 1 × 2 array in the first row of the first column on your Building Facts Chart.
- Fill in the second, third, and fourth columns of the chart for the fact modeled by the array.
- Draw the same array on an index card. Write the multiplication sentence, or fact, below the array.
- Turn the card over, and write only the multiplication problem (without the product).
- Repeat this process for the remaining rows in the chart. When you are finished, you will have ten flash cards of the 2-facts that you can use to begin memorizing the basic multiplication facts.
- When all your flash cards are made for the 2-facts, use a new Building Facts Chart to make ten flash cards for the 3-facts.
- Complete the 2-Facts and 3-Facts columns of the Multiplication Table. Write the number sentences and products in order, beginning with 1 × 2 = 2.

Monitoring Student Progress

| If . . . students finish making their flash cards before others, | Then . . . have them begin reviewing their facts, alone or with a partner. |

Building Blocks For additional practice with 2 and 3 multiplication facts, students should complete **Building Blocks** Function Machine 2.

 MathTools Use the Multiplication Table to demonstrate and explore 2 and 3 multiplication facts.

Using Student Pages

Have students complete **Workbook,** pages 22–23, on their own.

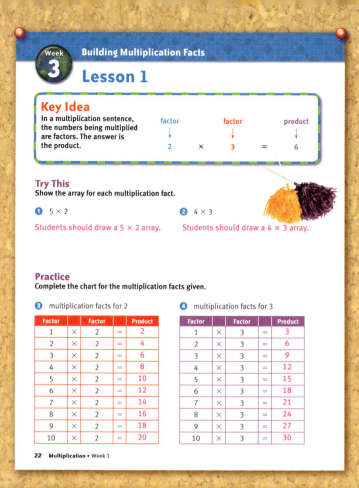

Building Multiplication Facts

Lesson 1

Key Idea

In a multiplication sentence, the numbers being multiplied are factors. The answer is the product.

	factor		factor		product
	↓		↓		↓
	2	×	3	=	6

Try This

Show the array for each multiplication fact.

1 5 × 2

Students should draw a 5 × 2 array.

2 4 × 3

Students should draw a 4 × 3 array.

Practice

Complete the chart for the multiplication facts given.

3 multiplication facts for 2

Factor		Factor		Product
1	×	2	=	2
2	×	2	=	4
3	×	2	=	6
4	×	2	=	8
5	×	2	=	10
6	×	2	=	12
7	×	2	=	14
8	×	2	=	16
9	×	2	=	18
10	×	2	=	20

4 multiplication facts for 3

Factor		Factor		Product
1	×	3	=	3
2	×	3	=	6
3	×	3	=	9
4	×	3	=	12
5	×	3	=	15
6	×	3	=	18
7	×	3	=	21
8	×	3	=	24
9	×	3	=	27
10	×	3	=	30

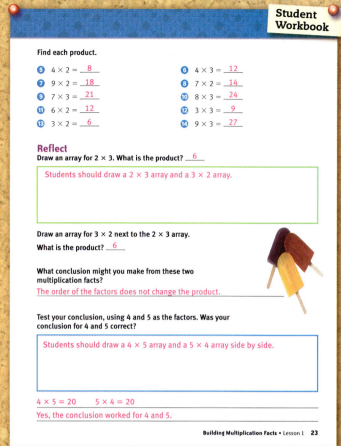

Find each product.

5 4 × 2 = 8

6 4 × 3 = 12

7 9 × 2 = 18

8 7 × 2 = 14

9 7 × 3 = 21

10 8 × 3 = 24

11 6 × 2 = 12

12 3 × 3 = 9

13 3 × 2 = 6

14 9 × 3 = 27

Reflect

Draw an array for 2 × 3. What is the product? 6

Students should draw a 2 × 3 array and a 3 × 2 array.

Draw an array for 3 × 2 next to the 2 × 3 array.
What is the product? 6

What conclusion might you make from these two multiplication facts?

The order of the factors does not change the product.

Test your conclusion, using 4 and 5 as the factors. Was your conclusion for 4 and 5 correct?

Students should draw a 4 × 5 array and a 5 × 4 array side by side.

4 × 5 = 20   5 × 4 = 20
Yes, the conclusion worked for 4 and 5.

---

3 Reflect   10

Extended Response REASONING

Review students' answers to the Reflect prompt at the bottom of student page 23.

Discuss that this is the Commutative Property of multiplication and that they have seen this relationship throughout the Multiplication unit. Guide students to understand that because of this property, as they progress through the facts this week, they will actually be learning fewer and fewer "new" facts in each set.

- **How would you explain this to someone who does not understand this concept?**
- **Can you think of other times outside school when you would use this skill?**

Real-World Application APPLYING

The products of the 2 multiplication facts are probably the most commonly used numbers when people skip count in everyday life. Have students name situations in which people recite the answers to two multiplication facts. Situations will vary, but one example is counting a pile of pennies. Most people count by twos.

4 Assess

Informal Assessment

Use the Student Assessment Record, **Assessment,** page 100, to record informal observations.

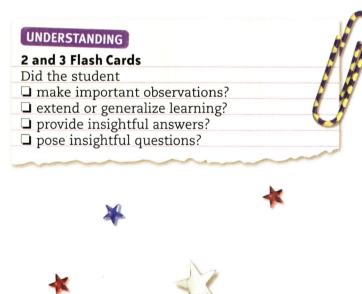

UNDERSTANDING

2 and 3 Flash Cards

Did the student
- ❏ make important observations?
- ❏ extend or generalize learning?
- ❏ provide insightful answers?
- ❏ pose insightful questions?

Lesson 2

Objective

Students know multiplication facts for fours and fives.

Materials

Program Materials
- Building Facts Charts, p. B17
- Multiplication Table, p. B18
- Multiplication Memory Game Cards, pp. B19–B20

Additional Materials
Index cards

Access Vocabulary

match Two cards that are a pair
winner The player who achieves the goal of the game first

Creating Context

Some students may not have experience with playing board games. Review with them the rules, how to win, taking turns, and what to do in case of a tie.

1 Warm Up · 5

Concept Building COMPUTING

Review multiplication facts from Lesson 1. Be sure to include 4×2, 4×3, 5×2, and 5×3.

2 Engage · 30

Skill Building

Students need all the materials listed for the Skill-Building Activity.

4 and 5 Flash Cards UNDERSTANDING

- For making the 4 and 5 flash cards, follow the same process as you did for making the flash cards in Lesson 1.
- Complete the 4-facts and 5-facts columns of the Multiplication Table. Write the number sentence and the products in order, beginning with $1 \times 4 = 4$.

Strategy Building

Multiplication Memory REASONING

Each group needs a set of cut-out Multiplication Memory Game Cards.

- Mix the cards, and place them facedown in four columns of nine.
- The object is to match the multiplication problems with their products. During each turn, a student turns over two cards. If the cards show the problem and its matching product, that student takes the cards and continues his or her turn. If the cards do not match, his or her turn is finished.

The game is finished when all cards are matched. The winner is the player with the most cards.

Monitoring Student Progress

If . . . students need additional practice with the 5-facts,

Then . . . provide nickels for them to place in groups and count.

Building Blocks For additional practice with 4 and 5 multiplication facts, students should complete **Building Blocks** Function Machine 2.

MathTools Use the Multiplication Table to demonstrate and explore 4 and 5 multiplication facts.

Using Student Pages

Have students complete **Workbook,** pages 24–25, on their own.

Lesson 2

Key Idea

In a multiplication sentence, the numbers being multiplied are factors. The answer is the product.

factor		factor		product
↓		↓		↓
4	×	5	=	20

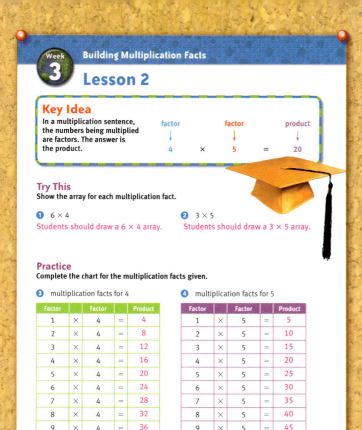

Try This

Show the array for each multiplication fact.

1 6 × 4
Students should draw a 6 × 4 array.

2 3 × 5
Students should draw a 3 × 5 array.

Practice

Complete the chart for the multiplication facts given.

3 multiplication facts for 4

Factor		Factor		Product
1	×	4	=	4
2	×	4	=	8
3	×	4	=	12
4	×	4	=	16
5	×	4	=	20
6	×	4	=	24
7	×	4	=	28
8	×	4	=	32
9	×	4	=	36
10	×	4	=	40

4 multiplication facts for 5

Factor		Factor		Product
1	×	5	=	5
2	×	5	=	10
3	×	5	=	15
4	×	5	=	20
5	×	5	=	25
6	×	5	=	30
7	×	5	=	35
8	×	5	=	40
9	×	5	=	45
10	×	5	=	50

Find each product.

5 5 × 2 = 10
6 8 × 4 = 32
7 10 × 5 = 50
8 2 × 4 = 8
9 7 × 4 = 28
10 9 × 5 = 45
11 6 × 5 = 30
12 7 × 2 = 14
13 3 × 2 = 6
14 6 × 3 = 18
15 1 × 3 = 3
16 9 × 4 = 36
17 4 × 4 = 16
18 9 × 2 = 18
19 1 × 5 = 5
20 8 × 5 = 40
21 4 × 3 = 12
22 10 × 4 = 40

Reflect

Think about the facts for 2 and 3 in Lesson 1. Knowing that the order of the factors does not matter, which multiplication facts for 4 are not new to you?

2 × 4 = 8 and 3 × 4 = 12

Which multiplication facts for 5 are not new to you when you compare the facts for 2, 3, and 4?

2 × 5 × 10, 3 × 5 = 15, and 4 × 5 = 20

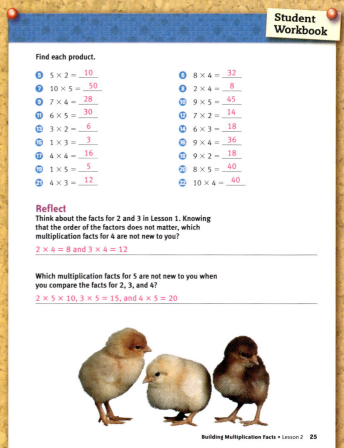

---

3 Reflect   10

Extended Response   REASONING

Review students' answers to the Reflect prompt at the bottom of student page 25.

You may want to list the facts for 2 and 3 on the board. Then write the facts for 4 and 5, but tell students to stop you before you write down a repeated fact.

- **Do you see a pattern in these lists?** Students should see that each list is shorter by one fact.
- **How do you think this will affect the next set of facts?** It will be shorter by one fact.

Real-World Application   APPLYING

The products of the 5-facts are commonly used in everyday life. Name a common use for skip counting by fives. Possible answers: counting nickels, tally mark groups, number of fingers, and so on

4 Assess

Informal Assessment

Use the Student Assessment Record, **Assessment,** page 100, to record informal observations.

UNDERSTANDING	REASONING
4 and 5 Flash Cards	**Multiplication Memory**
Did the student	Did the student
❑ make important observations?	❑ provide a clear explanation?
❑ extend or generalize learning?	❑ communicate reasons and strategies?
❑ provide insightful answers?	❑ choose appropriate strategies?
❑ pose insightful questions?	❑ argue logically?

Lesson 3

Objective

Students know multiplication facts for sixes and sevens.

Materials

Program Materials
- Building Facts Charts, p. B17
- Multiplication Table, p. B18

Additional Materials
Index cards

Access Vocabulary

Duck, Duck, Goose A circle game in which a selected player chases another player around the circle

Creating Context

In English we often talk about a set of objects without being specific about the number. For example, we may use a *lot*, a *bunch*, or *many*. Start a chart to collect some of these terms for collective nouns such as a *bunch* of grapes, a *school* of fish, a *batch* of cookies, and a *class* of students.

1 Warm Up 5

Concept Building COMPUTING

- **Name multiplication facts from Lessons 1 and 2 that have one factor that is a 6 or 7.** $6 \times 2 = 12$, $6 \times 3 = 18$, $6 \times 4 = 24$, $6 \times 5 = 30$, $7 \times 2 = 14$, $7 \times 3 = 21$, $7 \times 4 = 28$, $7 \times 5 = 35$

2 Engage 30

Skill Building

Students need all the materials listed for the Skill-Building Activity.

6 and 7 Flash Cards UNDERSTANDING

- For making the 6 and 7 flash cards, follow the same process as you did for making the flash cards in Lessons 1 and 2.
- Complete the 6-facts and 7-facts columns of the Multiplication Table.

Strategy Building

Factor, Factor, Product ENGAGING

Arrange students in a circle.

- This is a variation of the game Duck, Duck, Goose.
- The student who is "It" walks around the circle, taps someone on the head, and says a number between one and ten. The student continues to walk around the circle, taps another person, and says a number between two and seven. The student keeps walking around the circle until he or she taps the final person and says "product."
- If the selected student gives a correct product, then that student becomes "It."

Monitoring Student Progress

If . . . students want another variation of **Factor, Factor, Product,**

Then . . . modify the rules so that with the first tap the product is said. With the second tap, one of the factors is said. The third person tapped says the entire number sentence.

 Building Blocks For additional practice with 6 and 7 multiplication facts, students should complete **Building Blocks** Function Machine 2.

MathTools Use the Multiplication Table to demonstrate and explore 6 and 7 multiplication facts.

Using Student Pages

Have students complete **Workbook,** pages 26–27, on their own.

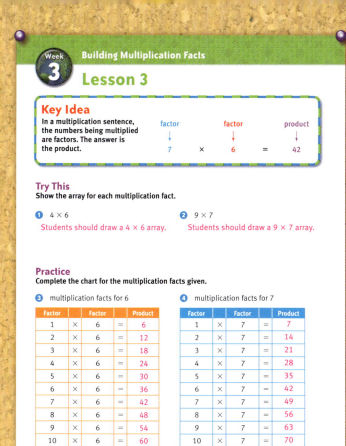

Building Multiplication Facts

Week 3 — Lesson 3

Key Idea

In a multiplication sentence, the numbers being multiplied are factors. The answer is the product.

factor × factor = product

7 × 6 = 42

Try This

Show the array for each multiplication fact.

1 4 × 6

Students should draw a 4 × 6 array.

2 9 × 7

Students should draw a 9 × 7 array.

Practice

Complete the chart for the multiplication facts given.

3 multiplication facts for 6

Factor		Factor		Product
1	×	6	=	6
2	×	6	=	12
3	×	6	=	18
4	×	6	=	24
5	×	6	=	30
6	×	6	=	36
7	×	6	=	42
8	×	6	=	48
9	×	6	=	54
10	×	6	=	60

4 multiplication facts for 7

Factor		Factor		Product
1	×	7	=	7
2	×	7	=	14
3	×	7	=	21
4	×	7	=	28
5	×	7	=	35
6	×	7	=	42
7	×	7	=	49
8	×	7	=	56
9	×	7	=	63
10	×	7	=	70

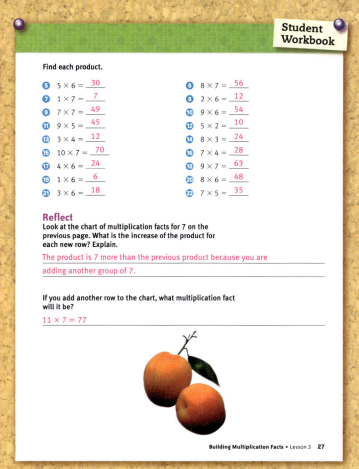

Find each product.

5 5 × 6 = 30
6 8 × 7 = 56
7 1 × 7 = 7
8 2 × 6 = 12
9 7 × 7 = 49
10 9 × 6 = 54
11 9 × 5 = 45
12 5 × 2 = 10
13 3 × 4 = 12
14 8 × 3 = 24
15 10 × 7 = 70
16 7 × 4 = 28
17 4 × 6 = 24
18 9 × 7 = 63
19 1 × 6 = 6
20 8 × 6 = 48
21 3 × 6 = 18
22 7 × 5 = 35

Reflect

Look at the chart of multiplication facts for 7 on the previous page. What is the increase of the product for each new row? Explain.

The product is 7 more than the previous product because you are adding another group of 7.

If you add another row to the chart, what multiplication fact will it be?

11 × 7 = 77

---

3 Reflect 10

Extended Response REASONING

Review students' answers to the Reflect prompt at the bottom of student page 27.

Guide students to see that the product column of the 7 multiplication facts contains the sums of repeated addition problems with 7 as the addend. Using repeated addition, have the class work together to name the products of 7 × 11, 7 × 12, 7 × 13, 7 × 14, and 7 × 15.

■ **Did this activity help you do anything you could not do before?**

■ **Did you find the correct solution the first time, or did you have to try more than once?**

Real-World Application APPLYING

Many items are packaged in groups of six. A group of six has a special name—a half dozen.

■ **Because a group of six is called a half dozen, how many are in a dozen?** 12

■ **Name a multiplication fact of 6 that has a product of 12.** 6 × 2 = 12

4 Assess

Informal Assessment

Use the Student Assessment Record, **Assessment,** page 100, to record informal observations.

UNDERSTANDING	ENGAGING
6 and 7 Flash Cards	**Factor, Factor, Product**
Did the student	Did the student
❑ make important observations?	❑ pay attention to the contributions of others?
❑ extend or generalize learning?	❑ contribute information and ideas?
❑ provide insightful answers?	❑ improve on a strategy?
❑ pose insightful questions?	❑ reflect on and check accuracy of work?

Lesson 4

Objective

Students know multiplication facts for eights and nines.

Materials

Program Materials
- Building Facts Charts, p. B17
- Multiplication Table, p. B18
- Product Bingo Cards, pp. B21–B22

Additional Materials
Index cards

Access Vocabulary

pattern A predictable repeating design made of numbers, colors, or pictures

Creating Context

English Learners who are new to school in the United States may not be familiar with Bingo. Explain that Bingo is a game in which players have a card with random numbers and a caller selects numbers and calls them out to the group. The first person to complete a straight line, diagonal line, or outer edge of the card wins.

1 Warm Up 5

Concept Building COMPUTING

- **Name any fact you know that has one factor that is 8 or 9.** Keep accepting answers until all are said. $2 \times 8 = 16$, $3 \times 8 = 24$, $4 \times 8 = 32$, $5 \times 8 = 40$, $6 \times 8 = 48$, $7 \times 8 = 56$, $2 \times 9 = 18$, $3 \times 9 = 27$, $4 \times 9 = 36$, $5 \times 9 = 45$, $6 \times 9 = 54$, $7 \times 9 = 63$

2 Engage 30

Skill Building

Students need all the materials listed for the Skill-Building Activity.

8 and 9 Flash Cards UNDERSTANDING

- For making the 8 and 9 flash cards, follow the same process as you did for making the flash cards in Lessons 1, 2, and 3.
- Complete the 8-facts and 9-facts columns of the Multiplication Table.

Strategy Building

Product Bingo COMPUTING

Distribute Bingo cards to students.

- The game is played the same as regular Bingo except instead of numbers being called, multiplication problems will be called. If you have a square on your card that has the product of the problem, mark that square on your card. When you get one row, one column, or a diagonal marked out, say "Bingo."

Use the Multiplication Table to randomly select problems to be called.

Monitoring Student Progress

If . . . students are doing well and you want to offer variety,

Then . . . change what makes a Bingo. (Possible changes: four corners, box of six squares)

Building Blocks For additional practice with 8 and 9 multiplication facts, students should complete **Building Blocks** Function Machine 2.

MathTools Use the Multiplication Table to demonstrate and explore 8 and 9 multiplication facts.

Using Student Pages

Have students complete **Workbook,** pages 28–29, on their own.

Lesson 4

Key Idea

In a multiplication sentence, the numbers being multiplied are factors. The answer is the product.

	factor		factor		product
	↓		↓		↓
	8	×	9	=	72

Try This

Show the array for each multiplication fact.

1. 3 × 9

   Students should draw a 3 × 9 array.

2. 8 × 8

   Students should draw an 8 × 8 array.

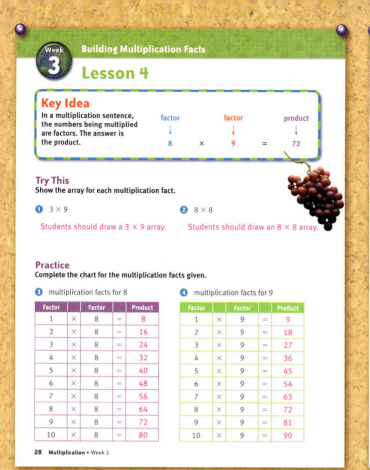

Practice

Complete the chart for the multiplication facts given.

3. multiplication facts for 8

Factor		Factor		Product
1	×	8	=	8
2	×	8	=	16
3	×	8	=	24
4	×	8	=	32
5	×	8	=	40
6	×	8	=	48
7	×	8	=	56
8	×	8	=	64
9	×	8	=	72
10	×	8	=	80

4. multiplication facts for 9

Factor		Factor		Product
1	×	9	=	9
2	×	9	=	18
3	×	9	=	27
4	×	9	=	36
5	×	9	=	45
6	×	9	=	54
7	×	9	=	63
8	×	9	=	72
9	×	9	=	81
10	×	9	=	90

Find each product.

5. 3 × 8 = 24
6. 9 × 9 = 81
7. 10 × 9 = 90
8. 5 × 8 = 40
9. 5 × 7 = 35
10. 9 × 6 = 54
11. 9 × 8 = 72
12. 3 × 9 = 27
13. 5 × 4 = 20
14. 8 × 8 = 64
15. 2 × 9 = 18
16. 6 × 5 = 30
17. 4 × 9 = 36
18. 7 × 8 = 56
19. 1 × 6 = 6
20. 5 × 3 = 15
21. 8 × 2 = 16
22. 7 × 9 = 63

Reflect

Other than facts that include a 1 or a 10 as a factor, what multiplication facts are included in only one list? Explain.

2 × 2 = 4, 3 × 3 = 9, 4 × 4 = 16, 5 × 5 = 25, 6 × 6 = 36, 7 × 7 = 49, 8 × 8 = 64, and 9 × 9 = 81. All other facts for any number are also included in the list of facts for the other factor.

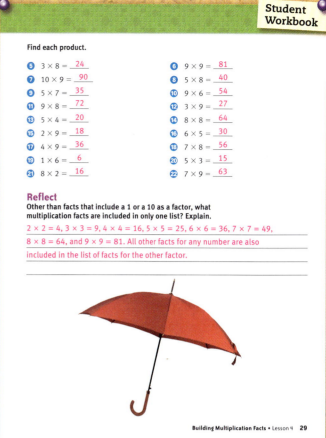

3 Reflect 10

Extended Response REASONING

Review students' answers to the Reflect prompt at the bottom of student page 29.

After one of the facts is named, students will see the pattern and should be quick to name the rest of the facts. Tell students that these products are called perfect square numbers.

- **Consider the arrays of these facts. Why do you think they are called perfect square numbers?** Answers will vary, but students should notice that their arrays are squares.

Real-World Application APPLYING

Now that students have spent three weeks learning and modeling basic multiplication facts, it is time for students to understand that the facts in their multiplication table contain information they will use throughout their lives, both professionally and personally. Discuss times in all adults' lives when they are called on to recall a basic multiplication fact.

4 Assess

Informal Assessment

Use the Student Assessment Record, **Assessment,** page 100, to record informal observations.

UNDERSTANDING

8 and 9 Flash Cards

Did the student
- ❏ make important observations?
- ❏ extend or generalize learning?
- ❏ provide insightful answers?
- ❏ pose insightful questions?

COMPUTING

Product Bingo

Did the student
- ❏ respond accurately?
- ❏ respond quickly?
- ❏ respond with confidence?
- ❏ self-correct?

Lesson 5 Review

Objective

Students review skills learned this week and complete the weekly assessment.

Materials

Review materials will be selected from those used in previous activities.

Creating Context

Often, when we talk about multiplying many different factors by a number such as 2, we put an *-s* on the end of it to make it plural. Explain to English Learners that *twos* refers to all the facts in the two times table.

1 Warm Up 5

Concept Building COMPUTING

Organize students into pairs. Provide time for students to review flash cards for facts from all four days.

2 Engage 20

Skill Building

Free-Choice Activity ENGAGING

For the last day of the week, allow students to choose a game activity from the previous lessons. These games can be adapted to include all the basic multiplication facts. The games they may choose include the following:

- **Multiplication Memory**
- **Factor, Factor, Product**
- **Product Bingo**

Make a note of the games children select. Do they prefer easy or challenging activities? If you believe students would benefit from extra practice on specific facts, choose an activity for them.

Monitoring Student Progress

If . . . students like to play one game in particular,

Then . . . encourage them to record the rules and to make copies of any game boards needed so that they can play at home with a caregiver and friends.

Using Student Pages

Have students complete **Workbook,** pages 30–31, on their own.

3 Reflect 10

Extended Response REASONING

Review students' answers to the Reflect prompts at the bottom of student pages 30–31.

Discuss the answers with the group to reinforce Week 3 concepts.

Week 3 — Building Multiplication Facts

Lesson 5 Review

This week you learned multiplication facts for 2s, 3s, 4s, 5s, 6s, 7s, 8s, and 9s. You have eighty flash cards to help you practice learning these multiplication facts.

Lesson 1 Find each product.

1. $5 \times 2 = \underline{10}$
2. $4 \times 3 = \underline{12}$
3. $8 \times 3 = \underline{24}$
4. $10 \times 2 = \underline{20}$

Lesson 2 Find each product.

5. $7 \times 4 = \underline{28}$
6. $6 \times 4 = \underline{24}$
7. $9 \times 5 = \underline{45}$
8. $2 \times 5 = \underline{10}$
9. $3 \times 5 = \underline{15}$
10. $3 \times 4 = \underline{12}$
11. $8 \times 4 = \underline{32}$
12. $7 \times 5 = \underline{35}$
13. $8 \times 5 = \underline{40}$
14. $10 \times 5 = \underline{50}$

Reflect

For which factors are the multiplication facts easier to recall? Explain your answer.

Answers will vary but will probably mention twos or fives because it is more common to skip count by twos and fives.

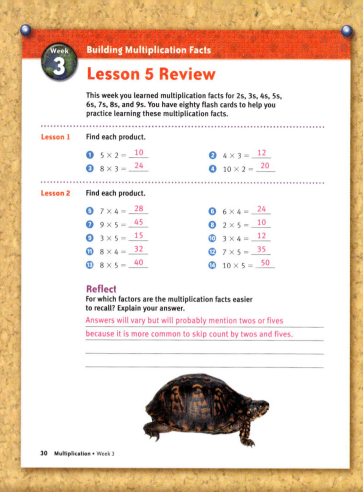

Lesson 3 Find each product.

15. $5 \times 6 = \underline{30}$
16. $1 \times 6 = \underline{6}$
17. $4 \times 7 = \underline{28}$
18. $3 \times 6 = \underline{18}$
19. $3 \times 7 = \underline{21}$
20. $10 \times 7 = \underline{70}$
21. $8 \times 6 = \underline{48}$
22. $7 \times 6 = \underline{42}$
23. $7 \times 7 = \underline{49}$
24. $9 \times 6 = \underline{54}$

Lesson 4 Find each product.

25. $2 \times 9 = \underline{18}$
26. $4 \times 8 = \underline{32}$
27. $6 \times 8 = \underline{48}$
28. $1 \times 8 = \underline{8}$
29. $8 \times 8 = \underline{64}$
30. $9 \times 8 = \underline{72}$
31. $9 \times 9 = \underline{81}$
32. $5 \times 9 = \underline{45}$
33. $6 \times 9 = \underline{54}$
34. $10 \times 9 = \underline{90}$

Reflect

Suppose that you are given the multiplication problem 7×8 and you do not know the product. You can remember that $5 \times 8 = 40$. Explain how you can use that fact and your understanding of multiplication to find the product of 7×8.

Because $5 \times 8 = 40$, the product of 6×8 is 8 more than 40, or 48. Then you can use 48 to find the product of 7×8 by adding 8 to 48. So the product of 7×8 is 56.

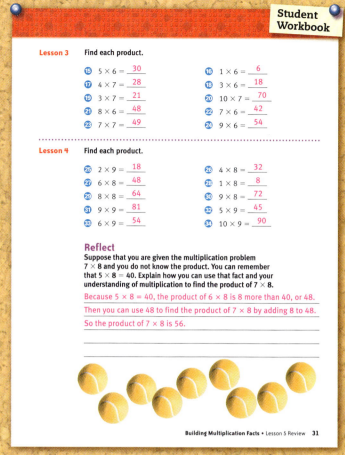

 Assess 10

A Gather Evidence

Formal Assessment

Have students complete the weekly test on **Assessment,** pages 52–53. Record progress on the Student Assessment Record, **Assessment,** page 100.

B Summarize Findings

Determine whether students have Minimal, Basic, or Secure understanding of the concepts presented in Week 3.

C Differentiate Instruction

Based on your observations, use these teaching strategies next week to follow up.

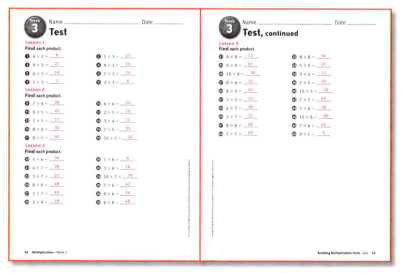

Assessment, pp. 52–53

Minimal Understanding

- Repeat the Warm-Up and Engage activities to develop Week 3 concepts.
- Use **Building Blocks** Function Machine 2 and **eMathTools** Multiplication Table to develop and reinforce this week's concepts.

Basic Understanding

- Repeat Engage activities in subsequent weeks.
- Use **Building Blocks** Function Machine 2 and **eMathTools** Multiplication Table to reinforce this week's concepts.

Secure Understanding

- Use **Building Blocks** Function Machine 2 and **eMathTools** to reinforce Week 3 concepts.
- Use variations of the weekly Warm-Up and Engage activities, using higher numbers and multiple steps.

Building Multiplication Facts • Lesson 5 Review **30–31**

Beyond the Basic Facts

Week at a Glance

This week, students continue *Number Worlds,* Level F, Multiplication. Students explore thinking strategies to use with multiplication and are introduced to division. Students should also learn shortcuts and methods for doing multiplication by using mental math.

Math Background

Finding and using patterns greatly simplifies the task of learning multiplication combinations. Moreover, it is one of the very essences of mathematics. Thus, approaching multiplication learning as pattern finding greatly simplifies the task and constitutes a core mathematical approach.

—Kilpatrick, Jeremy, W. Gary Martin, and Deborah Schifter. 2003. *A Research Companion to Principles and Standards for School Mathematics.* National Council of Teachers of Mathematics. Page 77.

How Students Learn

Array drawings are powerful models of multiplication for single-digit and double-digit factors. The arrays show the quantities, and these models provide visuals for understanding the effects of multiplying.

Teaching for Understanding

Observe closely while evaluating the assigned tasks this week to see whether students can demonstrate the following understandings.

Benchmark after Lesson 2: Students can multiply by 10 using mental math, and use 10 with the Distributive Property to compute multiplication problems beyond the basic facts.

Benchmark after Lesson 3: Students can use arrays to multiply combinations of factors for larger numbers and learn that division is the inverse of multiplication.

Benchmark after Lesson 4: Students can model division using counters and arrays.

Skills Focus

- Multiply by 10, using mental math
- Use the Distributive Property to multiply numbers that are greater than 10
- Understand that a product can have multiple pairs of factors
- Understand that multiplication and division are inverse operations

Math at Home

Give one copy of the Letter to Home, page A16, to each student. Encourage students to share and complete the activity with their caregivers.

Letter to Home, Teacher Edition, p. A16

PACING	LESSON	LEARNING GOALS	NCTM	MATERIALS	TECHNOLOGY
DAY 1	pages 32–33	Students multiply by 10 using mental math.	• Number and Operations • Problem Solving • Communication	Building Facts Chart, p. B17	**B**uilding**B**locks Function Machine 2 **e** MathTools 100 Table
DAY 2	pages 34–35	Students use the Distributive Property to find products when factors are greater than 10.	• Number and Operations • Problem Solving • Communication • Representation	Graph paper	**B**uilding**B**locks Arrays in Area **e** MathTools Array Tool
DAY 3	pages 36–37	Students use their knowledge of the basic facts and the Distributive Property to multiply and divide numbers.	• Number and Operations • Problem Solving • Communication • Connections • Representation	• Graph paper • Scissors • Counters	**B**uilding**B**locks Arrays in Area **e** MathTools Array Tool
DAY 4	pages 38–39	Students are introduced to division and play a game to practice their basic multiplication facts.	• Number and Operations • Problem Solving • Communication • Connections	• Graph paper • Cover Up game board, p. B23 • Number 1–6 Cubes, 2 • Colored pencils • Counters • Paper bags	**B**uilding**B**locks Arrays in Area **e** MathTools Multiplication Table
DAY 5	**Review and Assess** pages 40–41	Students review skills learned this week and complete the weekly assessment.	• Number and Operations • Problem Solving • Communication • Connections • Representation	Materials will be selected from Lessons 1–4.	**B**uilding**B**locks Review previous activities

Math Vocabulary

basic multiplication fact A basic number sentence that has two one-digit factors, a × symbol, and an answer

rectangular array An arrangement of tiles or objects in rows and columns

Distributive Property A property stating that one factor of a product can be written as a sum so that each addend is multiplied by the other factor and the product remains the same

English Learners

SPANISH COGNATES

English	Spanish
multiplication	multiplicación
factor	factor

ALTERNATE VOCABULARY

shortcut A way to arrive at the correct answer with fewer steps

Lesson 1

Objective

Students multiply by 10 and powers of 10, using mental math.

Materials

Program Materials
Building Facts Chart, p. B17

Access Vocabulary

mental math Solving math problems by thinking through the steps, without writing them on paper

Creating Context

In English, there are many words that have more than one meaning. This can be quite confusing, especially to English Learners. In this lesson, we learn the term *powers of ten*. Students may know the word *power* to mean "strength" or "special ability." Explain the meaning of the word *power* in mathematics, and encourage students to ask questions when they are uncertain of a word's meaning.

1 Warm Up   5

Concept Building   COMPUTING

On the board draw a 2 × 10 array.

- **What number sentence is modeled?** 2 × 10 = 20

Add another column of 10 to the array.

- **What number sentence is modeled?** 3 × 10 = 30
- **What pattern do you notice? What happens when you multiply by 10?**

Guide students to look closely at each product and to notice that it is the number of groups with a zero on the end.

2 Engage   30

Strategy Building

Times 10   UNDERSTANDING

- **Skip count by tens for a count of two. At what number are you?** 20
- **Write the number sentence for two groups of ten.** 2 × 10 = 20
- **Skip count by tens for a count of four. At what number are you?** 40
- **What did you model? Write the number sentence.** 4 groups of 10; 4 × 10 = 40
- **Skip count by tens for a count of six. At what number are you?** 60
- **What did you model? Write the number sentence.** 6 groups of 10: 6 × 10 = 60
- **Look at the factor that is not the 10 in each sentence and the product in each sentence. Do you see a pattern?** The product is that factor with a zero at the end.
- **Restate your conclusion for multiplying by 10 and 100 so that it tells how to multiply any whole number by any power of 10.** Count every zero in the factor that is a power of 10. For the product, write the other factor followed by zeros that match the number of zeros you counted.
- **Some people refer to this a rule as the "zero trick." Knowing this rule means that you can multiply by ten in your head and can use 10 as a way to find products when one factor is greater than 10.**

Building Blocks For additional practice multiplying by 10 and powers of 10, students should complete **Building Blocks** Function Machine 2.

 MathTools Use the 100 Table to demonstrate and explore multiplying by 10 and powers of 10.

Using Student Pages

Have students complete **Workbook,** pages 32–33, on their own.

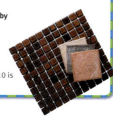

Lesson 1

Key Idea

Multiplying by 10 and multiples of 10 can be done by using mental math.

$$2 \times 10 = 20$$
$$2 \times 100 = 200$$
$$2 \times 1,000 = 2,000$$

The number of 0s in the factor that is a multiple of 10 is the same as the number of 0s in the product.

Try This

Write how many zeros will be in each product, and solve the equation.

❶ 5×100
  2   500

❷ 4×10
  1   40

❸ $8 \times 1,000$
  3   8,000

Practice

Complete the chart.

❹
Factor		Factor		Product
3	×	1	=	3
3	×	10	=	30
3	×	100	=	300
3	×	1,000	=	3,000
3	×	10,000	=	30,000
3	×	100,000	=	300,000
3	×	1,000,000	=	3,000,000
3	×	10,000,000	=	30,000,000
3	×	100,000,000	=	300,000,000

Find each product.

❺ $7 \times 10 = $ 70
❻ $3 \times 1,000 = $ 3,000
❼ $9 \times 1,000 = $ 9,000
❽ $6 \times 100 = $ 600
❾ $2 \times 100 = $ 200
❿ $4 \times 10 = $ 40
⓫ $5 \times 100 = $ 500
⓬ $6 \times 10,000 = $ 60,000
⓭ $8 \times 1,000 = $ 8,000
⓮ $4 \times 100,000 = $ 400,000

Reflect

Suppose that you are helping a friend who has been absent from school for a few days and who needs help learning the shortcut to multiplying by 10 and multiples of 10. Use the problems 7×10, 8×100, and $4 \times 1,000$ to explain how to use mental math to multiply.

Look at the factor that is 10 or a multiple of 10. Count how many zeros are in that factor. The product is the number that is the other factor followed by the number of zeros you counted.

7×10; There is one 0. The other factor is 7. The product is 7 followed by one 0, or 70.

8×100; There are two 0s. The other factor is 8. The product is 8 followed by two 0s, or 800.

$4 \times 1,000$; There are three 0s. The other factor is 4. The product is 4 followed by three 0s, or 4,000.

---

3 Reflect   10

Extended Response [REASONING]

Review students' answers to the Reflect prompt at the bottom of student page 33.

- **Did you find the correct solution the first time, or did you have to try more than once?**
- **How would you explain your method to someone who does not understand this concept?** Pair students. Have them "teach" the lesson written as an answer to the question. Have them combine the best parts of both their responses and write a single response to the question.

Real-World Application [APPLYING]

Multiplying by 10 and 100 are common skills that people use to estimate costs while shopping.

- **If you want to buy four beach towels that are $9.59 each, what multiplication problem could you create to estimate the total cost of the four towels?** $4 \times \$10 = \$40$

4 Assess

Informal Assessment

Use the Student Assessment Record, **Assessment,** page 100, to record informal observations.

UNDERSTANDING

Times 10

Did the student
❏ make important observations?
❏ extend or generalize learning?
❏ provide insightful answers?
❏ pose insightful questions?

Lesson 2

Objective

Students use the Distributive Property to find products when factors are greater than 10.

Materials

Additional Materials

Graph paper

Access Vocabulary

rewrite To write numbers again in a different way

Creating Context

Distributive Property is an academic term that English Learners may find difficult because the individual words are unfamiliar. Suggest that students examine the root of the word *distributive*. Show students an example of the Distributive Property and how it distributes the larger numbers into smaller, more manageable units.

1 Warm Up 5

Concept Building COMPUTING

- Rewrite each of these numbers as a sum of 10 and another number: 11, 14, and 18.
  10 + 1, 10 + 4, 10 + 8
- Rewrite each of these numbers as a sum of three numbers: 22 and 27. Use as many of the number 10 as possible. 10 + 10 + 2, 10 + 10 + 7

2 Engage 30

Skill Building

Distributive Property COMPUTING

- Let's look at the problem 3 × 12. This is not included in your list of basic facts. Let's try to rewrite it so that you can use your list of basic facts to find the product.
- We are going to use arrays to discover the answer. Because you know the 3-facts, you will create arrays that are three rows deep.
- You need to think of 12 as a sum of two numbers. A good number to use as an addend is 10. Why is 10 a good number to choose? Multiplying by 10 is something I can do in my head.
- Create an array that is 3 by 10 squares. Write the multiplication problem under the array.
- What number do you add to 10 to get 12? 2
- What other array do you add to create a 3 × 12 array? 3 × 2
- Set the arrays next to each other to see that they are the same height. Write the multiplication problems with a plus sign between. (3 × 10) + (3 × 2)
- Write each product, and then add them. 30 + 6 = 36
- What is the product of 3 × 12? 36

Monitoring Student Progress

If . . . students need to see another example,

Then . . . use 3 × 18, and ask the same questions as those posed above. Encourage students to show each step.

 Building Blocks For additional practice using the Distributive Property, students should complete **Building Blocks** Arrays in Area.

MathTools Use the Array Tool to demonstrate and explore the Distributive Property.

Using Student Pages

Have students complete **Workbook,** pages 34–35, on their own.

Lesson 2

Key Idea

You can use the Distributive Property to break a difficult problem into two easier problems and then add the results.

$$4 \times 18$$
$$4 \times (10 \times 8)$$
$$(4 \times 10) + (4 \times 8)$$
$$40 + 32 = 72$$

Try This

Fill in the blanks to name the multiplication problem shown in the arrays. Find the product.

1.

$$\underline{10} \times \underline{3} + \underline{4} \times \underline{3}$$
$$\underline{30} \qquad + \underline{12}$$
$$\underline{42}$$

Break the second factor of each problem into a sum of 10 + another number.

2. 2×18
$(2 \times 10) + (2 \times \underline{8})$

3. 6×14
$(6 \times \underline{10}) + (6 \times \underline{4})$

4. 4×19
$(4 \times \underline{10}) + (4 \times \underline{9})$

5. 3×13
$(3 \times \underline{10}) + (3 \times \underline{3})$

Practice
Fill in the blanks to find each product.

6. 3×15
$(\underline{3} \times \underline{10}) + (\underline{3} \times \underline{5})$
$\underline{30} + \underline{15}$
$\underline{45}$

7. 8×17
$(\underline{8} \times \underline{10}) + (\underline{8} \times \underline{7})$
$\underline{80} + \underline{56}$
$\underline{136}$

8. 9×18
$(\underline{9} \times \underline{10}) + (\underline{9} \times \underline{8})$
$\underline{90} + \underline{72}$
$\underline{162}$

9. 4×13
$(\underline{4} \times \underline{10}) + (\underline{4} \times \underline{3})$
$\underline{40} + \underline{12}$
$\underline{52}$

Find each product.

10. $5 \times 18 = \underline{90}$
11. $8 \times 18 = \underline{144}$
12. $9 \times 12 = \underline{108}$
13. $3 \times 16 = \underline{48}$
14. $6 \times 18 = \underline{108}$
15. $9 \times 16 = \underline{144}$

Reflect
Division is the inverse, or opposite, of multiplication. In multiplication we find the product of two factors, in division we find the missing factor if the other factor and the product are known. The example below shows that division undoes multiplication and multiplication undoes division.

$$5 \times 4 = 20 \qquad 20 \div 4 = 5$$

Choose three multiplication sentences from Problems 10–15 and rewrite them as division sentences.

Possible answers:
$90 \div 18 = 5 \qquad 144 \div 18 = 8 \qquad 108 \div 12 = 9$
$48 \div 16 = 3 \qquad 108 \div 18 = 6 \qquad 144 \div 16 = 9$

3 Reflect   10

Extended Response   REASONING

Review students' answers to the Reflect prompt at the bottom of student page 35.

Invite student volunteers to the board to display their problems. Allow each student time to explain the problem and the work to the class.

- **Is there more than one way to reverse a multiplication problem?** yes
- **Did anyone solve these problems differently?**
$90 \div 5 = 18, 144 \div 8 = 18, 108 \div 9 = 12,$
$48 \div 3 = 16, 108 \div 6 = 18, 144 \div 9 = 16$

Real-World Application   APPLYING

- **Many people use the Distributive Property to find the products of multiplication problems containing double-digit numbers without using paper and pencil or a calculator.**
- **For example: Solve 25×6. Think of 25 as 20 + 5. Multiply 20×6. Think $2 \times 6 = 12 \times 10 = 120$. Now multiply 5×6 to get 30. Add 120 and 30 to get the product 150.**

4 Assess

Informal Assessment

Use the Student Assessment Record, **Assessment,** page 100, to record informal observations.

COMPUTING

Distributive Property
Did the student
- ❏ respond accurately?
- ❏ respond quickly?
- ❏ respond with confidence?
- ❏ self-correct?

Lesson 3

Objective

Students use their knowledge of the basic facts and the Distributive Property to multiply and divide numbers.

Materials

Additional Materials
- Graph paper
- Scissors
- Counters

Access Vocabulary

operation sign The symbol that tells us which operation to use in a number sentence

Creating Context

Using visuals is an excellent strategy to use with students to aid comprehension and to check understanding. On the student pages are arrays that demonstrate multiplication sentences. Students can create arrays with graph paper and then write the multiplication sentences that go with them.

1 Warm Up 5

Concept Building COMPUTING

Draw a 4 × 3 array of circles on the board.
- **What multiplication sentence describes the array?** 4 × 3 or 3 × 4

Give each student 15 counters and ask him or her to make an array with three equal rows. Write the multiplication sentence on the board: 3 × ? = 15
- **Division is the inverse, or opposite, of multiplication, so the sentence can be rewritten as 15 ÷ 3 = ?** Count the number of counters in each row to find the missing number.

2 Engage 30

Skill Building

How Many Ways? ENGAGING

- How many ways can you show 100 using rectangular arrays? many
- On a sheet of graph paper, mark a 10 by 10 square.
- Cut two 5 by 5 arrays. Place these arrays on top of your 10 by 10 square.
- Cut at least two more arrays that can fit inside your 10 by 10 square. Put them in place on top of the 10 by 10 square.
- Are there still squares of your 100 that need to be covered? If yes, create more arrays.
- When your 100 squares are covered by arrays, place them in a line next to each other. Write the multiplication problem modeled by each array.
- What operation sign needs to be placed between the multiplication problems? +

Invite students to the board to share the ways they showed 100 by using rectangular arrays. Possible answer: (5 × 5) + (5 × 5) + (2 × 10) + (3 × 10) = 25 + 25 + 20 + 30 = 100

Repeat the activity using 24 as the product so students see that many numbers have more than one set of factors. Have students write an equivalent division sentence for each multiplication sentence they derive for the number 24.

Building Blocks For additional practice with different combinations of numbers that have the same product, students should complete **Building Blocks** Arrays in Area.

MathTools Use the Array Tool to demonstrate and explore different combinations of numbers that have the same product.

Using Student Pages

Have students complete **Workbook,** pages 36–37, on their own.

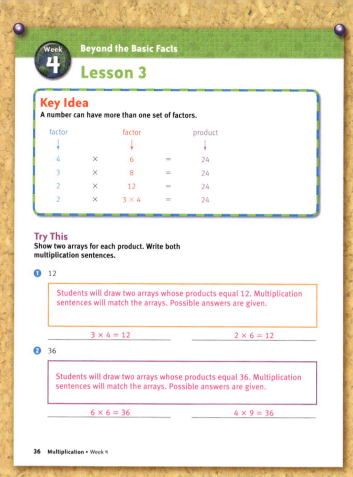

Week 4 — Beyond the Basic Facts

Lesson 3

Key Idea
A number can have more than one set of factors.

factor		factor		product
4	×	6	=	24
3	×	8	=	24
2	×	12	=	24
2	×	3 × 4	=	24

Try This
Show two arrays for each product. Write both multiplication sentences.

1. 12

Students will draw two arrays whose products equal 12. Multiplication sentences will match the arrays. Possible answers are given.

3 × 4 = 12   2 × 6 = 12

2. 36

Students will draw two arrays whose products equal 36. Multiplication sentences will match the arrays. Possible answers are given.

6 × 6 = 36   4 × 9 = 36

36   Multiplication • Week 4

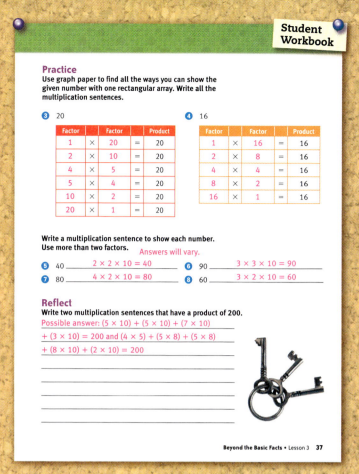

Practice
Use graph paper to find all the ways you can show the given number with one rectangular array. Write all the multiplication sentences.

3. 20

Factor		Factor		Product
1	×	20	=	20
2	×	10	=	20
4	×	5	=	20
5	×	4	=	20
10	×	2	=	20
20	×	1	=	20

4. 16

Factor		Factor		Product
1	×	16	=	16
2	×	8	=	16
4	×	4	=	16
8	×	2	=	16
16	×	1	=	16

Write a multiplication sentence to show each number. Use more than two factors.   Answers will vary.

5. 40   2 × 2 × 10 = 40   6. 90   3 × 3 × 10 = 90

7. 80   4 × 2 × 10 = 80   8. 60   3 × 2 × 10 = 60

Reflect
Write two multiplication sentences that have a product of 200.

Possible answer: (5 × 10) + (5 × 10) + (7 × 10) + (3 × 10) = 200 and (4 × 5) + (5 × 8) + (5 × 8) + (8 × 10) + (2 × 10) = 200

Beyond the Basic Facts • Lesson 3   37

---

3 Reflect   10

Extended Response   REASONING

Review students' answers to the Reflect prompt at the bottom of student page 37.

- **Can you think of another way to do this?**
Discuss the strategies used, and guide students to see that they could have used their results from **How Many Ways?** to simplify the task of naming all ways. From each list, a single factor needed to be doubled.

Real-World Application   APPLYING

- **Say any number less than 200.** Listen to the responses from students, and write a few on the board.

- **How many numbers did you hear others say that were between 1 and 10?** Students will probably say very few.

Point out that numbers used in the context of real-world situations are often greater than 10. Explain that this means that people need strategies for dealing with larger numbers.

4 Assess

Informal Assessment

Use the Student Assessment Record, **Assessment,** page 100, to record informal observations.

ENGAGING

How Many Ways?
Did the student
❏ pay attention to the contributions of others?
❏ contribute information and ideas?
❏ improve on a strategy?
❏ reflect on and check accuracy of work?

Lesson 4

Objective

Students are introduced to division and play a game to practice their basic multiplication facts.

Materials

Program Materials
- Cover Up game board, p. B23
- Number 1–6 Cubes, 2

Additional Materials
- Graph paper
- Colored pencils
- Counters
- Paper bags

Access Vocabulary

cover up To hide information from others

Creating Context

Review with English Learners game routines such as deciding who goes first, taking turns, and what to do in case of a tie.

1 Warm Up — 5

Concept Building COMPUTING

- **If you are making a gift bag and you want to make sure each person has the same number of items, what would you do?** Answers will vary. Possible answer: Put one item in each bag before adding the next item.

Set out 5 paper bags. Ask a student volunteer to put 10 counters in the bags so that each bag has the same number of counters in it.
- **This is division. The equation that describes what just happened is 10 divided by 5 equals 2.**

Write the equation on the board. Have other student volunteers take turns modeling different division problems. Write each equation on the board.

2 Engage — 30

Skill Building

Multiplication and Division REASONING

Give students graph paper and counters.
- **How many groups of 6 can you make with 18?** 3 **How would you find the answer?** Answers will vary.
- **Is 6 a factor of 18?** yes **What is the other factor you use to equal 18?** 3 **Draw an array or use counters if you need help.**
- **Division is the inverse or opposite of multiplication. Each multiplication fact has a division fact that uses the same numbers.**

On the board, write $10 \div 2 = ?$
- **How could you solve this problem? Show your solution with counters or by making an array.** Answers will vary.
- **What multiplication fact is related to $10 \div 2 = 5$?** $2 \times 5 = 10$

Repeat with different problems using single-digit factors.

Monitoring Student Progress

If . . . students want a more challenging version of the game,	Then . . . alter the rules so that each group has only one game board and players color their arrays with different-colored pencils. This variation of the game introduces a different kind of strategy.

Building Blocks For additional practice with basic multiplication facts, students should complete **Building Blocks** Arrays in Area.

MathTools Use the Multiplication Table to demonstrate and explore basic multiplication facts.

Using Student Pages

Have students complete **Workbook,** pages 38–39, on their own.

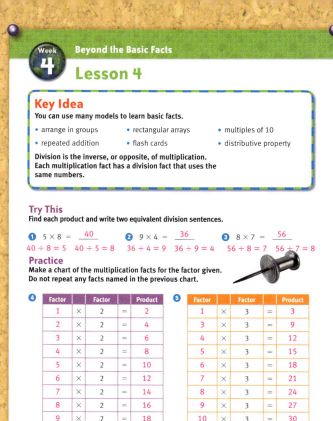

Week 4 — Beyond the Basic Facts

Lesson 4

Key Idea

You can use many models to learn basic facts.
- arrange in groups
- repeated addition
- rectangular arrays
- flash cards
- multiples of 10
- distributive property

Division is the inverse, or opposite, of multiplication. Each multiplication fact has a division fact that uses the same numbers.

Try This

Find each product and write two equivalent division sentences.

❶ $5 \times 8 = \underline{40}$
$40 \div 8 = 5 \quad 40 \div 5 = 8$

❷ $9 \times 4 = \underline{36}$
$36 \div 4 = 9 \quad 36 \div 9 = 4$

❸ $8 \times 7 = \underline{56}$
$56 \div 8 = 7 \quad 56 \div 7 = 8$

Practice

Make a chart of the multiplication facts for the factor given. Do not repeat any facts named in the previous chart.

❹
Factor		Factor		Product
1	×	2	=	2
2	×	2	=	4
3	×	2	=	6
4	×	2	=	8
5	×	2	=	10
6	×	2	=	12
7	×	2	=	14
8	×	2	=	16
9	×	2	=	18
10	×	2	=	20

❺
Factor		Factor		Product
1	×	3	=	3
3	×	3	=	9
4	×	3	=	12
5	×	3	=	15
6	×	3	=	18
7	×	3	=	21
8	×	3	=	24
9	×	3	=	27
10	×	3	=	30

❻
Factor		Factor		Product
1	×	4	=	4
4	×	4	=	16
5	×	4	=	20
6	×	4	=	24
7	×	4	=	28
8	×	4	=	32
9	×	4	=	36
10	×	4	=	40

❼
Factor		Factor		Product
1	×	5	=	5
5	×	5	=	25
6	×	5	=	30
7	×	5	=	35
8	×	5	=	40
9	×	5	=	45
10	×	5	=	50

❽
Factor		Factor		Product
1	×	6	=	6
6	×	6	=	36
7	×	6	=	42
8	×	6	=	48
9	×	6	=	54
10	×	6	=	60

❾
Factor		Factor		Product
1	×	7	=	7
7	×	7	=	49
8	×	7	=	56
9	×	7	=	63
10	×	7	=	70

❿
Factor		Factor		Product
1	×	8	=	8
8	×	8	=	64
9	×	8	=	72
10	×	8	=	80

⓫
Factor		Factor		Product
1	×	9	=	9
9	×	9	=	81
10	×	9	=	90

Reflect

Think of a game idea that will provide practice for the basic multiplication facts. Describe your idea. You should use words, and you may use drawings.

Answers will vary.

3 Reflect — 10

Extended Response — REASONING

Review students' answers to the Reflect prompt at the bottom of student page 39.

Have students share their game ideas with the class. Discuss how much work would be involved in creating some of the games. As a class, select the one favorite game idea.

- **What would some of the rules of this game be?**
- **How would this game help someone with his or her math skills?**

Real-World Application — APPLYING

- **Do you play board games at home with family and friends?**
- **Do you use strategies when playing any of these games?**
- **Name any game in which you use strategies similar to those used in Cover Up.**

4 Assess

Informal Assessment

Use the Student Assessment Record, **Assessment**, page 100, to record informal observations.

REASONING

Multiplication and Division

Did the student
- ❏ provide a clear explanation?
- ❏ communicate reason and strategies?
- ❏ choose appropriate strategies?
- ❏ argue logically?

Lesson 5 Review

Objective

Students review skills learned this week and complete the weekly assessment.

Materials

Review materials will be selected from those used in previous activities.

Creating Context

An excellent strategy for English Learners is to look for cognates. These are words that share their roots across languages. For speakers of Spanish, words that end in *-or* in English are often spelled the same way and mean exactly the same thing in Spanish. The use of cognates makes it easier to acquire new vocabulary.

Unit Assessment

Students should complete the Multiplication Test found on **Assessment,** pages 78–79. Using the key on **Assessment,** page 99, identify incorrect responses. Reteach and review the Warm-Up and Engage activities to reinforce concept understanding.

1 Warm Up 5

Concept Building COMPUTING

Pair students. Review flash cards made in Week 3, and ask students to modify problems by using the Distributive Property of multiplication.

2 Engage 20

Skill Building

Free-Choice Activity ENGAGING

For the last day of the week, allow students to choose an activity from the previous lessons. Some activities they may choose include the following:

- **Times 10**
- **Distributive Property**
- **How Many Ways?**
- **Cover Up**

If any games were made as an extension of Lesson 4, you may include them in the list above.

Make a note of the games children select. Do they prefer easy or challenging activities? If you believe your students would benefit from extra practice on specific facts, choose an activity for them.

Monitoring Student Progress

If . . . students need more practice learning the basic multiplication facts,

Then . . . they should choose **Cover Up** for their **Free-Choice** activity.

Using Student Pages

Have students complete **Workbook,** pages 40–41, on their own.

3 Reflect 10

Extended Response REASONING

Review students' answers to the Reflect prompts at the bottom of student pages 40–41.

Discuss the answers with the group to reinforce Week 4 concepts.

Lesson 5 Review

This week you learned more about multiplication facts. You learned a shortcut when one factor is a multiple of 10. You also learned how to use the Distributive Property to break a difficult problem into an easier problem.

Lesson 1 Find each product.

1. $5 \times 100 = \underline{500}$
2. $7 \times 100 = \underline{700}$
3. $8 \times 10 = \underline{80}$
4. $9 \times 10,000 = \underline{90,000}$
5. $4 \times 1,000 = \underline{4,000}$
6. $6 \times 10 = \underline{60}$

Lesson 2 Use the Distributive Property to find each product.

7. $7 \times 12 = \underline{84}$
8. $6 \times 14 = \underline{84}$
9. $9 \times 13 = \underline{117}$
10. $4 \times 12 = \underline{48}$
11. $5 \times 12 = \underline{60}$
12. $3 \times 20 = \underline{60}$
13. $6 \times 16 = \underline{96}$
14. $7 \times 15 = \underline{105}$

Reflect

In the above problems in which the Distributive Property was used to find the product, explain why you typically rewrite one factor using 10.

Renaming a number using 10 is easy to do because the other number is the digit that is in the ones place. Then, the product of a multiplication problem that has one factor that is a multiple of 10 is usually done by using mental math.

Lesson 3 Use graph paper to find all the ways you can show the given number with a rectangular array. Write all the multiplication sentences.

15. 18

Factor		Factor		Product
1	×	18	=	18
2	×	9	=	18
3	×	6	=	18
6	×	3	=	18
9	×	2	=	18
18	×	1	=	18

16. 40

Factor		Factor		Product
1	×	40	=	40
2	×	20	=	40
4	×	10	=	40
5	×	8	=	40
8	×	5	=	40
10	×	4	=	40
20	×	2	=	40
40	×	1	=	40

Lesson 4 Find each product or difference.

17. $9 \times 8 = \underline{72}$
18. $48 \div 6 = \underline{8}$
19. $4 \times 7 = \underline{28}$
20. $12 \div 2 = \underline{6}$
21. $5 \times 4 = \underline{20}$
22. $24 \div 3 = \underline{8}$
23. $10 \times 9 = \underline{90}$
24. $63 \div 7 = \underline{9}$
25. $3 \times 13 = \underline{39}$
26. $64 \div 4 = \underline{16}$

Reflect

Think of all the charts and flash cards you made during the past weeks of practicing multiplication and division facts. List ten facts you have trouble remembering. Then use any method of modeling to find the products. Show your work.

Answers will vary.

Assess 10

A Gather Evidence

Formal Assessment

Have students complete the weekly test, **Assessment,** pages 54–55. Record progress on the Student Assessment Record, **Assessment,** page 100.

B Summarize Findings

Determine whether students have Minimal, Basic, or Secure understanding of the concepts presented in Week 4.

C Differentiate Instruction

Based on your observations, use these teaching strategies next week to follow up.

Minimal Understanding
- Repeat the Warm-Up and Engage activities to develop Week 4 concepts.
- Use **Building Blocks** Function Machine 2 and **eMathTools** 100 Table to develop and reinforce this week's concepts.

Basic Understanding
- Repeat Engage activities in subsequent weeks.
- Use **Building Blocks** Arrays in Area and **eMathTools** Array tool to reinforce this week's concepts.

Secure Understanding
- Use **Building Blocks** Arrays in Area and **eMathTools** Array Tool to reinforce this week's concept.
- Use variations of the weekly Warm-Up and Engage activities, using higher numbers and multiple steps.

Assessment, pp. 54–55

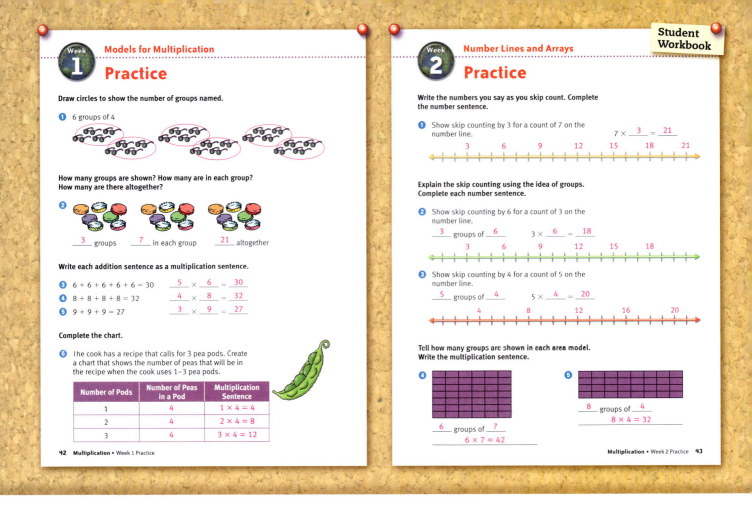

Week 1 Models for Multiplication

Practice

Draw circles to show the number of groups named.

1 6 groups of 4

How many groups are shown? How many are in each group? How many are there altogether?

2

____3____ groups   ____7____ in each group   ____21____ altogether

Write each addition sentence as a multiplication sentence.

3 6 + 6 + 6 + 6 + 6 = 30   ____5____ × ____6____ = ____30____

4 8 + 8 + 8 + 8 = 32   ____4____ × ____8____ = ____32____

5 9 + 9 + 9 = 27   ____3____ × ____9____ = ____27____

Complete the chart.

6 The cook has a recipe that calls for 3 pea pods. Create a chart that shows the number of peas that will be in the recipe when the cook uses 1–3 pea pods.

Number of Pods	Number of Peas in a Pod	Multiplication Sentence
1	4	1 × 4 = 4
2	4	2 × 4 = 8
3	4	3 × 4 = 12

Week 2 Number Lines and Arrays

Student Workbook

Practice

Write the numbers you say as you skip count. Complete the number sentence.

1 Show skip counting by 3 for a count of 7 on the number line.

7 × ____3____ = ____21____

3   6   9   12   15   18   21

Explain the skip counting using the idea of groups. Complete each number sentence.

2 Show skip counting by 6 for a count of 3 on the number line.

____3____ groups of ____6____   3 × ____6____ = ____18____

3   6   9   12   15   18

3 Show skip counting by 4 for a count of 5 on the number line.

____5____ groups of ____4____   5 × ____4____ = ____20____

4   8   12   16   20

Tell how many groups are shown in each area model. Write the multiplication sentence.

4

____6____ groups of ____7____

6 × 7 = 42

5

____8____ groups of ____4____

8 × 4 = 32

Unit 4
Multiplication
Practice Pages
Weeks 1 and 2

1 Assign the Practice pages, *Workbook,* pages 42–43, at the end of Weeks 1 and 2. Students should complete these pages independently.

2 Check student answers using the annotated pages above.

3 If students have difficulty with a Practice page, review the activities completed throughout the week, and have students complete the weekly practice again before they complete the weekly test.

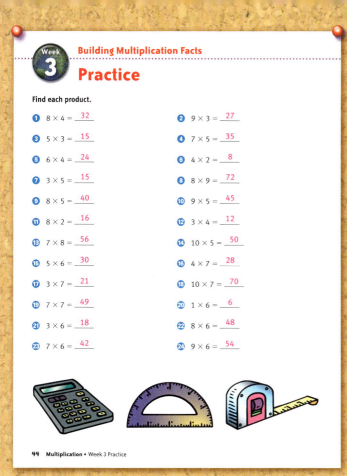

Practice

Find each product.

1. 8 × 4 = 32
2. 9 × 3 = 27
3. 5 × 3 = 15
4. 7 × 5 = 35
5. 6 × 4 = 24
6. 4 × 2 = 8
7. 3 × 5 = 15
8. 8 × 9 = 72
9. 8 × 5 = 40
10. 9 × 5 = 45
11. 8 × 2 = 16
12. 3 × 4 = 12
13. 7 × 8 = 56
14. 10 × 5 = 50
15. 5 × 6 = 30
16. 4 × 7 = 28
17. 3 × 7 = 21
18. 10 × 7 = 70
19. 7 × 7 = 49
20. 1 × 6 = 6
21. 3 × 6 = 18
22. 8 × 6 = 48
23. 7 × 6 = 42
24. 9 × 6 = 54

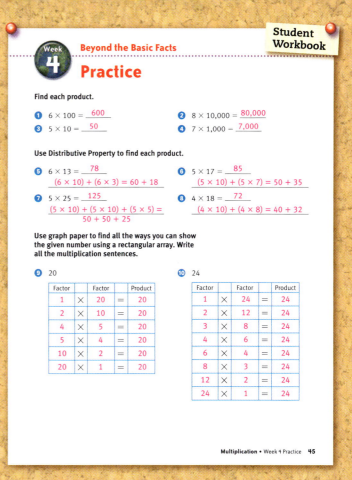

Student Workbook

Practice

Find each product.

1. 6 × 100 = 600
2. 8 × 10,000 = 80,000
3. 5 × 10 = 50
4. 7 × 1,000 = 7,000

Use Distributive Property to find each product.

5. 6 × 13 = 78
   (6 × 10) + (6 × 3) = 60 + 18
6. 5 × 17 = 85
   (5 × 10) + (5 × 7) = 50 + 35
7. 5 × 25 = 125
   (5 × 10) + (5 × 10) + (5 × 5) = 50 + 50 + 25
8. 4 × 18 = 72
   (4 × 10) + (4 × 8) = 40 + 32

Use graph paper to find all the ways you can show the given number using a rectangular array. Write all the multiplication sentences.

9. 20

Factor		Factor		Product
1	×	20	=	20
2	×	10	=	20
4	×	5	=	20
5	×	4	=	20
10	×	2	=	20
20	×	1	=	20

10. 24

Factor		Factor		Product
1	×	24	=	24
2	×	12	=	24
3	×	8	=	24
4	×	6	=	24
6	×	4	=	24
8	×	3	=	24
12	×	2	=	24
24	×	1	=	24

Unit 4
Multiplication Practice Pages

Weeks 3 and 4

1 Assign the Practice pages, *Workbook,* pages 44–45, at the end of Weeks 3 and 4. Students should complete these pages independently.

2 Check student answers using the annotated pages above.

3 If students have difficulty with a Practice page, review the activities completed throughout the week, and have students complete the weekly practice again before they complete the weekly test.

Investigating Shapes

Week at a Glance

This week, students begin **Number Worlds,** Level F, Geometry and Measurement, by investigating shapes. Students will recognize and construct shapes. They will also be able to describe properties of shapes and to categorize them by name.

Math Background

The reasoning skills that students develop in grades 3–5 allow them to investigate geometric problems of increasing complexity and to study geometric properties. As they move from grade 3 to grade 5, they should develop clarity and precision in describing the properties of geometric objects and then classifying them by these properties into categories.

—*Principles and Standards for School Mathematics.* 2000. National Council of Teachers of Mathematics. Page 165.

How Students Learn

Students benefit most from "doing" geometry. Their visualization and thinking skills develop with hands-on experiences in sorting, building, drawing, manipulating, tracing, and measuring geometric shapes. As students begin to analyze and compare the properties of shapes, they make conjectures about relationships among shapes and develop mathematical arguments to support their conjectures.

Teaching for Understanding

Observe closely while evaluating the assigned tasks this week to see whether students can demonstrate the following understandings:

Benchmark after Lesson 2: Students can describe characteristics of shapes.

Benchmark after Lesson 3: Students can name quadrilaterals based on their characteristics.

Benchmark after Lesson 4: Students can name shapes based on the number of sides.

Skills Focus

- Analyze characteristics and properties of two-dimensional shapes
- Identify quadrilaterals and other shapes based on the number of sides and other characteristics

Math at Home

Give one copy of the Letter to Home, page A17, to each student. Encourage students to share and complete the activity with their caregivers.

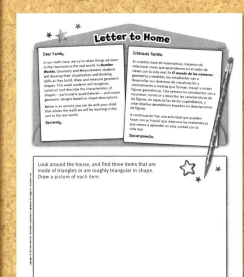

Letter to Home, Teacher Edition, p. A17

Week 1 Planner   Investigating Shapes

PACING	LESSON	LEARNING GOALS	NCTM	MATERIALS	TECHNOLOGY
DAY 1	pages 2–3	Students can combine triangles to create new shapes.	• Geometry • Communication • Connections • Representation	• Right isosceles triangles, p. B25 • Dot paper	**Building Blocks** Piece Puzzler **MathTools** Shape Tool
DAY 2	pages 4–5	Students identify the properties of shapes.	• Geometry • Communication • Connections • Representation	• Right isosceles triangles, p. B25 • Flex straws • Pattern blocks*	**Building Blocks** Geometry Snapshots **MathTools** Shape Tool
DAY 3	pages 6–7	Students identify shapes according to their properties.	• Geometry • Communication • Connections • Representation	Right isosceles triangles, p. B25	**Building Blocks** Geometry Snapshots **MathTools** Shape Tool
DAY 4	pages 8–9	Students name shapes by the number of sides.	• Geometry • Communication • Connections • Representation	Pattern blocks*	**Building Blocks** Super Shape **MathTools** Shape Tool
DAY 5	**Review and Assess** pages 10–11	Students review skills learned this week and complete the weekly assessment.	• Geometry • Communication • Connections • Representation	Materials will be selected from Lessons 1–4.	**Building Blocks** Review previous activities

Math Vocabulary

triangle A plane figure with three sides and three angles

right angle An angle with a measure of 90°

congruent sides Sides of a figure that have the same lengths

parallel sides Opposite sides of a figure at a continuous distance from each other that never touch, even when the sides are extended

English Learners

SPANISH COGNATES

English	Spanish
characteristics	características
geometry	geometría
triangles	triángulos
construct	construir

ALTERNATE VOCABULARY
composition How something is put together

---

* Available from SRA

Lesson 1

Objective

Students can combine triangles to create new shapes.

Materials

Program Materials
Right isosceles triangles cut from heavy card stock (8–10 per group of students), p. B25

Additional Materials
Dot paper

Access Vocabulary

match The edges meet exactly; they go together because they are the same

Creating Context

In order to build English Learners' academic language, practice describing why each figure in problems 1 through 4 is an example of correct or incorrect composition. When using terminology such as *edge* and *corner,* be sure to point to the appropriate places in the picture to reinforce understanding.

1 Warm Up · 5

Concept Building UNDERSTANDING

Display two right isosceles triangles on a table or overhead projector.

Demonstrate how these triangles can be lined up to create a new shape. Create a square, an isosceles triangle, and a parallelogram by using two triangles.

As each shape is named, move the triangles to show it on the overhead projector.

2 Engage · 30

Skill Building

Organize students into groups of two. Each group needs eight right isosceles triangles.

Composing Shapes UNDERSTANDING

Invite students to create as many shapes as they can with the triangles. They should record each shape on dot paper.

- Challenge students to make several new shapes with three triangles. Allow time for students to work.
- Challenge students to make several new shapes with four triangles. Allow time for students to work.
- Challenge students to make at least five new shapes with six triangles. Allow time for students to work.
- Challenge students to make at least five new shapes using eight triangles. Allow time for students to work.

As a class, discuss the shapes the groups made. Make sure that all unique shapes are shared.

Monitoring Student Progress

If . . . students struggle to create new shapes,

Then . . . encourage them to make a known shape and alter the arrangement of one triangle.

Building Blocks For additional practice with constructing shapes, students should complete **Building Blocks** activity Piece Puzzler.

MathTools Use the Shape Tool to demonstrate and explore geometric shapes.

Using Student Pages

Have students complete **Workbook,** pages 2–3, on their own. Provide students with additional dot paper, if needed.

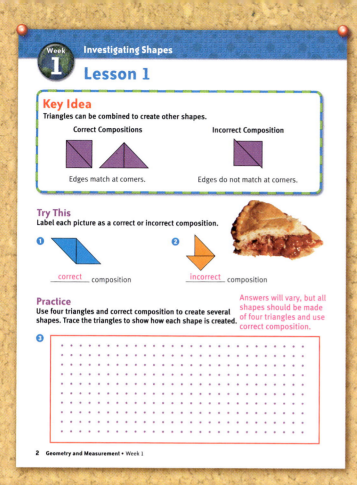

Week 1 Investigating Shapes

Lesson 1

Key Idea
Triangles can be combined to create other shapes.

Correct Compositions   **Incorrect Composition**

Edges match at corners.   Edges do not match at corners.

Try This
Label each picture as a correct or incorrect composition.

❶ <u>correct</u> composition

❷ <u>incorrect</u> composition

Practice
Use four triangles and correct composition to create several shapes. Trace the triangles to show how each shape is created.

Answers will vary, but all shapes should be made of four triangles and use correct composition.

❸

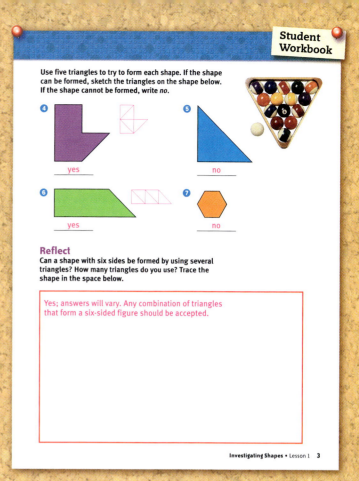

Use five triangles to try to form each shape. If the shape can be formed, sketch the triangles on the shape below. If the shape cannot be formed, write *no*.

❹ <u>yes</u>

❺ <u>no</u>

❻ <u>yes</u>

❼ <u>no</u>

Reflect
Can a shape with six sides be formed by using several triangles? How many triangles do you use? Trace the shape in the space below.

Yes; answers will vary. Any combination of triangles that form a six-sided figure should be accepted.

---

③ Reflect 10

Extended Response REASONING
Review students' answers to the Reflect prompt at the bottom of student page 3.

Invite students to share their shapes by drawing them on the board.
- **Was this activity easy or difficult?**
- **Is there more than one way to solve this problem?**

Real-World Application APPLYING
Figures used in architecture are often composed of shapes that are composed of other shapes. Ask students to look around their classroom or home and find examples of this concept. In particular, they may want to look for examples of triangles being used. Have students draw an example of a building that uses combined shapes in its design.

④ Assess

Informal Assessment
Use the Student Assessment Record, *Assessment,* page 100, to record informal observations.

UNDERSTANDING

Composing Shapes
Did the student
- ❏ make important observations?
- ❏ extend or generalize learning?
- ❏ provide insightful answers?
- ❏ pose insightful questions?

Lesson 2

Objective

Students identify the properties of shapes.

Materials

Program Materials

Right isosceles triangles cut from heavy card stock, p. B25

Additional Materials

- Flex straws
- Pattern blocks

Access Vocabulary

right angle An angle measuring 90 degrees, such as the corner of a piece of paper

Creating Context

In English, words often have more than one meaning. When we listen or read, we have to find clues to give us a context for the word we are not sure about. For example, in the context of this lesson, when you are instructed to make a table, *table* means "an organized display of information," not a piece of furniture.

1 Warm Up 5

Concept Building UNDERSTANDING

Have students use flex straws to create angles of different sizes. Demonstrate and describe a right angle. Demonstrate and describe an acute angle as an angle that has a measure less than that of a right angle. Demonstrate and describe an obtuse angle as an angle greater than that of a right angle. Ask students to use straws to practice making and naming angles.

2 Engage 30

Skill Building

Display a square on the board or overhead projector. Point out the top and bottom of the square.

- **If these two lines continued on, would they ever touch?** No
- **These are parallel sides or lines. Can you point out a shape in the classroom that has parallel sides?** Possible answer: the edges of the board

Point out that the four sides of the square are the same length.

- **When a shape has two or more sides that are the same length, those sides are called** *congruent*.
- **How can you tell whether two sides are congruent?** You can measure the sides.

Recognizing Properties UNDERSTANDING

Students need pattern blocks.

- **Make a table with these headings: Right Angle, Congruent Sides, Parallel Sides.**
- **Under each heading, place any of the pattern blocks that have that characteristic.** Allow students time to do this and then review. Under the Right Angle column is the square. Under the Congruent Sides column are the triangle, square, parallelogram, trapezoid, and hexagon. Under the Parallel Sides column are the square, parallelogram, trapezoid, and hexagon.

Monitoring Student Progress

| **If . . .** students are struggling to place the shapes under the correct headings, | **Then . . .** encourage them either to use the straw to compare right angles or to measure the sides with a ruler. |

Building Blocks For additional practice describing shapes, students should complete *Building Blocks* Geometry Snapshots.

MathTools Use the Shape Tool to demonstrate and explore geometric shapes.

Using Student Pages

Have students complete *Workbook,* pages 4–5, on their own.

Lesson 2

Key Idea

Symbols are used to indicate the different properties of quadrilaterals.

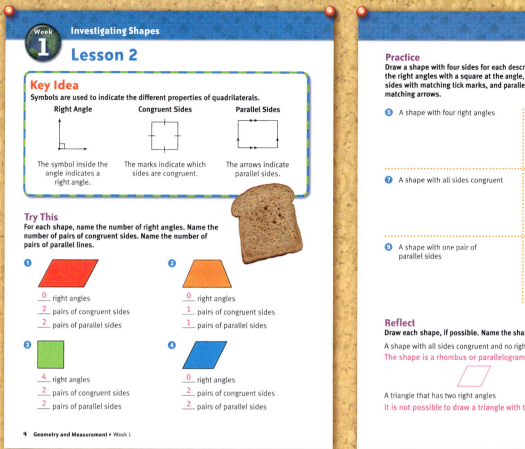

Right Angle	Congruent Sides	Parallel Sides
The symbol inside the angle indicates a right angle.	The marks indicate which sides are congruent.	The arrows indicate parallel sides.

Try This

For each shape, name the number of right angles. Name the number of pairs of congruent sides. Name the number of pairs of parallel lines.

1
<u>0</u> right angles
<u>2</u> pairs of congruent sides
<u>2</u> pairs of parallel sides

2
<u>0</u> right angles
<u>1</u> pairs of congruent sides
<u>1</u> pairs of parallel sides

3
<u>4</u> right angles
<u>2</u> pairs of congruent sides
<u>2</u> pairs of parallel sides

4
<u>0</u> right angles
<u>2</u> pairs of congruent sides
<u>2</u> pairs of parallel sides

Practice

Draw a shape with four sides for each description. Mark the right angles with a square at the angle, congruent sides with matching tick marks, and parallel sides with matching arrows.

Answers will vary.

5 A shape with four right angles

6 A shape with no right angles and two pairs of parallel sides

7 A shape with all sides congruent

8 A shape with two sides congruent and one pair of parallel lines

9 A shape with one pair of parallel sides

10 A shape with two sets of parallel sides

Reflect

Draw each shape, if possible. Name the shape.

A shape with all sides congruent and no right angles
The shape is a rhombus or parallelogram.

A triangle that has two right angles
It is not possible to draw a triangle with two right angles.

3 Reflect   10

Extended Response   REASONING

Review students' answers to the Reflect prompt at the bottom of student page 5.

- **How did you solve this problem?**

Discuss that there are specific names for shapes with certain properties. Guide students to the generalization that if a figure has two or more right angles, it must have at least four sides.

- **How would you explain this to someone who has never done this before?**

Real-World Application   APPLYING

Find five different shapes in the classroom. Describe the characteristics of the shapes.

- **Do the shapes have any right angles?**
- **How many sides are parallel?**
- **Are any of the sides congruent?**
- **How many lines of symmetry can be drawn through the shape?**

4 Assess

Informal Assessment

Use the Student Assessment Record, *Assessment,* page 100, to record informal observations.

UNDERSTANDING

Recognizing Properties
Did the student
- ❑ make important observations?
- ❑ extend or generalize learning?
- ❑ provide insightful answers?
- ❑ pose insightful questions?

Investigating Shapes • Lesson 2   4–5

Lesson 3

Objective

Students identify shapes according to their properties.

Materials

Program Materials

Right isosceles triangles cut from heavy card stock, p. B25

Access Vocabulary

have in common Share, have similar qualities

Creating Context

An excellent strategy to help English Learners acquire new academic vocabulary is to find real-world examples of the concept. In this lesson show examples, or have students draw examples, of traffic signs that are printed on different shapes.

1 Warm Up   5

Concept Building UNDERSTANDING

Draw these four shapes on the board.

- **What shape has four right angles and four congruent sides?** square
- **What shape has four right angles and two pairs of parallel sides?** rectangle or square
- **What shape has only one pair of parallel sides?** trapezoid
- **What shape has two pairs of parallel sides?** parallelogram or rectangle, or square

Discuss what the four shapes have in common. Students should notice that all are four-sided figures.

- **Any four-sided figure is a quadrilateral. Squares, rectangles, trapezoids, and parallelograms are special quadrilaterals.**

2 Engage   30

Skill Building

Students need 8–10 right isosceles triangles. As students create each shape, ask them to describe the properties of the shape, as well as naming each one.

Introduction to Quadrilaterals UNDERSTANDING

- **Use triangles to create a parallelogram. What names can you give this shape?** quadrilateral and parallelogram
- **If a parallelogram has four congruent sides and no right angles, it is called a rhombus.**
- **Use triangles to create a rectangle. What names can you give this shape?** quadrilateral, rectangle, and parallelogram
- **Use triangles to create a square. What names can you give this shape?** quadrilateral, square, rectangle, and parallelogram
- **Use triangles to create a trapezoid. What names can you give this shape?** quadrilateral and trapezoid
- **Use triangles to create a quadrilateral that cannot be classified as a square, rectangle, trapezoid, or parallelogram. This quadrilateral would not have any characteristics in common with the four named quadrilaterals.**

Monitoring Student Progress

| **If . . .** students need more practice classifying shapes, | **Then . . .** provide dot paper for students to practice drawing more quadrilaterals. |

Building Blocks For additional practice categorizing quadrilaterals, students should complete **Building Blocks** Geometry Snapshots.

MathTools Use the Shape Tool to demonstrate and explore geometric shapes.

Using Student Pages

Have students complete **Workbook,** pages 6–7, on their own.

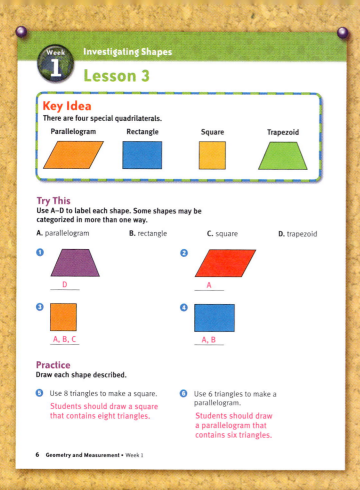

Lesson 3

Key Idea
There are four special quadrilaterals.

| Parallelogram | Rectangle | Square | Trapezoid |

Try This
Use A–D to label each shape. Some shapes may be categorized in more than one way.

A. parallelogram   **B.** rectangle   **C.** square   **D.** trapezoid

1
D

2
A

3
A, B, C

4
A, B

Practice
Draw each shape described.

5 Use 8 triangles to make a square.
Students should draw a square that contains eight triangles.

6 Use 6 triangles to make a parallelogram.
Students should draw a parallelogram that contains six triangles.

Use triangles to create each shape named.
Trace the triangles to show the shape.

7 square
Drawings will vary.

8 rectangle
Drawings will vary.

9 parallelogram
Drawings will vary.

10 trapezoid
Drawings will vary.

Reflect
What do all the shapes in this lesson have in common?
All are four-sided figures, so all are quadrilaterals.

What are the differences between a parallelogram and a square? How are they alike?
Parallelograms do not have to have right angles, but a square does.
Parallelograms have two pairs of parallel sides, but not all sides have to be congruent. A square has four congruent sides and four right angles.
They are alike because they both have two pairs of parallel sides.

3 Reflect   10

Extended Response   REASONING
Review students' answers to the Reflect prompt at the bottom of student page 7.

Discuss the characteristics of quadrilaterals. Remind students that all quadrilaterals have four sides.

- **Are there other quadrilaterals that do not look like the four special quadrilaterals?**
- **How would you explain this to someone who has never done this before?**

Real-World Application   APPLYING
Road signs often have different shapes, but many are quadrilaterals.

- **Name a road sign that is a quadrilateral.**
  Sample answers: speed limit, reserved parking, deer crossing, hospital, do not enter, no U-turn
- **What type of quadrilateral is a speed limit sign?** rectangle

4 Assess

Informal Assessment
Use the Student Assessment Record, *Assessment,* page 100, to record informal observations.

UNDERSTANDING
Introduction to Quadrilaterals
Did the student
- ❏ make important observations?
- ❏ extend or generalize learning?
- ❏ provide insightful answers?
- ❏ pose insightful questions?

Lesson 4

Objective

Students name shapes by the number of sides.

Materials

Additional Materials
Pattern blocks

Creating Context

Creating a hands-on model is an excellent strategy for English Learners because it allows them to work in steps and experiment with different possibilities. It also allows the teacher to check students' progress.

 1 Warm Up 5

Concept Building UNDERSTANDING

Review the following prefixes:

3	tri-	4	quad-	5	pent-
6	hex-	7	hept-	8	oct-

- Shapes are named for their number of sides.
- How many sides does a triangle have? three
- How many wheels does a tricycle have? three
- The prefixes of the names of shapes often tell you the number of sides.
- What prefix means "3"? tri-

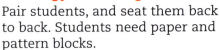 **2 Engage** 30

Strategy Building

Pair students, and seat them back to back. Students need paper and pattern blocks.

Build My Design ENGAGING

Arrange pattern blocks into a shape that the students cannot see. Describe your shape, and ask students to re-create it as you describe it. Use the names of the shapes as much as possible in your description. Have students compare their shapes with yours.

Without their partner seeing the design, students should create a simple design with no more than ten pattern blocks. Students should describe their design while their partners try to re-create the design. Students take turns in each role.

Ask each pair to create a design and write instructions for making their design. The instructions should contain only words—no pictures. Have pairs switch instructions. Each pair should then try to re-create the new design.

Monitoring Student Progress

If . . . students need more of a challenge in this activity,	Then . . . tell them that they cannot use color in their descriptions. They must refer to shapes by name only.

Building Blocks For additional practice composing shapes, students should complete **Building Blocks** activity Super Shape.

MathTools Use the Shape Tool to demonstrate and explore geometric shapes.

Using Student Pages

Have students complete **Workbook,** pages 8–9, on their own.

Lesson 4

Key Idea
Shapes are named based on the number of sides they have.

Name	Triangle	Quadrilateral	Pentagon	Hexagon	Heptagon	Octagon
Number of Sides	3	4	5	6	7	8

Try This
Name each figure.

① 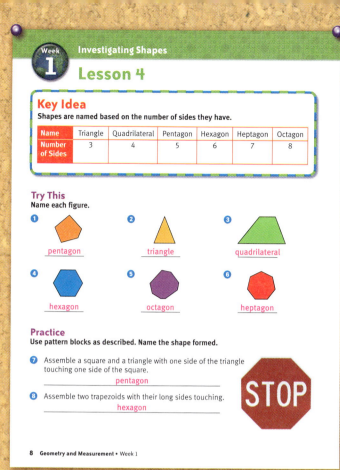 pentagon

② triangle

③ quadrilateral

④ hexagon

⑤ octagon

⑥ heptagon

Practice
Use pattern blocks as described. Name the shape formed.

⑦ Assemble a square and a triangle with one side of the triangle touching one side of the square.
pentagon

⑧ Assemble two trapezoids with their long sides touching.
hexagon

Use pattern blocks. Follow the directions given to create the design.

⑨
- Put a square in the middle.
- Put a triangle on top of the square so that the sides fit exactly.
- Put a hexagon on the bottom of the square so that the sides fit exactly.
- Put a triangle on the right side of the square so that the sides fit exactly.
- Put a triangle on the left side of the square so that the sides fit exactly.
Check student's patterns for accuracy.

⑩
- Put a hexagon in the middle.
- Put a triangle on the top and bottom of the hexagon.
- Put a square against each of the 4 sides of the hexagon that are not being touched by a triangle.
Check student's patterns for accuracy.

⑪ Sketch or trace each of the six pattern blocks in order of increasing number of sides. Write the name of each figure below it.

| Triangle | Rhombus | Square | Trapezoid | Hexagon |

Reflect
Is the word *quadrilateral* the best word to use in a design description? Why or why not?
No; a quadrilateral could be a parallelogram, a rectangle, a square, a rhombus, or another four-sided shape.

---

3 Reflect 10

Extended Response REASONING
Review students' answers to the Reflect prompt at the bottom of student page 9.

Have students share their reasons with the class.

- **What words other than *quadrilateral* can be used to describe a four-sided shape?** rhombus, square, parallelogram, rectangle, trapezoid

- **Can you think of other times outside school when you would use this skill?**

Real-World Application APPLYING
Discuss the shapes of road signs with which the students are familiar.

- **Not all road signs are quadrilaterals.**

- **What road sign is in the shape of a pentagon?** school zone

- **What shape is the most recognizable road sign? What sign is it?** octagon; stop sign

4 Assess

Informal Assessment
Use the Student Assessment Record, **Assessment,** page 100, to record informal observations.

ENGAGING

Build My Design
Did the student
- ❏ pay attention to the contributions of others?
- ❏ contribute information and ideas?
- ❏ improve on a strategy?
- ❏ reflect on and check accuracy of work?

Lesson 5 Review

Objective

Students review skills learned this week and complete the weekly assessment.

Materials

Review materials will be selected from those used in previous activities.

Creating Context

When multiple-step directions are given, take one step at a time so that the task is less overwhelming.

1 Warm Up 5

Concept Building APPLYING

Allow students time to use pattern blocks or triangles to explore and create shapes at their desks.

2 Engage 20

Skill Building

Free-Choice Activity ENGAGING

For the last day of the week, allow students to choose an activity from the following:

- **Composing Shapes**
- **Recognizing Properties**
- **Introduction to Quadrilaterals**
- **Build My Design**

Make a note of the activities children select. Do they prefer easy or challenging activities? If you believe your students would benefit from extra practice on specific facts, choose an activity for them.

Monitoring Student Progress

If . . . students need more practice with sorting and classifying shapes,

Then . . . they should choose Introduction to Quadrilaterals for their Free-Choice activity.

Using Student Pages

Have students complete **Workbook,** pages 10–11, on their own.

3 Reflect 10

Extended Response REASONING

Review students' answers to the Reflect prompts at the bottom of student pages 10 and 11.

Discuss the answers with the group to reinforce Week 1 concepts.

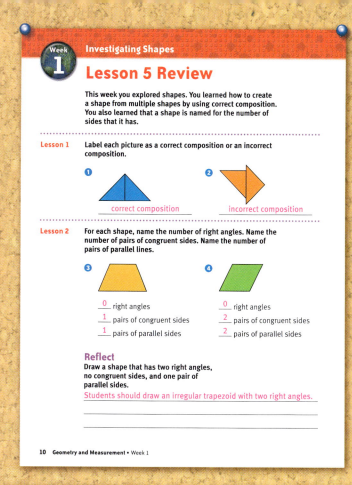

Lesson 5 Review

This week you explored shapes. You learned how to create a shape from multiple shapes by using correct composition. You also learned that a shape is named for the number of sides that it has.

Lesson 1 Label each picture as a correct composition or an incorrect composition.

❶ correct composition

❷ incorrect composition

Lesson 2 For each shape, name the number of right angles. Name the number of pairs of congruent sides. Name the number of pairs of parallel lines.

❸
0 right angles
1 pairs of congruent sides
1 pairs of parallel lines

❹
0 right angles
2 pairs of congruent sides
2 pairs of parallel lines

Reflect
Draw a shape that has two right angles, no congruent sides, and one pair of parallel lines.
Students should draw an irregular trapezoid with two right angles.

Lesson 3 Use A–D to match each shape with its description. Some descriptions may fit more than one shape.

A. parallelogram   B. rectangle   C. square   D. trapezoid

❺ a four-sided shape with four right angles
B, C

❻ a four-sided shape with one pair of parallel sides
D

❼ a four-sided shape with four right angles and four congruent sides
C

❽ a four-sided shape with two pairs of parallel sides
A, B, C

Lesson 4 Name each shape with the given number of sides.

❾ 6 sides — hexagon
❿ 7 sides — heptagon
⓫ 8 sides — octagon
⓬ 3 sides — triangle
⓭ 5 sides — pentagon
⓮ 4 sides — quadrilateral, parallelogram, rectangle, rhombus, square, or trapezoid

Reflect
What is another name for the shape described in 8 above?
rhombus

 ## Assess 10

A Gather Evidence

Formal Assessment
Have students complete the weekly test, **Assessment,** pages 56–57. Record progress on the Student Assessment Record, **Assessment,** page 100.

B Summarize Findings
Determine whether students have Minimal, Basic, or Secure understanding of the concepts presented in Week 1.

C Differentiate Instruction
On the basis of your observations, use these teaching strategies next week to follow up.

Assessment, pp. 56–57

Minimal Understanding
- Repeat the Warm-Up and Engage activities to develop Week 1 concepts.
- Use **Building Blocks** computer activities beginning with Geometry Snapshots to develop and reinforce this week's concepts.

Basic Understanding
- Repeat Engage activities in subsequent weeks.
- Use **Building Blocks** computer activities beginning with Piece Puzzler and **eMathTools** Shape Tool to reinforce Week 1 concepts.

Secure Understanding
- Use **Building Blocks** Super Shape and **eMathTools** Shape Tool to reinforce Week 1 concepts.
- Use variations of the weekly Warm-Up and Engage activities, using higher numbers and multiple steps.

Week 2 Exploring Congruence

Week at a Glance

This week, students continue with Geometry and Measurement, Level F, Exploring Congruence, by constructing figures and using transformations (slides, flips, and turns) to answer the question, "How can we tell whether two shapes are congruent?"

Math Background

When shapes have the same corresponding angles and corresponding sides of the same length, they are *congruent*. They may have different colors or be in different locations or orientations, but they are still congruent. Slides, flips, and turns are rigid transformations, actions that move a shape from one location and orientation to another without changing the shape itself.

How Students Learn

Visualization is a mental skill, similar to performing mental arithmetic. Students developing spatial sense can be nurtured by opportunities to construct, draw, visualize, compare, and transform geometric shapes. As students work through the activities during this week, it is important to connect what they are saying (slide, flip, turn) with what they are doing. Students need plenty of practice with verbalizing what they visualize and visualizing what is verbalized.

Teaching for Understanding

Observe closely while evaluating the assigned tasks this week to see whether students can demonstrate the following understandings:

Benchmark after Lesson 2: Students can identify congruent shapes.

Benchmark after Lesson 3: Students can perform simple shape transformations.

Benchmark after Lesson 4: Students can perform shape transformations involving three or more steps.

Skills Focus

- Identify congruent shapes
- Recognize, describe, and apply transformations
- Use visualization, spatial reasoning, and geometric modeling to solve problems

Math at Home

Give one copy of the Letter to Home, page A18, to each student. Encourage students to share and complete the activity with their caregivers.

Letter to Home, Teacher Edition, p. A18

PACING	LESSON	LEARNING GOALS	NCTM	MATERIALS	TECHNOLOGY
DAY 1	pages 12–13	Students determine whether shapes are congruent.	• Geometry • Communication • Connections • Representation	• Square tiles • Graph paper • Scissors	**Building Blocks** Piece Puzzler **MathTools** Shape Tool
DAY 2	pages 14–15	Students recognize, describe, and apply the simple transformations of flips, turns, and slides.	• Geometry • Communication • Connections • Representation	• Square tiles • Transformation Arrows, p. B26 • Tetrominoes, p. B26 • Graph paper • Scissors	**Building Blocks** Super Shape **MathTools** Shape Tool
DAY 3	pages 16–17	Students recognize, describe, and apply combinations of flips, turns, and slides.	• Geometry • Communication • Connections • Representation	• Transformation Arrows, p. B26 • Tetrominoes, p. B26 • Graph paper • Dot paper • Scissors	**Building Blocks** Super Shape **MathTools** Shape Tool
DAY 4	pages 18–19	Students recognize, describe, and apply equivalent sets of transformations.	• Geometry • Communication • Connections • Representation	• Transformation Arrows, p. B26 • Tetrominoes, p. B26 • Graph paper • Dot paper • Scissors	**Building Blocks** Super Shape **MathTools** Shape Tool
DAY 5	**Review and Assess** pages 20–21	Students review skills learned this week and complete the weekly assessment.	• Geometry • Communication • Connections • Representation	Materials will be selected from Lessons 1–4.	**Building Blocks** Review previous activities

Math Vocabulary

tetromino A shape made from 4 squares

congruence Shapes that fit exactly on top of one another when flipped, turned, or slid

slide To be able to move a shape in any direction

flip To reflect a shape over an invisible line

turn To rotate a shape clockwise or counterclockwise

English Learners

SPANISH COGNATES

English	Spanish
congruence	congruencia
direction	dirección
reflections	reflexiones

ALTERNATE VOCABULARY

clockwise Moving in the same direction as the hands of an analog clock

counterclockwise Moving in the opposite direction of the hands of an analog clock

Lesson 1

Objective

Students determine whether shapes are congruent.

Materials

Additional Materials
- Square tiles
- Graph paper

Access Vocabulary

clockwise Moving in the same direction as the hands of an analog clock

congruence Shapes that fit exactly on top of one another when flipped, turned, or slid

tetromino A shape made from four squares

Creating Context

Using graphic organizers with English Learners is an excellent way to help them develop critical thinking skills. Create a two-column chart or a Venn diagram to compare and contrast congruent and similar figures.

1 Warm Up 5

Concept Building UNDERSTANDING

Give each student four square tiles, or squares cut from cardboard, and graph paper.

Place four square tiles on the overhead projector. Create one of the tetrominoes below by using the four square tiles. Have students re-create the tetromino at their desks.

- A tetromino is a shape made out of four squares.

2 Engage 30

Skill Building

Creating Tetrominoes APPLYING

Each student should use four square tiles to create tetrominoes. Students should then sketch their tetrominoes on graph paper.

- **Create as many tetrominoes as you can with four square tiles.**
- **Sketch your tetrominoes on the graph paper. How many different tetrominoes did you make?**

Demonstrate with cut-out tetronimoes that you can flip and turn to try to fit them on top of one another.

- **Are there any arrangements you sketched that could fit exactly on top of one another and match each square?**

Explain congruence as a situation in which shapes fit exactly on top of one another when flipped, turned, or slid.

Define slides, flips, and turns for students as they are defined on page 14A. Have students cut out their tetrominoes.

Have students use flips, turns, and slides to determine whether any of the tetrominoes are congruent.

Have students find all five unique tetrominoes.

Monitoring Student Progress

If . . . students are missing some of the tetrominoes,

Then . . . suggest that they share tetrominoes with other students to see which ones are missing.

Building Blocks For additional practice finding congruence, students should complete **Building Blocks** Piece Puzzler.

MathTools Use the Shape Tool to demonstrate and explore geometric shapes.

Using Student Pages

Have students complete **Workbook,** pages 12–13, on their own.

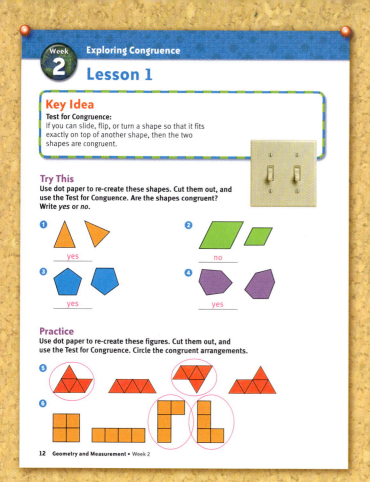

Week 2 — Exploring Congruence
Lesson 1

Key Idea

Test for Congruence:
If you can slide, flip, or turn a shape so that it fits exactly on top of another shape, then the two shapes are congruent.

Try This

Use dot paper to re-create these shapes. Cut them out, and use the Test for Congruence. Are the shapes congruent? Write *yes* or *no*.

1. yes
2. no
3. yes
4. yes

Practice

Use dot paper to re-create these figures. Cut them out, and use the Test for Congruence. Circle the congruent arrangements.

5.
6.

12 Geometry and Measurement • Week 2

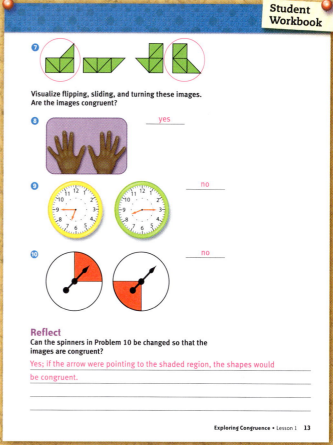

7.

Visualize flipping, sliding, and turning these images. Are the images congruent?

8. yes

9. no

10. no

Reflect

Can the spinners in Problem 10 be changed so that the images are congruent?

Yes; if the arrow were pointing to the shaded region, the shapes would be congruent.

---

3 Reflect   10

Extended Response   REASONING

Review students' answers to the Reflect prompt at the bottom of student page 13.

Invite a student to the board to illustrate what needs to be done to make the spinners congruent. Discuss the difference between similar and congruent.

Real-World Application   APPLYING

Many designs are made of congruent shapes. For example, wallpaper may have a pattern made of congruent shapes. Have students list five items in their homes or in the classroom that are made of congruent shapes and then list which congruent shapes make up the items.

4 Assess

Informal Assessment

Use the Student Assessment Record, **Assessment,** page 100, to record informal observations.

APPLYING

Creating Tetrominoes

Did the student
- ❏ apply learning to new situations?
- ❏ contribute concepts?
- ❏ contribute answers?
- ❏ connect mathematics to the real world?

Lesson 2

Objective

Students recognize the simple transformations of flips, turns, and slides.

Materials

Program Materials
- Transformation Arrows, p. B26
- Tetrominoes, p. B26

Additional Materials
- Square tiles
- Graph paper
- Scissors

Access Vocabulary

counterclockwise Moving in the opposite direction of the hands of an analog clock

transformation Turning a shape to a new position

Creating Context

English Learners may benefit from clarification of some common phrases and words that proficient English speakers probably know. Occasionally words have more than one meaning or are used in potentially puzzling idiomatic expressions. Before or during the lesson, be sure to clarify words that may be confusing.

1 Warm Up   5

Concept Building   COMPUTING

Cut out transformation arrows. Explain and demonstrate what each symbol means.
- A one-sided arrow means a slide.
- A horizontal double-sided arrow means a horizontal flip.
- A vertical double-sided arrow means a vertical flip.
- A curved arrow pointing right means a right turn.
- A curved arrow pointing left means a left turn.

2 Engage   30

Skill Building

Distribute a tetromino and transformation arrow worksheet to each student. Have students cut out the five unique tetrominoes and the five transformation arrows.

Transformations   UNDERSTANDING

Pair students. One student should choose a tetromino and a transformation arrow and perform the transformation.
- **Does the transformation change the size of the tetromino?** no

Have students switch roles and repeat the process, using a different tetromino each time.

When students understand the idea of applying transformations, challenge them to use two transformation arrows for each problem.

Monitoring Student Progress

If . . . students are having trouble determining how far to turn a shape,

Then . . . have them stand up and practice making quarter turns to the right and to the left.

 Building Blocks For additional practice creating a transformation, students should complete **Building Blocks** activity Super Shape.

ⓔMathTools Use the Shape Tool to demonstrate and explore geometric shapes.

Using Student Pages

Have students complete **Workbook,** pages 14–15, on their own.

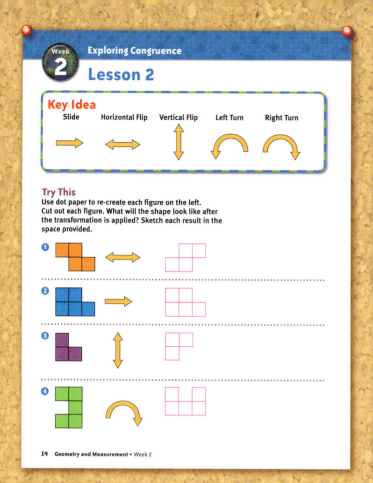

Lesson 2

Key Idea

Slide   Horizontal Flip   Vertical Flip   Left Turn   Right Turn

Try This

Use dot paper to re-create each figure on the left. Cut out each figure. What will the shape look like after the transformation is applied? Sketch each result in the space provided.

1
2
3
4

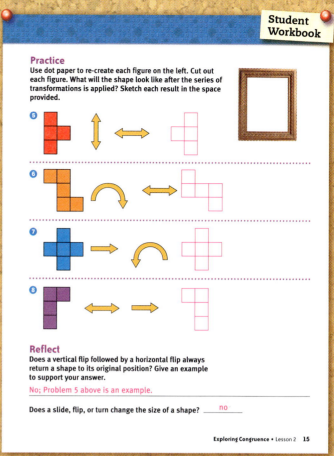

Practice

Use dot paper to re-create each figure on the left. Cut out each figure. What will the shape look like after the series of transformations is applied? Sketch each result in the space provided.

5
6
7
8

Reflect

Does a vertical flip followed by a horizontal flip always return a shape to its original position? Give an example to support your answer.

No; Problem 5 above is an example.

Does a slide, flip, or turn change the size of a shape? ___ no

3 Reflect   10

Extended Response   REASONING

Review students' answers to the Reflect prompt at the bottom of student page 15.

Have students use their tetrominoes to support their answers.

- **Was this activity easy or difficult?**
- **How would you explain your method to someone who does not understand this concept?**

Real-World Application   APPLYING

Students may have seen arrows on street and building signs. Have students name examples of signs that are similar in meaning to the turn arrow. Have students name examples of signs that are similar in meaning to the slide arrow. For example, a one-way street sign has the same form as the slide arrow.

4 Assess

Informal Assessment

Use the Student Assessment Record, **Assessment,** page 100, to record informal observations.

UNDERSTANDING

Transformations

Did the student
- ❏ make important observations?
- ❏ extend or generalize learning?
- ❏ provide insightful answers?
- ❏ pose insightful questions?

Lesson 3

Objective

Students recognize, describe, and apply combinations of flips, turns, and slides.

Materials

Program Materials
- Transformation Arrows, p. B26
- Tetrominoes, p. B26

Additional Materials
Graph paper

Access Vocabulary

transformation Turning a shape to a new position

Creating Context

The word *flip* is used in a number of contexts meaning "to turn over." You flip a pancake, do a flip in gymnastics, and "flip your lid" when you get mad and yell about something. Interview some adults at your school to see whether you can find other ways the word *flip* is used.

1 Warm Up 5

Concept Building REASONING

Select two copies of the same tetromino, and place them on the overhead projector in different orientations.

Ask a volunteer to place one or more transformation arrows between the shapes to show the sequence of changes.

- **Are there any other combinations of arrows that could cause this change?**

Allow volunteers to show other transformations.

2 Engage 30

Skill Building

Pair students. Give each pair tetrominoes and transformation arrows.

Multistep Transformations UNDERSTANDING

One student should choose a tetromino and two transformation arrows and record this information on graph paper. The other student should perform the transformations and record the ending position on the graph paper.

Students should trade roles and repeat several times. Each pair should share a multistep transformation with the class.

How Did That Happen? REASONING

Students continue to work with their partners. Each student should select a tetromino and two transformation arrows without his or her partner knowing which arrows were chosen. They should then trace the starting position of the tetromino on the graph paper. Then perform the two transformations and record the ending position of the tetromino. When they have finished, they should trade papers with their partner who must decide what two transformations were used to result in the ending position.

Students may also adapt this activity using **MathTools** Shape Tool.

Monitoring Student Progress

If . . . students are unable to guess the missing transformations,

Then . . . have the partner reveal *one* of the transformation arrows used.

Building Blocks For additional practice performing multistep transformations, students should complete **Building Blocks** activity Super Shape.

MathTools Use the Shape Tool to demonstrate and explore geometric shapes.

Using Student Pages

Have students complete **Workbook,** pages 16–17, on their own.

Week 2 — Exploring Congruence

Lesson 3

Key Idea
A combination of transformations can be used to change the position of a shape.

Try This
Will the given combination of transformations result in the ending position? Use pattern blocks to help you decide. Write *yes* or *no*.

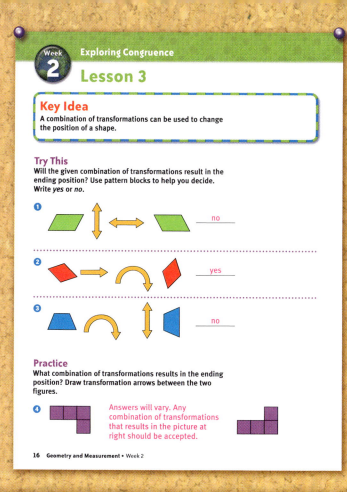

① ___no___

② ___yes___

③ ___no___

Practice
What combination of transformations results in the ending position? Draw transformation arrows between the two figures.

④ Answers will vary. Any combination of transformations that results in the picture at right should be accepted.

16 Geometry and Measurement • Week 2

⑤ Answers will vary. Any combination of transformations that results in the picture at right should be accepted.

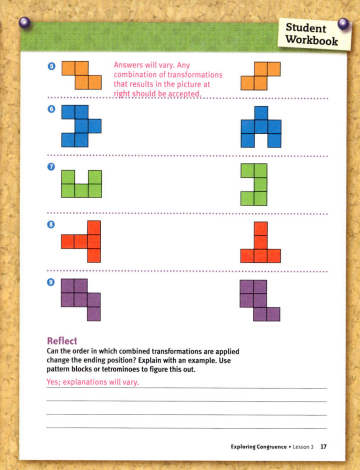

⑥

⑦

⑧

⑨

Reflect
Can the order in which combined transformations are applied change the ending position? Explain with an example. Use pattern blocks or tetrominoes to figure this out.

Yes; explanations will vary.

_____

_____

_____

Exploring Congruence • Lesson 3   17

---

3 Reflect   10

Extended Response   REASONING

Review students' answers to the Reflect prompt at the bottom of student page 17.

Have students create an example in which the order of transformations changes the ending position. Have students create their examples on graph paper. Display students' graph papers around the room.

- **How did you figure that out?**
- **Was this activity easy or difficult?**

Real-World Application   APPLYING

Architects often use turns, flips, or slides to design houses. Have students find examples of houses or parts of houses that use turns, flips, or slides.

4 Assess

Informal Assessment

Use the Student Assessment Record, **Assessment,** page 100, to record informal observations.

UNDERSTANDING	REASONING
Multistep Transformations	**How Did That Happen?**
Did the student	Did the student
❏ make important observations?	❏ provide a clear explanation?
❏ extend or generalize learning?	❏ communicate reasons and strategies?
❏ provide insightful answers?	❏ choose appropriate strategies?
❏ pose insightful questions?	❏ argue logically?

Lesson 4

Objective

Students recognize, describe, and apply equivalent sets of transformations.

Materials

Program Materials
- Transformation Arrows, p. B26
- Tetrominoes, p. B26

Additional Materials
Graph paper

Access Vocabulary

rotate To turn a shape clockwise or counterclockwise

Creating Context

In this lesson, we ask students to reflect on how a flip and a turn are similar. English Learners at early levels of English proficiency may need to draw, make a model, or discuss these ideas in their primary language to answer the question.

1 Warm Up   5

Concept Building   REASONING

Place two congruent tetrominoes on the overhead projector. Refer to the leftmost tetromino as the starting position. Place the second tetromino to the right of the first tetromino. Rotate the second tetromino 90° clockwise.

- **How can I use transformation arrows to transform the shape from the starting position to the ending position in exactly three moves?** Answers will vary. Any combination of three transformations that results in the final position should be accepted.

2 Engage   30

Strategy Building

Pair students. Give each pair tetrominoes and transformation arrows.

Step by Step   ENGAGING

Have each student choose a tetromino and create a transformation without his or her partner seeing it. Students should then sketch the starting position and ending position of the tetromino on graph paper and switch graph paper with his or her partner.

- **What is the fewest number of transformations you can use to transform the shape from the starting position to the ending position?** Answers will vary.
- **Can you transform the shape from the starting position to the ending position in exactly three moves? Draw the transformation arrows on the graph paper.** Answers will vary.
- **Can you transform the shape from the starting position to the ending position in exactly four moves?** Draw the transformation arrows on the graph paper. Answers will vary.

Monitoring Student Progress

If . . . a student is unable to perform the transformation in the given number of moves,

Then . . . suggest that they use a slide as one of the steps.

Building Blocks For additional practice performing equivalent transformations, students should complete **Building Blocks** activity Super Shape.

eMathTools Use the Shape Tool to demonstrate and explore geometric shapes.

Using Student Pages

Have students complete **Workbook,** pages 18–19, on their own.

Lesson 4

Key Idea
Different combinations of transformations can change a shape from its starting position to its ending position.

Try This
Use dot paper to re-create each figure on the left. Cut out each figure. Transform the shape from the starting position to the ending position in exactly five moves. Record the symbol of each transformation used. *Answers will vary.*

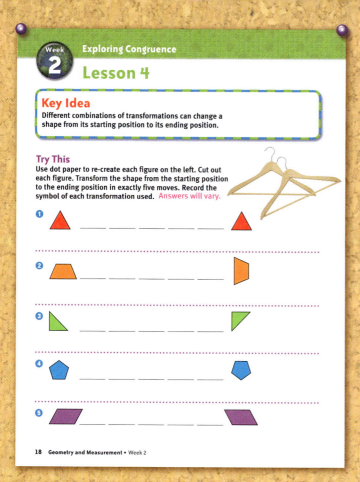

Practice
Transform the shape from the starting position to the ending position in as few transformations as possible. Record the symbol of the transformations used.

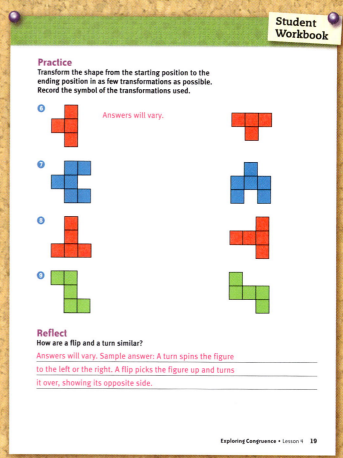

6 *Answers will vary.*

Reflect
How are a flip and a turn similar?

Answers will vary. Sample answer: A turn spins the figure to the left or the right. A flip picks the figure up and turns it over, showing its opposite side.

---

3 Reflect   10

Extended Response   REASONING
Review students' answers to the Reflect prompt at the bottom of student page 19.

Discuss ways that students can confirm their answers.

- **What if you applied two right turns and then two left turns to a tetromino?**
- **How would that compare with a flip?**

Real-World Application   APPLYING
There are many ways to reach a destination. A map can show alternate ways to travel to the same place. Describe three ways to get from your home to the school.

4 Assess

Informal Assessment
Use the Student Assessment Record, **Assessment,** page 100, to record informal observations.

ENGAGING

Step By Step
Did the student
- ❑ pay attention to the contributions of others?
- ❑ communicate reasons and strategies?
- ❑ choose appropriate strategies?
- ❑ reflect on and check accuracy of work?

Lesson 5 Review

Objective

Students review skills learned this week and complete the weekly assessment.

Materials

Review materials will be selected from those used in previous activities.

Creating Context

In *Workbook,* page 20, Problems 1–3 ask students to determine whether congruence exists in the selected shapes and to indicate by writing *yes* or *no*. Work with students to explain why each shape is or is not congruent.

---

2 Engage   20

Skill Building

Free-Choice Activity   ENGAGING

For the last day of the week, allow students to choose an activity from the previous lessons. Some activities they may choose include the following:

- **Creating Tetrominoes**
- **Transformations**
- **Multistep Transformations**
- **How Did That Happen?**
- **Step by Step**

Make a note of the activities children select. Do they prefer easy or challenging activities? If you believe your students would benefit from extra practice on specific transformations, choose an activity for them.

Monitoring Student Progress

If . . . students need more practice to understand basic transformations,

Then . . . they should choose Transformations and then Multistep Transformations for their Free-Choice activities.

Using Student Pages

Have students complete **Workbook,** pages 20–21, on their own.

---

1 Warm Up   5

Concept Building   APPLYING

Pair students. Provide time for students to practice using tetrominoes and the transformation arrows to perform multistep transformations.

---

3 Reflect   10

Extended Response   REASONING

Review students' answers to the Reflect prompts at the bottom of student pages 20 and 21.

Discuss the answers with the group to reinforce Week 2 concepts.

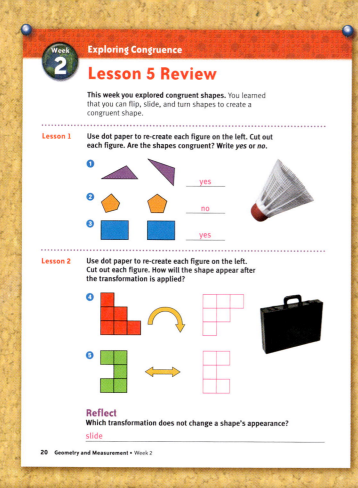

Lesson 5 Review

This week you explored congruent shapes. You learned that you can flip, slide, and turn shapes to create a congruent shape.

Lesson 1 Use dot paper to re-create each figure on the left. Cut out each figure. Are the shapes congruent? Write *yes* or *no*.

❶ yes

❷ no

❸ yes

Lesson 2 Use dot paper to re-create each figure on the left. Cut out each figure. How will the shape appear after the transformation is applied?

❹

❺

Reflect
Which transformation does not change a shape's appearance?

slide

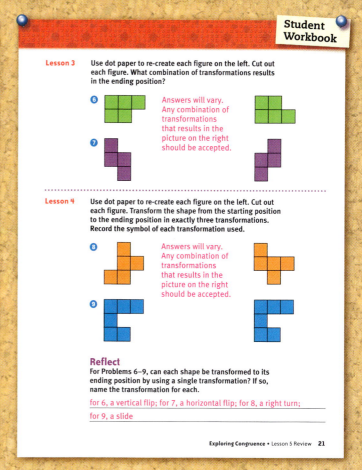

Lesson 3 Use dot paper to re-create each figure on the left. Cut out each figure. What combination of transformations results in the ending position?

❻
❼

Answers will vary. Any combination of transformations that results in the picture on the right should be accepted.

Lesson 4 Use dot paper to re-create each figure on the left. Cut out each figure. Transform the shape from the starting position to the ending position in exactly three transformations. Record the symbol of each transformation used.

❽
❾

Answers will vary. Any combination of transformations that results in the picture on the right should be accepted.

Reflect
For Problems 6–9, can each shape be transformed to its ending position by using a single transformation? If so, name the transformation for each.

for 6, a vertical flip; for 7, a horizontal flip; for 8, a right turn; for 9, a slide

4 Assess 10

A Gather Evidence

Formal Assessment

Have students complete the weekly test, *Assessment,* pages 58–59. Record progress on the Student Assessment Record, *Assessment,* page 100.

B Summarize Findings

Determine whether students have Minimal, Basic, or Secure understanding of the concepts presented in Week 2.

C Differentiate Instruction

On the basis of your observations, use these teaching strategies next week to follow up.

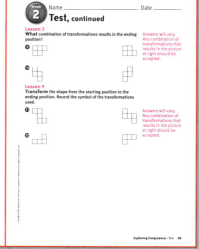

Assessment, pp. 58–59

Minimal Understanding
- Repeat the Warm-Up and Engage activities to develop Week 2 concepts.
- Use *Building Blocks* computer activities beginning with Piece Puzzler and *eMathTools* Shape Tool to develop and reinforce this week's concepts.

Basic Understanding
- Repeat Engage activities in subsequent weeks.
- Use *Building Blocks* computer activities beginning with Super Shape and *eMathTools* Shape Tool to reinforce Week 2 concepts.

Secure Understanding
- Use subsequent *Building Blocks* Super Shape and *eMathTools* Shape Tool.
- Use variations of the weekly Warm-Up and Engage activities.

Week 3 Measuring Area

Week at a Glance

This week, students continue with Geometry and Measurement, Level F, Measuring Area. Students will define area and determine the appropriate units used to measure area.

Math Background

To solve problems that involve area, students must understand the following:

- Area is an expression of how much surface is covered, not a measurement of length.
- Some shapes cover the plane more completely than others.
- The size of the unit used to designate how much surface is covered determines the number of units.
- Areas of regular and irregular shapes can be determined by counting square units or using formulas.

—Chapin, Suzanne, and Art Johnson, 2000. *Math Matters Grades K–6: Understanding the Math You Teach*. Math Solutions Publications. Pages 184–185.

How Students Learn

As students begin measuring area, they will have opportunities to look for patterns and relationships that lead to the informal discovery of formulas. As students use the strategy of counting squares and recording the information in tables, they will begin to see a pattern and realize that it is not necessary to count all the squares when you know the length and the width of the rectangle.

Teaching for Understanding

Observe closely while evaluating the assigned tasks this week to see whether students can demonstrate the following understandings:

Benchmark after Lesson 2: Students can identify the appropriate units of measurement for area.

Benchmark after Lesson 3: Students can calculate the area of a rectangle.

Benchmark after Lesson 4: Students can calculate the area of shapes consisting of multiple rectangles.

Math at Home

Give one copy of the Letter to Home, page A19, to each student. Encourage students to share and complete the activity with their caregivers.

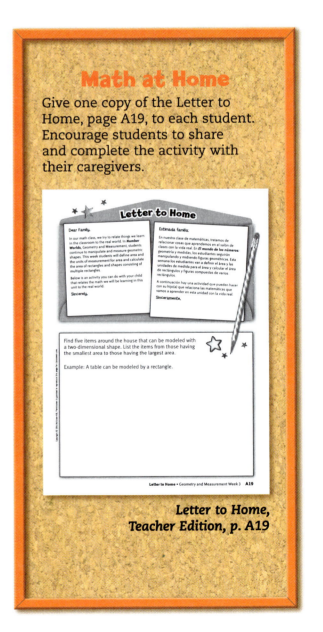

Letter to Home, Teacher Edition, p. A19

PACING	LESSON	LEARNING GOALS	NCTM	MATERIALS	TECHNOLOGY
DAY 1	pages 22–23	Students define area and its attributes.	• Geometry • Communication • Connections • Representation	• Scissors • Graph paper • Straightedge/ruler	**Building Blocks** Arrays in Area **MathTools** Shape Tool
DAY 2	pages 24–25	Students identify appropriate units of measurement for area.	• Geometry • Communication • Connections • Representation	• Rulers • Scissors	**Building Blocks** Arrays in Area **MathTools** Shape Tool
DAY 3	pages 26–27	Students find the area of a rectangle by using the standard formula.	• Geometry • Communication • Connections • Representation	• Index cards, 3 × 5 or 4 × 6 • Scissors • Inch ruler	**Building Blocks** Arrays in Area
DAY 4	pages 28–29	Students determine the area of irregular shapes consisting of several rectangles.	• Geometry • Communication • Connections • Representation	Graph paper	**Building Blocks** Arrays in Area
DAY 5	**Review and Assess** pages 30–31	Students review skills learned this week and complete the weekly assessment.	• Geometry • Communication • Connections • Representation	Materials will be selected from Lessons 1–4.	**Building Blocks** **MathTools** Review previous activities

Math Vocabulary

area The measure of a two-dimensional shape, in square units

rectangle A quadrilateral that has opposite parallel sides and all right angles

rectilinear shape A shape consisting of two or more rectangles

chunking Dividing a rectilinear shape into individual rectangles

English Learners

SPANISH COGNATES

English	Spanish
area	área
attributes	atributos
units	unidades
parallel	paralelo

ALTERNATE VOCABULARY

opposite Having attributes that are directly contrary to each other

irregular Not regular

Lesson 1

Objective

Students define area and its attributes.

Materials

Additional Materials
- Scissors
- Graph paper
- Straightedge/ruler

Access Vocabulary

area The measure of a two-dimensional shape in square units

rectangle A quadrilateral that has opposite parallel sides and all right angles

Creating Context

English Learners may benefit from a review of the abbreviations used for standard units of measure. Abbreviations usually rely on the consonant sounds and eliminate the vowels to shorten a word. Review the following: *inch* = in., *foot* = ft, *yard* = yd, and *mile* = mi.

---

1 Warm Up | 5

Concept Building UNDERSTANDING

Select a variety of items from the room.

- **How do we measure these objects?**

Students can brainstorm several ways to measure, such as length, width, height, weight, volume, and area.

- **Area is the measure of a two-dimensional shape. It is measured in square units.**
- **Name some items that have area.** Answers will vary.

---

2 Engage | 30

Skill Building

Have students measure and cut twenty or more one-inch squares from graph paper. Organize students in groups of three or four.

Using Squares COMPUTING

Have students select one of their notebooks or folders.

- **Look only at the cover. How can we find the area of the cover?** Answers will vary.
- **Place your paper squares side by side and end to end so that the squares cover the entire cover of the item. Count the number of squares needed to cover the item. This is the area of the cover in square inches.**

Rectangle Graphs APPLYING

- **Use graph paper to draw three rectangles of various sizes.**
- **Find the area of each rectangle by counting the number of squares inside the rectangle.**
- **In this case you are measuring area in square inches. What is the area of each of your rectangles?** Answers will vary but should be expressed in square inches.

Monitoring Student Progress

If . . . students are having trouble counting the number of squares,	Then . . . have them number each square counted inside the rectangle. The largest number is the area of the rectangle.

Building Blocks For additional practice finding area, students should complete **Building Blocks** Arrays in Area.

ⓔ MathTools Use the Shape Tool to demonstrate and explore geometric shapes.

Using Student Pages

Have students complete **Workbook,** pages 22–23, on their own.

Measuring Area

Lesson 1

Key Idea
Area is the size of a two-dimensional shape.
Area is measured in square units.

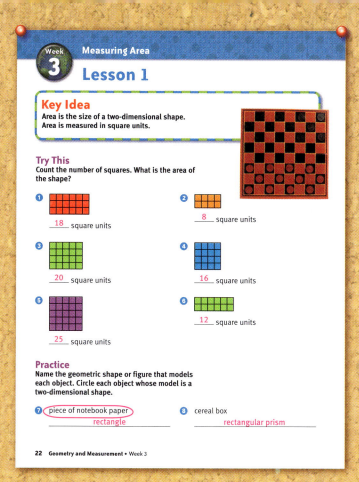

Try This
Count the number of squares. What is the area of the shape?

1 __18__ square units

2 __8__ square units

3 __20__ square units

4 __16__ square units

5 __25__ square units

6 __12__ square units

Practice
Name the geometric shape or figure that models each object. Circle each object whose model is a two-dimensional shape.

7 (piece of notebook paper)
__rectangle__

8 cereal box
__rectangular prism__

---

9 tissue box
__rectangular prism__

10 (bathroom floor)
__rectangle__

11 (a wall with wallpaper)
__rectangle__

12 (rug)
__rectangle__

Count the number of squares. Find the area of each shape.

13 __11__ square units

14 __26__ square units

15 __36__ square units

16 __29__ square units

Reflect
Name four things in your home that can be measured in square units.
Possible answers: rug, tile, floor, wallpaper,
carpet, picture, window

---

3 Reflect 10

Extended Response REASONING

Review students' answers to the Reflect prompt at the bottom of student page 23.

Discuss students' answers. Emphasize that area is the measure of two-dimensional shapes and that in the real world you compute the area of a surface.

- **Was this question easy or difficult?**
- **Are some items easier to measure in square units than other items? Why?**

Real-World Application APPLYING

A piece of graph paper shows its area because it is already marked in square units.

- **What is the area of a sheet of graph paper in square units?**
- **What is its perimeter?**
- **How is perimeter different from area?** Area is measured in square units.

---

4 Assess

Informal Assessment

Use the Student Assessment Record, *Assessment,* page 100, to record informal observations.

COMPUTING	APPLYING
Using Squares	**Rectangle Graphs**
Did the student	Did the student
❑ respond accurately?	❑ apply learning to new situations?
❑ respond quickly?	
❑ respond with confidence?	❑ contribute concepts?
❑ self-correct?	❑ contribute answers?
	❑ connect mathematics to the real world?

Lesson 2

Objective

Students identify appropriate units of measurement for area.

Materials

Additional Materials
- Rulers
- Scissors
- Graph paper

Access Vocabulary

unit of measurement Consistent, standard measurement in common use

Creating Context

Bring in an advertisement for carpet so that English Learners can see a real-world example of measurement in square feet. Have students measure an area of the classroom for carpet and figure the prices of different types of carpet. Which is cheapest? Which is most expensive?

1 Warm Up   5

Concept Building   UNDERSTANDING

Create a chart with seven columns on the board. Label the columns Square Inch, Square Foot, Square Yard, Square Mile, Square Centimeter, Square Meter, and Square Kilometer. Give students an idea of the size of these units.

■ **These are different units that can be used to measure area. The size of an object determines the appropriate unit of measurement for that object.**

2 Engage   30

Skill Building

Choosing Measurements   UNDERSTANDING

■ **Name some different objects having a surface area that can be measured.**

When students name an object, ask the class which square unit they think would be the most appropriate for measuring the item.

When the class has agreed on an appropriate unit of measure, write the name of the object in the appropriate column of the chart.

Some items may appear in more than one column. For example, a soccer field could be measured in square feet, square yards, or square meters.

Measuring Objects   APPLYING

Have students measure and cut several one-inch squares, called *square inches,* and one-foot squares, called *square feet,* from graph paper. Have them use these squares to measure items in the room. Have students record each object and its area on the board. Remind students to include the unit of measure used for each object.

Monitoring Student Progress

If . . . a student is having difficulty thinking of objects with surface area,

Then . . . suggest a few objects in the room, and have the student classify the specified object.

Building Blocks For additional practice finding area, students should complete **Building Blocks** Arrays in Area.

MathTools Use the Shape Tool to demonstrate and explore geometric shapes.

Using Student Pages

Have students complete **Workbook,** pages 24–25, on their own.

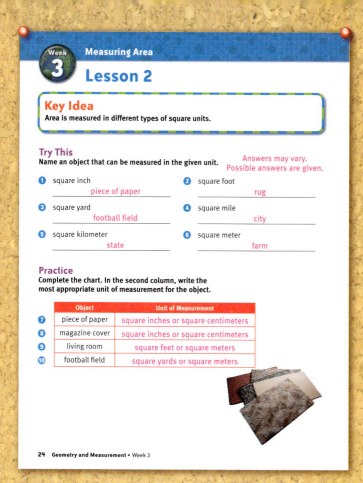

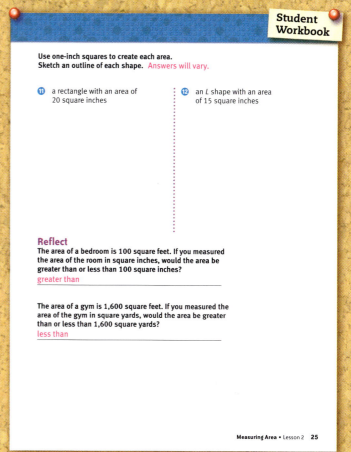

Student Workbook

Week 3 — Measuring Area
Lesson 2

Key Idea
Area is measured in different types of square units.

Try This
Name an object that can be measured in the given unit. Answers may vary. Possible answers are given.

1. square inch — piece of paper
2. square foot — rug
3. square yard — football field
4. square mile — city
5. square kilometer — state
6. square meter — farm

Practice
Complete the chart. In the second column, write the most appropriate unit of measurement for the object.

Object	Unit of Measurement
7. piece of paper	square inches or square centimeters
8. magazine cover	square inches or square centimeters
9. living room	square feet or square meters
10. football field	square yards or square meters

Use one-inch squares to create each area. Sketch an outline of each shape. Answers will vary.

11. a rectangle with an area of 20 square inches

12. an L shape with an area of 15 square inches

Reflect
The area of a bedroom is 100 square feet. If you measured the area of the room in square inches, would the area be greater than or less than 100 square inches?
greater than

The area of a gym is 1,600 square feet. If you measured the area of the gym in square yards, would the area be greater than or less than 1,600 square yards?
less than

---

3 — Reflect   10

Extended Response   REASONING
Review students' answers to the Reflect prompt at the bottom of student page 25.

- **How does the unit of measure used affect the numerical part of an area measurement?** Guide students to understand that the size of the unit used determines the number of units. Explain that if the same area is measured using small units and large units, the numerical part of the measurement using the small units is greater than the numerical part of the measurement using large units. Use a piece of paper that has an area of 1 square foot. Have students measure the paper in inches and find the area to be 144 square inches.

Real-World Application   APPLYING
Carpet is often sold by the square foot. Explore advertisements for carpeting. Students may notice that carpeting is sold by the square yard as well. Have students determine whether a square yard or a square foot is larger.

4 — Assess

Informal Assessment
Use the Student Assessment Record, *Assessment*, page 100, to record informal observations.

UNDERSTANDING	APPLYING
Choosing Measurements	**Measuring Objects**
Did the student	Did the student
❏ make important observations?	❏ apply learning to new situations?
❏ extend or generalize learning?	❏ contribute concepts?
❏ provide insightful answers?	❏ contribute answers?
❏ pose insightful questions?	❏ connect mathematics to the real world?

Lesson 3

Objective

Students find the area of a rectangle by using the standard formula.

Materials

Additional Materials
- Index cards, 3 × 5 or 4 × 6
- Scissors
- Inch ruler

Access Vocabulary

length The measure of how long an object is
width The measure of how wide an object is

Creating Context

Students who have lived outside the United States may be more familiar with the metric system. Help them see which customary units of measure are similar in size to the metric measures. If they know, for example, that a soccer field would be better measured by meters than by centimeters, they can understand that it is better measured by yards than by inches.

1 Warm Up • 5

Concept Building ENGAGING

- **How many one-inch squares do we need to cover the floor of this classroom?** Students may guess but will probably not answer correctly.

Guide students to determine that covering the floor with one-inch squares would not be an easy way to find the area of the classroom.

2 Engage • 30

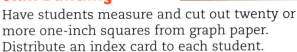

Skill Building

Have students measure and cut out twenty or more one-inch squares from graph paper. Distribute an index card to each student.

- **Find the area of the card by using the squares.** 15 square inches for a 3 × 5 card and 24 square inches for a 4 × 6 card

Have students measure the edges of the card, and discuss that the product of the length and width is the same as the number of square inches counted.

- **For larger areas, counting square units is not always possible or practical.**

Using the Area Formula APPLYING

Have students measure and calculate the area of rectangular objects around the classroom, or provide them with paper rectangles.

- **Measure the length and width of a rectangle by using a ruler.**
- **Use the formula** *Area of a Rectangle = length × width* **to find the area of the rectangle.**
- **Find the area of five other rectangles.**

Conclude this activity with the entire group working together to use the area formula to calculate the area of the classroom.

> ### Monitoring Student Progress
>
> **If . . .** a student is having difficulty finding area,
>
> **Then . . .** make sure students are not finding the perimeter instead.

 Building Blocks For additional practice finding area by using a formula, students should complete *Building Blocks* Arrays in Area.

Using Student Pages

Have students complete *Workbook,* pages 26–27, on their own.

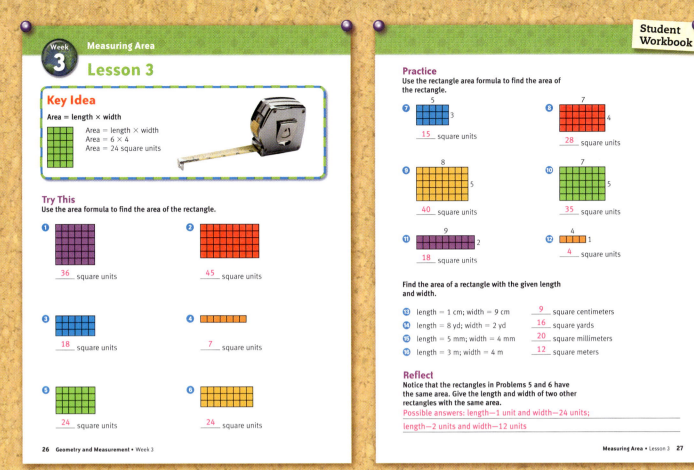

Lesson 3

Key Idea

Area = length × width

Area = length × width
Area = 6 × 4
Area = 24 square units

Try This

Use the area formula to find the area of the rectangle.

1 __36__ square units

2 __45__ square units

3 __18__ square units

4 __7__ square units

5 __24__ square units

6 __24__ square units

Practice

Use the rectangle area formula to find the area of the rectangle.

7 5 / 3 — __15__ square units

8 7 / 4 — __28__ square units

9 8 / 5 — __40__ square units

10 7 / 5 — __35__ square units

11 9 / 2 — __18__ square units

12 4 / 1 — __4__ square units

Find the area of a rectangle with the given length and width.

13 length = 1 cm; width = 9 cm   __9__ square centimeters

14 length = 8 yd; width = 2 yd   __16__ square yards

15 length = 5 mm; width = 4 mm   __20__ square millimeters

16 length = 3 m; width = 4 m   __12__ square meters

Reflect

Notice that the rectangles in Problems 5 and 6 have the same area. Give the length and width of two other rectangles with the same area.

Possible answers: length—1 unit and width—24 units; length—2 units and width—12 units

---

3 Reflect   10

Extended Response — REASONING

Review students' answers to the Reflect prompt at the bottom of student page 27.

Students may want to use 1-inch squares to construct rectangles. This may be helpful for visualizing the dimensions.

- If the rectangles have the same area, why do they look different?
- How would you explain this to someone who has never done this before?

Real-World Application — APPLYING

Area rugs come in many sizes. Locate area rugs on Internet shopping Web sites. Write the length and width of four different rugs. Find the area of the rugs, using the formula for area of a rectangle.

4 Assess

Informal Assessment

Use the Student Assessment Record, **Assessment,** page 100, to record informal observations.

APPLYING

Using the Area Formula

Did the student

- ❏ apply learning to new situations?
- ❏ contribute concepts?
- ❏ contribute answers?
- ❏ connect mathematics to the real world?

Measuring Area • Lesson 3   26–27

Lesson 4

Objective

Students determine the area of irregular shapes consisting of several rectangles.

Materials

Additional Materials
Graph paper

Access Vocabulary

irregular Not regular
chunking Dividing a shape into large units, or chunks

Creating Context

English Learners can accelerate their acquisition of English by breaking down compound words into their component meanings. In this lesson we talk about *footprint, handprint,* and *outline.* Look for other compound words in this book, and collect them on a list.

1 Warm Up   5

Concept Building   UNDERSTANDING

Draw an irregular rectilinear shape similar to those found on **Workbook,** page 28, on the overhead.

- **Can we find the area of this shape similar to those found on *Workbook,* page 28, by using the area of a rectangle formula?** no
- **We can divide this shape into smaller rectangles and find the area of each rectangle. This is called *chunking.***

Divide the shape into regular rectangles. Find the area of each smaller rectangle, and add the areas together.

- **When we add together the areas of the smaller rectangles that make up the larger shape, we get the area of the entire shape.**

2 Engage   30

Skill Building

Chunking Shapes   APPLYING

Distribute graph paper to each student.

- **Draw an irregular rectilinear shape on the graph paper.**
- **The shape you made can be divided into smaller rectangles.**
- **Divide the original shape into smaller rectangles.**

Allow students time to divide the shape.

- **Find the area of each rectangle.**
- **To find the area of the original shape, add together all the areas of the smaller rectangles that make up the shape.**

Instruct students to draw several more rectilinear shapes and follow the process again.

Remind students of the difference between *perimeter* and *shape* for each shape. Ask this question:

- **What is the perimeter of the rectangle?**

Monitoring Student Progress

If . . . a student is having trouble finding the area of the rectangles,

Then . . . have the student divide the shape into smaller rectangles with more manageable side lengths.

Building Blocks For additional practice with chunking, students should complete **Building Blocks** Arrays in Area.

Using Student Pages

Have students complete **Workbook,** pages 28–29, on their own.

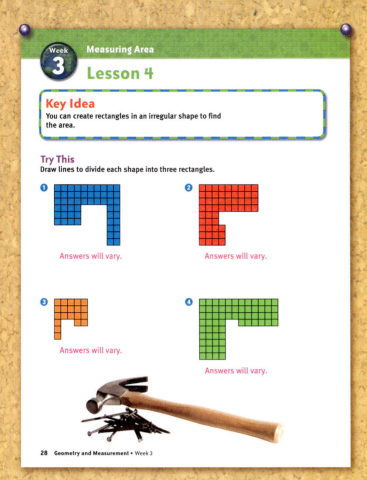

Lesson 4

Key Idea
You can create rectangles in an irregular shape to find the area.

Try This
Draw lines to divide each shape into three rectangles.

1 Answers will vary.

2 Answers will vary.

3 Answers will vary.

4 Answers will vary.

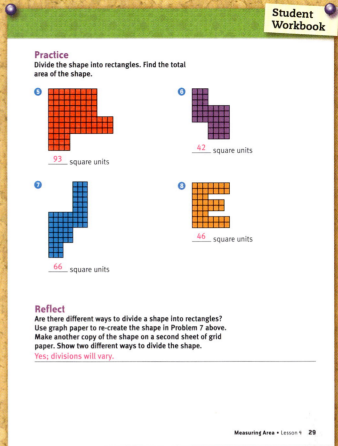

Practice
Divide the shape into rectangles. Find the total area of the shape.

5 93 square units

6 42 square units

7 66 square units

8 46 square units

Reflect
Are there different ways to divide a shape into rectangles? Use graph paper to re-create the shape in Problem 7 above. Make another copy of the shape on a second sheet of grid paper. Show two different ways to divide the shape.
Yes; divisions will vary.

3 Reflect 10

Extended Response REASONING
Review students' answers to the Reflect prompt at the bottom of student page 29.

■ **Did you come up with the same answer both times?**

Have students discuss the differences between their first and second attempts at chunking.

Real-World Application APPLYING
The layout of school hallways can resemble rectilinear shapes. Design a school hallway that is rectilinear. Label the hallway with its dimensions. Use chunking to divide the hallway into rectangles. Now find the area of the hallway. If time and availability allow, have students measure the school's hallway and compute the area.

4 Assess

Informal Assessment
Use the Student Assessment Record, *Assessment,* page 100, to record informal observations.

APPLYING

Chunking Shapes
Did the student
❏ apply learning to new situations?
❏ contribute concepts?
❏ contribute answers?
❏ connect mathematics to the real world?

Lesson 5 Review

Objective

Students review skills learned this week and complete the weekly assessment.

Materials

Review materials will be selected from those used in previous activities.

Creating Context

The reflect question on page 30 asks: *Which has a smaller area, a poster board or a postcard?* English Learners may not know these objects or what they are called in English. Bring in examples of each, or draw a picture to demonstrate them.

2 Engage   20

Skill Building

Free-Choice Activity ENGAGING

For the last day of the week, allow students to choose an activity from the previous lessons. Some activities they may choose include the following:

- Using Squares
- Rectangle Graphs
- Choosing Measurements
- Measuring Objects
- Using the Area Formula
- Chunking Shapes

Make a note of the activities children select. Do they prefer easy or challenging activities? If you believe your students would benefit from extra practice on specific facts, choose an activity for them.

Monitoring Student Progress

If . . . students need more practice with using the formula and finding areas of rectilinears,

Then . . . they should choose Areas of Chunks for their Free-Choice activity and should use the formula to find the area of each smaller rectangle.

Using Student Pages

Have students complete *Workbook,* pages 30–31, on their own.

1 Warm Up   5

Concept Building COMPUTING

Instruct students to draw several rectangles and rectilinear shapes on graph paper.

- **Compute the area of each figure.**

3 Reflect   10

Extended Response REASONING

Review students' answers to the Reflect prompts at the bottom of student pages 30 and 31.

Discuss the answers with the group to reinforce Week 3 concepts.

Week 3 — Measuring Area

Lesson 5 Review

Lesson 1 Count the number of squares. What is the area of the shape?

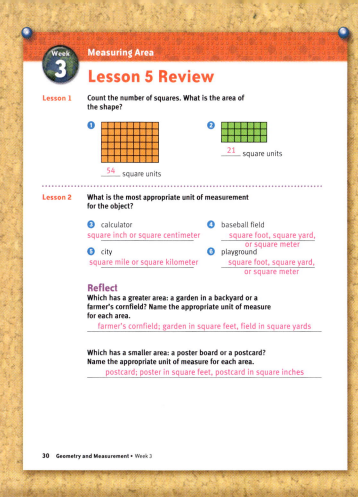

① _____54_____ square units

② _____21_____ square units

Lesson 2 What is the most appropriate unit of measurement for the object?

③ calculator
square inch or square centimeter

④ baseball field
square foot, square yard, or square meter

⑤ city
square mile or square kilometer

⑥ playground
square foot, square yard, or square meter

Reflect
Which has a greater area: a garden in a backyard or a farmer's cornfield? Name the appropriate unit of measure for each area.
farmer's cornfield; garden in square feet, field in square yards

Which has a smaller area: a poster board or a postcard? Name the appropriate unit of measure for each area.
postcard; poster in square feet, postcard in square inches

Lesson 3 Find the area of a rectangle with the given length and width.

⑦ length = 10 in.; width = 6 in.   ___60___ square inches

⑧ length = 5 cm; width = 3 cm   ___15___ square centimeters

⑨ length = 5 miles; width = 6 miles   ___30___ square miles

⑩ length = 8 km; width = 7 km   ___56___ square kilometers

Lesson 4 Divide the shape into rectangles. Find the total area of the shape.

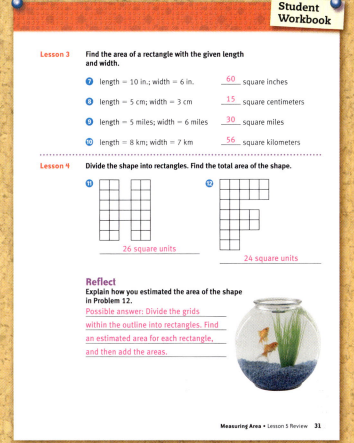

⑪ 26 square units

⑫ 24 square units

Reflect
Explain how you estimated the area of the shape in Problem 12.
Possible answer: Divide the grids within the outline into rectangles. Find an estimated area for each rectangle, and then add the areas.

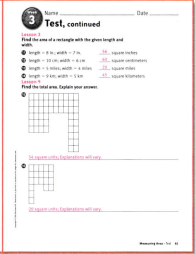

---

4 Assess   10

A Gather Evidence

Formal Assessment

Have students complete the weekly test on *Assessment,* pages 60–61. Record progress on the Student Assessment Record, *Assessment,* page 100.

B Summarize Findings

Determine whether students have Minimal, Basic, or Secure understanding of the concepts presented in Week 3.

C Differentiate Instruction

On the basis of your observations, use these teaching strategies next week to follow up.

Minimal Understanding
- Repeat the Warm-Up and Engage activities to develop Week 3 concepts.
- Use *Building Blocks* computer activities beginning with Arrays in Area to develop and reinforce this week's concepts.

Basic Understanding
- Repeat Engage activities in subsequent weeks.
- Use *eMathTools* Shape Tool to reinforce Week 3 concepts.

Secure Understanding
- Use *Building Blocks* Arrays in Area.
- Use variations of the weekly Warm-Up and Engage activities, using larger areas and multiple steps.

Assessment, pp. 60–61

Measuring Area • Lesson 5 Review   30–31

Measurement Conversions

Week at a Glance

This week, students continue with Geometry and Measurement, Level F, Measurement Conversions, by investigating the various types of conversions. Students should be able to convert measurements for length, weight, volume, and time by the end of the week.

Math Background

Measurement is a skill that spans the course of schooling. Students need to go beyond learning procedures to learn the mathematical relationships among units of measurement. Developing knowledge of effective procedures is a form of conceptual development, and developing concepts is helped by constructing and reflecting on ways to measure and applications of measurement.

—Kilpatrick, J., W.G. Martin, and D. Schifter (Eds.). 2003. *A Research Companion to Principles and Standards for School Mathematics.* National Council of Teachers of Mathematics. Page 190.

How Students Learn

Teachers should emphasize the standard units that are used in the United States (customary units). Students should become familiar with these and should form mental images to associate with each unit. Being able to express measurements in equivalent forms is essential to problem solving in the real world. Students need to use their knowledge of relationships between units and their understanding of the situation to make conversions.

Teaching for Understanding

Observe closely while evaluating the assigned tasks this week to see whether students can demonstrate the following understandings:

Benchmark after Lesson 2: Students can convert measurements for length and weight.

Benchmark after Lesson 3: Students can convert measurements for volume.

Benchmark after Lesson 4: Students can convert measurements for time.

Skills Focus

- Develop measurement sense for conversions
- Create and use representations to organize, record, and communicate mathematical ideas

Math at Home

Give one copy of the Letter to Home, page A20, to each student. Encourage students to share and complete the activity with their caregivers.

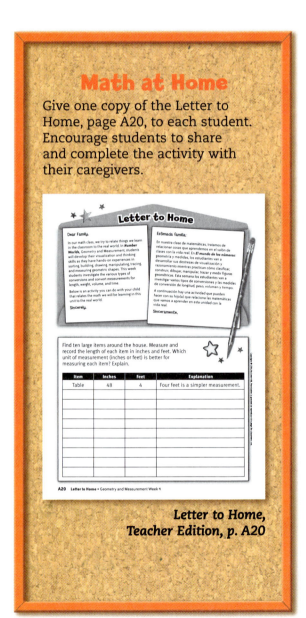

Letter to Home, Teacher Edition, p. A20

Week 4 Planner   Measurement Conversions

PACING	LESSON	LEARNING GOALS	NCTM	MATERIALS	TECHNOLOGY
DAY 1	pages 32–33	Students convert standard measurements for length.	• Geometry • Communication • Connections • Representation	Graph paper	**MathTools** Metric/Customary Conversion
DAY 2	pages 34–35	Students convert standard measurements for weight.	• Geometry • Communication • Connections • Representation	Graph paper	**MathTools** Metric/Customary Conversion
DAY 3	pages 36–37	Students convert standard measurements for volume.	• Geometry • Communication • Connections • Representation	Graph paper	**MathTools** Metric/Customary Conversion
DAY 4	pages 38–39	Students convert standard measurements for time.	• Geometry • Communication • Connections • Representation	• Graph paper • Calculator	**MathTools** Metric/Customary Conversion
DAY 5	**Review and Assess** pages 40–41	Students review skills learned this week and complete the weekly assessment.	• Geometry • Communication • Connections • Representation	Materials will be selected from Lessons 1–4.	**MathTools** Review previous activities

Math Vocabulary

length The distance between two points

weight The mass of a three-dimensional object

capacity The amount of matter held by a three-dimensional object

time The measurement that expresses the duration of events

English Learners

SPANISH COGNATES

English	Spanish
conversion	conversión
hour	hora
minutes	minutos
scale	escala
seconds	segundos

ALTERNATE VOCABULARY

convert To change from one standard to another, such as inches to feet or inches to centimeters

Lesson 1

Objective

Students convert standard measurements for length.

Materials

Additional Materials
Graph paper

Access Vocabulary

length The distance between two points
convert To change from one standard to another, such as inches to feet or inches to centimeters

Creating Context

English Learners who have attended school outside the United States may be more familiar with the metric system than the customary system of measurement. Ask students to determine their height in both systems. Have them use the measurement to compare the size of the units. Remind English Learners that the correct way to ask about height is to ask, *How tall are you?* not *How high are you?*

1 Warm Up 5

Concept Building UNDERSTANDING

Review basic units of length. If possible, reteach by using manipulatives.

- **Which is longer—a foot or an inch?** foot
- **How many inches are in a foot?** Twelve inches are in 1 foot.
- **Which is longer—a foot or a yard?** yard
- **How many feet are in a yard?** Three feet are in a yard.
- **We are going to use pictures to help us convert inches, feet, and yards.**

2 Engage 30

Skill Building

Display the following chart on the board.

Unit	Equivalent Measurements					
Inches	12	24	36	48	60	72
Feet	1	2	3	4	5	6
Yards			1			2

Charting Length COMPUTING

Have students copy and complete the chart at the top of a sheet of graph paper.

- **Outline twelve squares horizontally on your paper. Number them 1–12.**
- **Below this, write** 12 *inches* = 1 *foot.*
- **Outline twenty-four squares horizontally on your paper. Number them 1–24.**
- **Below this, write** 24 *inches* = 2 *feet.*
- **Outline thirty-six squares horizontally on your paper. Number them 1–36.**
- **Below this, write** 36 *inches* = 3 *feet,* 3 *feet* = 1 *yard,* and 36 *inches* = 1 *yard.*

Strategy Building

Length in Action APPLYING

Present the following problem to students.

- **Caitlin uses plastic string to make bead bracelets. She uses one foot of string for each bracelet. How many bracelets can she make with 72 inches of string?** 6 bracelets
- **How many feet of string did Caitlin have? How many yards?** 6 feet; 2 yards

Monitoring Student Progress

If . . . students notice a pattern,	**Then . . .** allow them to use repeated addition or multiplication to complete the charts.

 MathTools For additional practice with measurements, use *eMathTools* Metric/Customary Conversion.

Using Student Pages

Have students complete *Workbook,* pages 32–33, on their own.

Lesson 1

Key Idea
1 foot = 12 inches
1 yard = 3 feet

Try This
Fill in the blanks to convert each measurement.

1. 2 feet = 1 foot + 1 foot = __12__ inches + __12__ inches = __24__ inches

2. 4 feet = __1__ foot + __1__ foot + __1__ foot + __1__ foot = __12__ inches + __12__ inches + __12__ inches + __12__ inches = __48__ inches

3. 12 inches = __1__ foot

4. 36 inches = 12 inches + 12 inches + 12 inches = __1__ foot + __1__ foot + __1__ foot = __3__ feet

5. 18 feet = __3__ feet + __3__ feet + __3__ feet + __3__ feet + __3__ feet = __1__ yard + __1__ yard + __1__ yard + __1__ yard + __1__ yard + __1__ yard = __6__ yards

Practice
Convert each measurement.

6. 27 feet = __9__ yards
7. 6 feet = __2__ yards
8. 2 yards = __6__ feet
9. 24 inches = __2__ feet
10. 3 feet = __36__ inches
11. 30 feet = __10__ yards
12. 9 feet = __108__ inches
13. 96 inches = __8__ feet

Draw a picture to represent the problem. Solve the problem.

14. Sarah is 5 feet tall. How many inches tall is Sarah?
    60 inches

15. A toy rocket is 2 feet long. How many inches long is the rocket?
    24 inches

16. The length of a basketball court is 84 feet. How many yards long is the court?
    28 yards

17. The gondola of a hot air balloon is 10 feet long. How long is the gondola in inches?
    120 inches

18. Mr. Hernandez is 6 feet tall. How tall in inches is Mr. Hernandez?
    72 inches

19. A room is 360 inches wide. How wide in yards is the room?
    10 yards

Reflect
On a regulation baseball field, the distance between bases is 90 feet. How many inches are between bases? How many yards are between bases?
1,080 inches; 30 yards

---

3 Reflect 10 ▶

Extended Response REASONING
Review students' answers to the Reflect prompt at the bottom of student page 33.

■ **Why do we sometimes measure in inches and sometimes in feet or yards?**

Discuss that different units are appropriate depending on the size of the object being measured.

Real-World Application APPLYING
Trees can grow to be very tall. However, they all begin very small. They may be measured in inches initially but later may be measured in feet.

■ **Are the trees in the school yard measured in inches or feet?**

---

4 Assess

Informal Assessment
Use the Student Assessment Record, *Assessment,* page 100, to record informal observations.

COMPUTING	APPLYING
Charting Length	**Length in Action**
Did the student	Did the student
❏ respond accurately?	❏ apply learning to new situations?
❏ respond quickly?	❏ contribute concepts?
❏ respond with confidence?	❏ contribute answers?
❏ self-correct?	❏ connect mathematics to the real world?

Lesson 2

Objective

Students convert standard measurements for weight.

Materials

Additional Materials
Graph paper

Access Vocabulary

weight The measure of the heaviness of an object

scale A tool used to measure the weight of a person or object

Creating Context

In many places, a person's weight is considered personal information. Remind English Learners that not all people like to be asked about their weight.

1 Warm Up   5

Concept Building   COMPUTING

Review basic units of weight. If possible, reteach by using manipulatives.

- **Which weighs more, a pound or an ounce?** pound
- **How many ounces are in a pound?** 16 ounces are in 1 pound.
- **We are going to use pictures to help us convert between pounds and ounces.**

2 Engage   30

Skill Building

Display the following chart on the board.

Unit	Equivalent Measurements				
Ounces	16	32	48	64	80
Pounds	1	2	3	4	5

Charting Weight   COMPUTING

Have students copy and complete the chart at the top of a sheet of graph paper.

- **Outline sixteen squares horizontally on your paper. Number them 1–16.**
- **Below this, write** *16 ounces = 1 pound.*
- **Outline thirty-two squares horizontally on your paper. Number them 1–32.**
- **Below this, write** *32 ounces = 2 pounds.*

Continue this process for three, four, and five pounds.

Skill Building

Weighing the Issue   APPLYING

Present the following problem to students.

- **Sandra decided to recycle her old magazines. Her magazines weighed 9 pounds altogether. How many ounces of magazines did Sandra recycle? Draw a picture to solve the problem.** 144 ounces
- **Sandra then opened and heated a can of beef stew weighing 16 ounces. How many pounds of stew did she heat?** 1 pound

Monitoring Student Progress

| **If . . .** a student is having trouble circling groups of graph squares, | **Then . . .** suggest that the student use different-colored markers or crayons to mark sets of sixteen. |

 MathTools For additional practice converting weight, use *eMathTools* Metric/Customary Conversion.

Using Student Pages

Have students complete **Workbook,** pages 34–35, on their own.

Lesson 2

Key Idea
1 pound = 16 ounces

Try This
Fill in the blanks to convert each measurement.

1. 4 pounds = 1 pound + 1 pound + 1 pound + 1 pound = __16__ ounces + __16__ ounces + __16__ ounces + __16__ ounces = __64__ ounces

2. 8 pounds = 4 pounds + __4__ pounds = __64__ ounces + __64__ ounces = __128__ ounces

3. 16 ounces = __1__ pound

4. 32 ounces = 16 ounces + __16__ ounces = __1__ pound + __1__ pound = __2__ pounds

Convert each measurement.

5. 144 ounces = __9__ pounds
6. 2 pounds = __32__ ounces
7. 10 pounds = __160__ ounces
8. 48 ounces = __3__ pounds
9. 80 ounces = __5__ pounds
10. 64 ounces = __4__ pounds

Practice
Draw a picture to represent the problem. Solve the problem.

11. A baby weighs 9 pounds. How many ounces does the baby weigh?
   __144 ounces__

12. A bag of sugar weighs 2 pounds. How many ounces does the sugar weigh?
   __32 ounces__

13. A dog weighs 25 pounds. How many ounces does the dog weigh?
   __400 ounces__

14. A can of baked beans weighs 16 ounces. How many pounds does the can weigh?
   __1 pound__

Reflect
Andy says that a bag of dog food that weighs 160 ounces is heavier than a bag of dog food that weighs 11 pounds. Do you agree or disagree? Explain.

Disagree; 160 ounces is equal to 10 pounds. Because 10 pounds is less than 11 pounds, Andy's statement is incorrect.

---

3 Reflect 10

Extended Response REASONING

Review students' answers to the Reflect prompt at the bottom of student page 35.
Draw a picture to represent the Reflect prompt.

- **Why would a company list the weight of products in pounds instead of ounces?** to avoid using large numbers

Real-World Application APPLYING

- **Many food items are sold by their weight. Explore your kitchen. Find five food items that have their weights listed. Write the name of the item and its weight on graph paper. Draw a picture to represent the weights of the food items.**

---

4 Assess

Informal Assessment

Use the Student Assessment Record, **Assessment,** page 100, to record informal observations.

COMPUTING	APPLYING
Charting Weight	**Weighing the Issue**
Did the student	Did the student
❏ respond accurately?	❏ apply learning to new situations?
❏ respond quickly?	❏ contribute concepts?
❏ respond with confidence?	❏ contribute answers?
❏ self-correct?	❏ connect mathematics to the real world?

Measurement Conversions • Lesson 2   34–35

Lesson 3

Objective

Students convert standard measurements for volume.

Materials

Additional Materials
Graph paper

Access Vocabulary

capacity The amount of matter that a three-dimensional object can hold

Creating Context

If you were having a large party with 100 people, would you rather buy punch in gallons or pints? Is there an example of buying drinks for 100 people in which it is better to purchase in pints?

1 Warm Up 5

Concept Building COMPUTING

Review basic units of capacity. If possible, reteach by using manipulatives.

- **Which has a larger capacity, a quart or a pint?** quart
- **How many pints are in a quart?** two
- **Which has a larger capacity, a quart or a gallon?** gallon
- **How many quarts are in a gallon?** four
- **We are going to use pictures to help us convert pints, quarts, and gallons.**

2 Engage 30

Skill Building

Display the following chart on the board.

Unit	Equivalent Measurements							
Pint	1	2	3	4	5	6	7	8
Quart		1		2		3		4
Gallon								1

Charting Capacity COMPUTING

Have students copy and complete the chart at the top of a sheet of graph paper.

- **Outline two squares horizontally on your paper. Number them 1–2.**
- **Below this, write** *2 pints = 1 quart.*
- **Outline 8 squares horizontally on your paper. Number them 1–8.**
- **Below this, write** *8 pints = 4 quarts = 1 gallon.*

Strategy Building

Present the following problem to students.

Lemonade Stand APPLYING

- **Reggie is running a lemonade stand. A quart serves 5 people. He has 3 gallons of lemonade. How many people can he serve with this amount of lemonade?** 60 people
- **Reggie had one gallon of lemonade left over when he was done. How many pints did he have left? Draw a picture to solve the problem.** 8 pints

Monitoring Student Progress

If . . . students cannot grasp the concept of capacity,

Then . . . provide samples of a pint, a quart, and a gallon. Students should then be able to relate to a container they see regularly.

 MathTools For additional practice converting capacity, students should complete *eMathTools* Metric/Customary Conversion.

Using Student Pages

Have students complete **Workbook,** pages 36–37, on their own.

Lesson 3

Key Idea
1 quart = 2 pints
1 gallon = 4 quarts

Try This
Fill in the blanks to convert each measurement.

1 4 quarts = 1 quart + 1 quart + 1 quart + 1 quart = __2__ pints + __2__ pints + __2__ pints + __2__ pints = __8__ pints

2 6 pints = 2 pints + __2__ pints + __2__ pints = __1__ quart + __1__ quart + __1__ quart = __3__ quarts

3 5 gallons = __1__ gallon + __1__ gallon + __1__ gallon + __1__ gallon + __1__ gallon = __4__ quarts + __4__ quarts + __4__ quarts + __4__ quarts + __4__ quarts = __20__ quarts

4 8 quarts = __4__ quarts + __4__ quarts = __1__ gallon + __1__ gallon = __2__ gallons

Convert each measurement.

5 6 pints = __3__ quarts

6 18 pints = __9__ quarts

7 3 gallons = __12__ quarts

8 9 gallons = __36__ quarts

9 7 quarts = __14__ pints

10 16 quarts = __4__ gallons

Practice
Draw a picture to represent the problem. Solve the problem.

11 Mr. Diego purchased 9 quarts of lemonade. He gives each student in his class 1 pint of lemonade. How many students are in his class?
__18 students__

12 Rebecca has 5 gallons of orange juice. How many quarts of orange juice does she have?
__20 quarts__

13 The school cafeteria sells 240 pints of milk every day. How many gallons of milk are sold each day?
__30 gallons__

14 Wilson drank 1 gallon of sport drink throughout the marathon. How many pints of sport drink did he drink?
__8 pints__

Reflect
What is the relationship between pints and gallons?
Draw a picture to support your equation.
1 gallon = 8 pints; pictures will vary.

---

3 Reflect 10

Extended Response REASONING

Review students' answers to the Reflect prompt at the bottom of student page 37.

- **How did you figure out how to answer this question?** Discuss students' methods of determining the relationship between gallons and pints. Discuss how converting between pints and gallons might be beneficial.

Real-World Application APPLYING

Milk is often sold in pints, quarts, and gallons. Discuss why milk is sold in different sizes. Explore other beverages, such as lemonade and orange juice. Are they sold in pints, quarts, or gallons?

---

4 Assess

Informal Assessment

Use the Student Assessment Record, **Assessment,** page 100, to record informal observations.

COMPUTING	APPLYING
Charting Capacity	**Lemonade Stand**
Did the student	Did the student
❏ respond accurately?	❏ apply learning to new situations?
❏ respond quickly?	❏ contribute concepts?
❏ respond with confidence?	❏ contribute answers?
❏ self-correct?	❏ connect mathematics to the real world?

Lesson 4

Objective

Students convert standard measurements for time.

Materials

Additional Materials
- Graph paper
- Calculator

Access Vocabulary

time Measurement used to express the duration of events

Creating Context

Are measurements of time the same in countries that use the metric system as in those that use the customary system of measurement? The military uses a 24-hour clock instead of the 12-hour clock. What time is it when an army captain says "1900 hours"?

1 Warm Up 5

Concept Building COMPUTING

Review basic units of time. If possible, reteach by using manipulatives.

- **How many seconds are in one minute?** 60
- **How many minutes are in one hour?** 60
- **Which is longer, a week or a day?** a week
- **How many days are in a week?** 7
- **There are 365 days in one year. We are going to use this information to explore converting time.**

2 Engage 30

Display the following chart on the board.

60 seconds	60 minutes	24 hours	7 days	365 days
1 minute	1 hour	1 day	1 week	1 year

Strategy Building

Time of Day APPLYING

- **How many seconds are in an hour? Use the chart to help determine the number of seconds in an hour. You might need to do several conversions to find the answer. Draw a picture to solve the problem.** 360 seconds; Students can use graph squares to represent minutes. Because the numbers will be large, a calculator may be necessary. Guide students through the conversions if necessary.
- **How many hours are in one week? Draw a picture to solve the problem.** 168 hours

Skill Building

A Moment in Time COMPUTING

- **Liam has 9 weeks until his 10th birthday. How many days until Liam turns 10? Draw a picture to help solve the problem.** 63 days
- **Patrice must wait 42 days for a rebate on her DVD player. How many weeks must she wait for the rebate? Draw a picture to help solve the problem.** 6 weeks

Monitoring Student Progress

If . . . students are having difficulty counting the number of graph squares,

Then . . . suggest that they allow each square to represent more than one minute.

 MathTools For additional practice converting time, students should use *eMathTools* Metric/Customary Conversion.

Using Student Pages

Have students complete *Workbook,* pages 38–39, on their own.

Lesson 4

Key Idea
1 minute = 60 seconds
1 hour = 60 minutes
1 day = 24 hours
1 week = 7 days
1 year = 365 days

Try This
Fill in the blanks to convert each measurement.

1. 3 minutes = 1 minute + 1 minute + 1 minute = __60__ seconds + __60__ seconds + __60__ seconds = __180__ seconds

2. 2 hours = __1__ hour + __1__ hour = __60__ minutes + __60__ minutes = __120__ minutes

3. 4 days = __1__ day + __1__ day + __1__ day + __1__ day = __24__ hours + __24__ hours + __24__ hours + __24__ hours = __96__ hours

4. 2 weeks = __1__ week + __1__ week = __7__ days + __7__ days = __14__ days

Convert each measurement.

5. 5 days = __120__ hours
6. 2 years = __730__ days
7. 6 minutes = __360__ seconds
8. 168 hours = __7__ days
9. 9 weeks = __63__ days
10. 21 days = __3__ weeks
11. 1,095 days = __3__ years
12. 240 minutes = __4__ hours
13. 11 hours = __660__ minutes
14. 480 seconds = __8__ minutes

Practice
Draw a picture to represent the problem. Solve the problem.

15. Sally receives her water bill every 56 days. How many weeks are there between water bills?
8 weeks

16. Tasha was able to run a mile in 8 minutes. How many seconds did it take Tasha to run a mile?
480 seconds

17. In 4 years, Len will be able to get his driver's license. How many days until he can get his license?
1,460 days

18. At Brookforest Elementary School, students receive a 7-week summer break. How many days long is the summer break?
49 days

19. The instructions for a model airplane state that it takes 9 hours to complete the model kit. How many minutes is this?
540 minutes

20. Today is Parker's birthday. In 730 days, he will celebrate his tenth birthday. How old is Parker today?
8 years old

Reflect
Find the number of hours in a day. Find the number of minutes in a day. Find the number of seconds in a day.
24 hours; 1,440 minutes; 86,400 seconds

---

3 Reflect 10

Extended Response APPLYING

Review students' answers to the Reflect prompt at the bottom of student page 39.
Have students refer back to the first activity. Their charts should contain the information they need to calculate.

- **Are there other strategies that could be used to solve this problem?**
- **How can you check your answer?**

Real-World Application REASONING

- **Marathons are often measured in hours, minutes, and seconds. Find the results of a recent marathon on the Internet. Convert the winning time into seconds.**

---

4 Assess

Informal Assessment

Use the Student Assessment Record, *Assessment,* page 100, to record informal observations.

APPLYING	COMPUTING
Time of Day	**A Moment in Time**
Did the student	Did the student
❑ apply learning to new situations?	❑ respond accurately?
❑ contribute concepts?	❑ respond quickly?
❑ contribute answers?	❑ respond with confidence?
❑ connect mathematics to the real world?	❑ self-correct?

Measurement Conversions • Lesson 4 38–39

Lesson 5 Review

Objective

Students review skills learned this week and complete the weekly assessment.

Materials

Review materials will be selected from those used in previous activities.

Creating Context

English Learners who have studied outside the United States may know the metric system best. Help make a conversion chart to show which measures in the customary system of measurement are closest to the units of metric measurement. When working with the metric system, be sure to call on these students as experts who can help other students learn the units of measure and their conversions.

✓ Unit Assessment

Students should complete the Geometry and Measurement Test found on *Assessment,* pages 80–81. Use the key, *Assessment,* page 99, to identify incorrect responses. Reteach and review the Warm-Up and Engage activities to reinforce concept understanding.

1 Warm Up — 5

Concept Building COMPUTING

- Write five units of measure.
- Write an equivalent form for each measure.
- Look through the charts you made this week to check your work.

2 Engage — 20

Skill Building

Free-Choice Activity ENGAGING

For the last day of the week, allow students to choose an activity from the previous lessons. Some activities they may choose include the following:

- **Charting Length**
- **Length in Action**
- **Charting Weight**
- **Weighing the Issue**
- **Charting Capacity**
- **Lemonade Stand**
- **Time of Day**
- **A Moment in Time**

Make a note of the activities children select. Do they prefer easy or challenging activities? If you believe your students would benefit from extra practice on specific conversions, choose an activity for them.

Monitoring Student Progress

If . . . students need more practice with solving problems,

Then . . . provide alternative problems to replace the second activity in each day's lesson.

Using Student Pages

Have students complete **Workbook,** pages 40–41, on their own.

3 Reflect — 10

Extended Response REASONING

Review students' answers to the Reflect prompts at the bottom of student pages 40 and 41.

Discuss the answers with the group to reinforce Week 4 concepts.

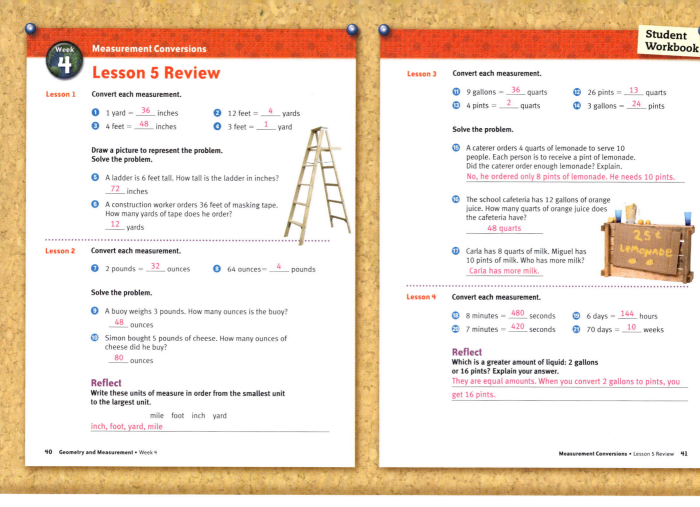

Week 4 — Measurement Conversions

Lesson 5 Review

Lesson 1

Convert each measurement.

1. 1 yard = 36 inches
2. 12 feet = 4 yards
3. 4 feet = 48 inches
4. 3 feet = 1 yard

Draw a picture to represent the problem. Solve the problem.

5. A ladder is 6 feet tall. How tall is the ladder in inches?
72 inches
6. A construction worker orders 36 feet of masking tape. How many yards of tape does he order?
12 yards

Lesson 2

Convert each measurement.

7. 2 pounds = 32 ounces
8. 64 ounces = 4 pounds

Solve the problem.

9. A buoy weighs 3 pounds. How many ounces is the buoy?
48 ounces
10. Simon bought 5 pounds of cheese. How many ounces of cheese did he buy?
80 ounces

Reflect

Write these units of measure in order from the smallest unit to the largest unit.

mile   foot   inch   yard

inch, foot, yard, mile

40   Geometry and Measurement • Week 4

Lesson 3

Convert each measurement.

11. 9 gallons = 36 quarts
12. 26 pints = 13 quarts
13. 4 pints = 2 quarts
14. 3 gallons = 24 pints

Solve the problem.

15. A caterer orders 4 quarts of lemonade to serve 10 people. Each person is to receive a pint of lemonade. Did the caterer order enough lemonade? Explain.
No, he ordered only 8 pints of lemonade. He needs 10 pints.

16. The school cafeteria has 12 gallons of orange juice. How many quarts of orange juice does the cafeteria have?
48 quarts

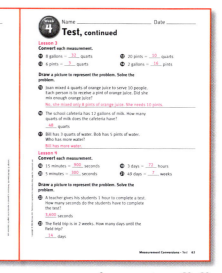

17. Carla has 8 quarts of milk. Miguel has 10 pints of milk. Who has more milk?
Carla has more milk.

Lesson 4

Convert each measurement.

18. 8 minutes = 480 seconds
19. 6 days = 144 hours
20. 7 minutes = 420 seconds
21. 70 days = 10 weeks

Reflect

Which is a greater amount of liquid: 2 gallons or 16 pints? Explain your answer.
They are equal amounts. When you convert 2 gallons to pints, you get 16 pints.

Measurement Conversions • Lesson 5 Review   41

---

4 Assess 10

A Gather Evidence

Formal Assessment

Have students complete the weekly test, *Assessment,* pages 62–63. Record progress on the Student Assessment Record, *Assessment,* page 100.

B Summarize Findings

Determine whether students have Minimal, Basic, or Secure understanding of the concepts presented in Week 4.

C Differentiate Instruction

On the basis of your observations, use these teaching strategies next week to follow up.

Minimal Understanding
- Repeat the Warm-Up and Engage activities to develop Week 4 concepts.
- Use *eMathTools* Metric/Customary Conversion to develop and reinforce this week's concepts.

Basic Understanding
- Repeat Engage activities in subsequent weeks.
- Use *eMathTools* Metric/Customary Conversion to reinforce Week 4 concepts.

Secure Understanding
- Use *eMathTools* Metric/Customary Conversion.
- Use variations of the weekly Warm-Up and Engage activities, using higher numbers and multiple steps.

Assessment, pp. 62–63

Measurement Conversions • Lesson 5 Review   40–41

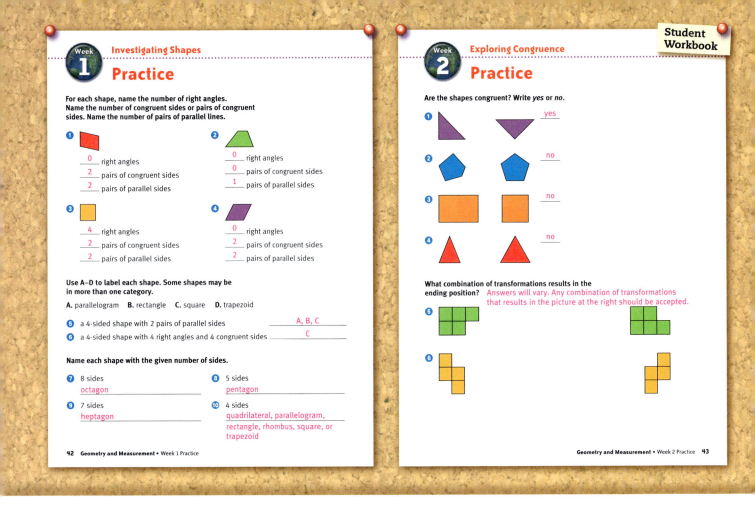

Week 1 — Investigating Shapes

Practice

For each shape, name the number of right angles. Name the number of congruent sides or pairs of congruent sides. Name the number of pairs of parallel lines.

1
___0___ right angles
___2___ pairs of congruent sides
___2___ pairs of parallel sides

2
___0___ right angles
___0___ pairs of congruent sides
___1___ pairs of parallel sides

3
___4___ right angles
___2___ pairs of congruent sides
___2___ pairs of parallel sides

4
___0___ right angles
___2___ pairs of congruent sides
___2___ pairs of parallel sides

Use A–D to label each shape. Some shapes may be in more than one category.

A. parallelogram   **B.** rectangle   **C.** square   **D.** trapezoid

5 a 4-sided shape with 2 pairs of parallel sides   ___A, B, C___

6 a 4-sided shape with 4 right angles and 4 congruent sides   ___C___

Name each shape with the given number of sides.

7 8 sides
octagon

8 5 sides
pentagon

9 7 sides
heptagon

10 4 sides
quadrilateral, parallelogram, rectangle, rhombus, square, or trapezoid

Week 2 — Exploring Congruence

Practice

Are the shapes congruent? Write *yes* or *no*.

1 ___yes___

2 ___no___

3 ___no___

4 ___no___

What combination of transformations results in the ending position?   Answers will vary. Any combination of transformations that results in the picture at the right should be accepted.

5

6

Unit 5
Geometry and Measurement Practice Pages
Weeks 1 and 2

1 Assign the Practice pages, *Workbook,* pages 42–43, at the end of Weeks 1 and 2. Students should complete these pages independently.

2 Check student answers using the annotated pages above.

3 If students have difficulty with a Practice page, review the activities completed throughout the week, and have students complete the weekly practice again before they complete the weekly test.

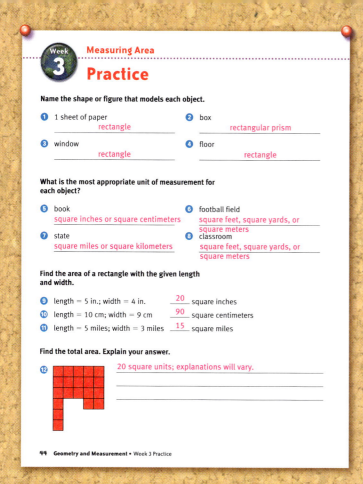

Practice

Name the shape or figure that models each object.

1. 1 sheet of paper — rectangle
2. box — rectangular prism
3. window — rectangle
4. floor — rectangle

What is the most appropriate unit of measurement for each object?

5. book — square inches or square centimeters
6. football field — square feet, square yards, or square meters
7. state — square miles or square kilometers
8. classroom — square feet, square yards, or square meters

Find the area of a rectangle with the given length and width.

9. length = 5 in.; width = 4 in. — 20 square inches
10. length = 10 cm; width = 9 cm — 90 square centimeters
11. length = 5 miles; width = 3 miles — 15 square miles

Find the total area. Explain your answer.

12. 20 square units; explanations will vary.

Practice

Convert each measurement.

1. 1 yard = 36 inches
2. 12 feet = 4 yards
3. 4 feet = 48 inches
4. 3 feet = 1 yards
5. 3 pounds = 48 ounces
6. 16 ounces = 1 pounds
7. 8 gallons = 32 quarts
8. 20 pints = 10 quarts
9. 6 pints = 3 quarts
10. 2 gallons = 16 pints
11. 15 minutes = 900 seconds
12. 3 days = 72 hours
13. 5 minutes = 300 seconds
14. 49 days = 7 weeks

Draw a picture to represent the problem. Solve the problem.

15. John mixed 4 quarts of orange juice to serve 10 people. Each person is to receive a pint of orange juice. Did he mix enough?

    No, he mixed only 8 pints of orange juice. He needs 10 pints.

16. The dairy farm sold fresh milk for $1 per quart. Joan bought 4 gallons. How many quarts did she buy?

    16 quarts

Unit 5
Geometry and Measurement
Practice Pages

Weeks 3 and 4

1. Assign the Practice pages, *Workbook,* pages 44–45, at the end of Weeks 3 and 4. Students should complete these pages independently.

2. Check student answers using the annotated pages above.

3. If students have difficulty with a Practice page, review the activities completed throughout the week, and have students complete the weekly practice again before they complete the weekly test.

Week 1 Examining Data

Week at a Glance

This week, students begin *Number Worlds,* Level F, Data Analysis and Applications. Students should explore different kinds of data and ways to classify data. They should also explore how to formulate appropriate questions and should draw conclusions from sets of data.

Math Background

The amount of data available to help make decisions in business, politics, research, and everyday life is staggering. Students need to know about data analysis and related aspects of probability in order to reason statistically—skills necessary to become informed citizens and intelligent consumers.

—*Principles and Standards for School Mathematics.* 2000. National Council of Teachers of Mathematics. Page 48.

How Students Learn

Students need to be exposed to many examples of data displays and to learn to interpret and analyze the information that is shown visually. Teachers need to help students become educated about the types of data, types of data displays, and the proper uses of both. Elementary students need to be introduced to many types of graphs. It is critical that students understand how to read and construct these graphs so that they can become aware of the misleading ways data can be presented.

Teaching for Understanding

Observe closely while evaluating the assigned tasks this week to see whether students can demonstrate the following understandings.

Benchmark after Lesson 2: Students can categorize data as being fact or opinion as well as being numerical or categorical.

Benchmark after Lesson 3: Students can sort and classify elements in a data set.

Benchmark after Lesson 4: Students can ask appropriate questions about a set of data and to draw conclusions.

Skills Focus

- Recognize different kinds of data
- Sort and categorize elements within a data set
- Ask appropriate questions to draw conclusions from a set of data

Math at Home

Give one copy of the Letter to Home, page A21, to each student. Encourage students to share and complete the activity with their caregivers.

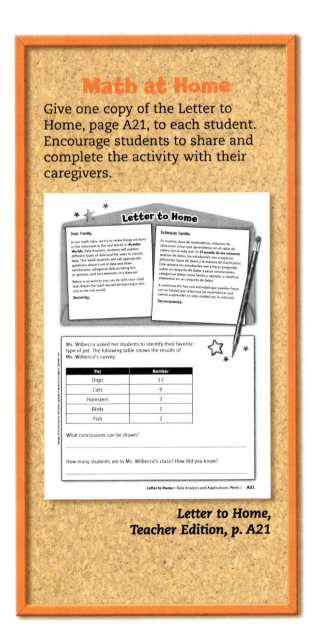

Letter to Home, Teacher Edition, p. A21

Week 1 Planner — Examining Data

PACING	LESSON	LEARNING GOALS	NCTM	MATERIALS	TECHNOLOGY
DAY 1	pages 2–3	Students identify questions used to conduct surveys and collect data.	• Data Analysis and Applications • Communication	No additional materials needed	**MathTools** Spreadsheet
DAY 2	pages 4–5	Students use multiple strategies for categorizing different types of data.	• Data Analysis and Applications • Communication • Connections	No additional materials needed	**MathTools** Spreadsheet
DAY 3	pages 6–7	Students sort and classify elements in a data set.	• Data Analysis and Applications • Problem Solving • Communication • Representation	Scissors	**MathTools** Spreadsheet
DAY 4	pages 8–9	Students ask appropriate questions to analyze data and draw conclusions.	• Data Analysis and Applications • Problem Solving • Communication • Connections	Magazines and newspapers	**MathTools** Spreadsheet
DAY 5	Review and Assess pages 10–11	Students review skills learned this week and complete the weekly assessment.	• Data Analysis and Applications • Problem Solving • Communication • Connections • Representation	Materials will be selected from Lessons 1–4.	**MathTools** Review previous activities

Math Vocabulary

survey A method of collecting information by asking people the same question

data Information that is collected

fact Data that that can be verified, such as age

opinion Data that involves a person's likes or dislikes

numerical data Data that can be described by numbers such as a phone number

categorical data Data that can be organized into non-overlapping groups, such as eye color

English Learners

SPANISH COGNATES

English	Spanish
data	datos
probability	probabilidad
analysis	análisis
conduct	conducer

ALTERNATE VOCABULARY

survey An organized list of questions designed to elicit the opinions and preferences of a large number of people

data set A collection of data that goes together and shares a topic

Lesson 1

Objective

Students identify questions used to conduct surveys and collect data.

Materials

No additional materials needed

Access Vocabulary

tally chart A display that can be used to organize data collected from a survey

draw conclusions To form a theory based on the answers collected

Creating Context

A survey is a data-gathering tool that can help English Learners practice asking questions. For English Learners in early proficiency levels, provide three potential answers that respondents may choose from rather than open-ended answers that may be difficult for English Learners to write.

1 Warm Up 5

Concept Building REASONING

On the board, write the names of groups of people whom students might find interesting. These might include athletes, actors, politicians, and school personnel. Choose one of the groups at random and ask students to think of a question they would like to ask that group. Have students explain why they would want to ask that particular question.

2 Engage 30

Skill Building

Explain that when conducting surveys, each person surveyed must be asked the same question.

Conduct a Survey APPLYING

Have students think of different survey questions they would ask their classmates. Some examples include the following: What is your favorite sport? What is your favorite dessert? What is your favorite weekend activity? How many hours of television do you watch each week? Choose one question from those the students provide.

Ask each student the survey question. Organize the results in a tally chart or a circle graph on the board. After the survey is completed, ask the students these questions.

- **What was the purpose of the survey question?**
- **What conclusions can be drawn from the data?**
- **Who might benefit from knowing the results of this survey?**

Monitoring Student Progress

| **If . . .** students are having difficulty drawing conclusions from the results of a survey, | **Then . . .** help them list some possible conclusions that could be drawn. Have them also use the process of elimination to determine which of those listed is most likely. |

MathTools Use the Spreadsheet tool to demonstrate and explore organizing data.

Using Student Pages

Have students complete **Workbook,** pages 2–3, on their own.

Lesson 1

Key Idea
When you look at the results of a survey, you should consider the following questions:
- What question was asked?
- What was the purpose of asking the question?
- What conclusions, if any, can be drawn?
- Who might benefit from having asked the question?

Try This
What question was asked in each survey?

❶

Favorite Pizza Topping		
Topping	Tally	Number
Pepperoni		9
Sausage		4
Green peppers		2
Mushrooms		8
Ground beef		4
Onions		3

What is your favorite pizza topping?

❷

Class President Voting

Suzie 30% · Pedro 25% · Michelle 45%

For whom did you vote to be class president?

❸

High Temperature	
Day	Temperature
Monday	78°F
Tuesday	81°F
Wednesday	80°F
Thursday	84°F
Friday	83°F
Saturday	80°F
Sunday	77°F

What is the daily high temperature?

❹

Favorite Movie Type		
Movie Type	Tally	Number
Comedy		7
Drama		3
Action		12
Suspense		8
Animated		6

What is your favorite type of movie?

Practice
For the following data display, tell:
- a. What question was asked?
- b. What conclusions, if any, can be drawn?
- c. Who might have conducted the survey and why?

❺

Favorite Cafeteria Food		
Food	Tally	Number
Lasagna		4
Chicken		7
Vegetable soup		3
Pizza		15
Fish		6
Meat loaf		5

Sample answers: a. What is the favorite cafeteria food among fourth graders? b. Pizza is the most favored food, and vegetable soup is the least favored food. c. The cafeteria staff might have conducted the survey to help them plan what meals to serve.

❻

Harrison Elementary School			
New Playground Construction			
	Third Graders	Fourth Graders	Fifth Graders
For	28	29	35
Against	3	1	2

Sample answers: a. Do you want a new playground built? b. Students at Harrison Elementary School really want a new playground built. c. School administrators might have conducted the survey to decide if they should build a new playground.

Reflect
Suppose a survey question was phrased, "Do you support the construction of a new state-of-the-art highway, or are you satisfied with the old, outdated system?" Are the constructors of the survey for or against the new highway?

for

---

3 Reflect   10 ▶

Extended Response REASONING
Review students' answers to the Reflect prompt at the bottom of student page 3. Discuss different examples of how one who conducts a survey can ask a biased question.

- **What effect might a biased question have on a survey?** It might trick people into answering a question differently than they otherwise would.

Real-World Application APPLYING
Have students think about surveys they encounter in everyday life. Give examples of surveys presented in newspapers, magazines, online, or on television. Answer the following questions for each survey discussed.

- **What was the purpose of the survey question?**
- **What conclusions are drawn or could be drawn?**
- **Who might benefit from the data?**

4 Assess

Informal Assessment
Use the Student Assessment Record, **Assessment,** page 100, to record informal observations.

APPLYING

Conduct a Survey
Did the student
- ❏ apply learning to new situations?
- ❏ contribute concepts?
- ❏ contribute answers?
- ❏ connect mathematics to the real world?

Lesson 2

Objective

Students use multiple strategies for categorizing different types of data.

Materials

No additional materials needed

Access Vocabulary

organize data To group the responses in ways that clarify the information

biased question A question whose answer is influenced by the wording, intonation, or another variable

Creating Context

Present students with examples of surveys from magazines or from the internet. Point out differences in the types of questions asked. Surveys can have yes/no questions, multiple-choice questions, or open-ended questions.

 1 Warm Up 5

Concept Building REASONING

Present the following two statements. Instruct students to say if the statement is a fact or an opinion.

- **A cheetah can run at speeds of up to 70 miles per hour.** fact
- **Chocolate cake is the best dessert.** opinion

Have students define *fact* and *opinion* in their own words.

 2 Engage 30

Strategy Building

Arrange students into groups of three or four. Students will need paper and pencils.

- **There are two types of data: numerical data, which involves numbers, and categorical data, which classifies data.**
- **Is the fact that the cheetah can run up to 70 miles per hour numerical or categorical data?** numerical data
- **Is the opinion that chocolate cake is the best dessert numerical or categorical data?** categorical data
- **If a survey showed that 11 students in a class preferred chocolate cake and 9 preferred yellow cake, would that data be numerical or categorical?** numerical
- **Facts can be numerical or categorical data, and opinions can be numerical or categorical data.**

Fact or Fiction? REASONING

Instruct each student to write a statement on a sheet of paper that involves a fact or an opinion. On the back of the sheet of paper, have each student write the words that identify the written statement as *fact* or *opinion* and *categorical* or *numerical*. Students should exchange papers and answer the following questions:

- **Does the statement represent a fact or an opinion?**
- **Does the statement describe data that are numerical or categorical?**
- **What survey question was likely asked to make this conclusion?**

Students can compare their answers to those on the back of the paper.

MathTools Use the Spreadsheet tool to demonstrate and explore organizing data.

Using Student Pages

Have students complete **Workbook,** pages 4–5, on their own.

Lesson 2

Key Idea

Information gathered or collected is *data*. Data can represent facts or opinions.

A *fact* is actual information such as your age, and an *opinion* involves how you feel about something.

Data can also be *numerical* or *categorical*. Numerical data can be described by numbers such as the number of votes or your zip code. Categorical data can be classified into nonoverlapping categories such as your eye color, gender, or city of birth.

Try This

Identify each statement as either *fact* or *opinion*. Circle your answer.

1. Three students in Mr. Nelson's homeroom were absent today.
   (fact)   opinion
2. Dogs make the best pets.
   fact   (opinion)
3. The most exciting sport to watch is soccer.
   fact   (opinion)
4. There are 68 students going on the field trip.
   (fact)   opinion
5. Malcolm ran a mile in 6 minutes 14 seconds.
   (fact)   opinion
6. Miss Hernandez is the best teacher at Lincoln Elementary School.
   fact   (opinion)

Practice

Write *categorical* or *numerical* in each blank to tell what type of data each survey question will collect.

7. How do you get to school each day?
   Bus   Car   Bicycle   Walk   Other
   __categorical__

8. How many days per week do you usually eat cafeteria food?
   __numerical__

9. How many books did you read last summer?
   __numerical__

10. What is your favorite ride at the amusement park?
    __categorical__

11. What is your favorite color?
    Red   Blue   White   Purple   Green   Other
    __categorical__

12. How often do you go to the mall each month?
    __numerical__

Reflect

Write a survey question that will collect factual numerical data. Write a survey question that will collect opinions and categorical data.
Sample answers:
How many sit-ups did you do during gym class? (numerical, fact)
What is your favorite after-school activity? (categorical, opinion)

 ## Reflect   10

Extended Response REASONING

Review students' answers to the Reflect prompt at the bottom of student page 5.

Discuss different statements to make sure students understand the differences between facts and opinions.

- **How would you explain the difference between a fact and an opinion?**
- **Can you give some examples of both?**

Real-World Application APPLYING

Have students brainstorm different ways that they could classify students in their grade level. Some possibilities might include students' gender, height, eye color, sports they play, homeroom teacher, and so on. Ask students to determine if each of these classifications describes numerical data or categorical data.

 ## Assess

Informal Assessment

Use the Student Assessment Record, **Assessment,** page 100, to record informal observations.

REASONING
Fact or Fiction?
Did the student
- ❏ provide a clear explanation?
- ❏ communicate reasons and strategies?
- ❏ choose appropriate strategies?
- ❏ argue logically?

Lesson 3

Objective

Students sort and classify elements in a data set.

Materials

Additional Materials

Scissors

Access Vocabulary

characteristic A feature that makes an object distinct from another object

category A group of ideas or things that have the same characteristic

Creating Context

The concept-building section of this lesson includes questions that allow students to participate by giving short answers, lists, and either/or questions. These are excellent examples of the types of questions to use to check for understanding with English Learners.

1   Warm Up   5

Concept Building   REASONING

Have students brainstorm ways they could classify the sports they play during Physical Education class. For example, they can classify each sport by the number of players on a team, the equipment used, whether points or goals are scored, and so on. Ask them to group the sports by asking questions such as the following:

- **Which sports use a ball?**
- **How many players are on each team?**
- **Which sports involve scoring goals?**

2   Engage   30

Skill Building

Have students brainstorm as many musical instruments as they can. Write the names of the instruments on the board.

Orchestra Activity   UNDERSTANDING

Discuss with students the four categories of instruments: strings, woodwinds, brass, and percussion. Write these headings on the board, and have the students determine which instruments belong in each group and why. For example, a guitar belongs in strings, a clarinet belongs in woodwinds, a saxophone in brass, and drums in percussion.

Have students think of other ways that the instruments can be classified, such as instruments that require your mouth to play (brass and woodwinds) or instruments that require your hands to play (strings and percussion).

Monitoring Student Progress

If . . . students are having difficulty classifying instruments,	**Then . . .** have them list the characteristics of each category. They may need to do some research to learn about musical instruments. Explain to students that they can match characteristics of a specific instrument to the characteristics of each category.

MathTools Use the Spreadsheet tool to demonstrate and explore organizing data.

Using Student Pages

Have students complete **Workbook,** pages 6–7, on their own.

Lesson 3

Key Idea
It is important to be able to sort and classify elements in a data set. Data can be organized into different categories and groups.

Try This
Copy the information onto separate cards, and use them to answer the Practice questions.

Name: Saltwater Crocodile **Type:** Reptile **Habitat:** Ocean **Size (weight):** 1,150 pounds	**Name:** Blue Whale **Type:** Mammal **Habitat:** Ocean **Size (weight):** 420,000 pounds	**Name:** Giraffe **Type:** Mammal **Habitat:** Grasslands/Jungle **Size (weight):** 3,000 pounds
Name: Whale Shark **Type:** Fish **Habitat:** Ocean **Size (weight):** 26,000 pounds	**Name:** Cheetah **Type:** Mammal **Habitat:** Grasslands/Jungle **Size (weight):** 125 pounds	**Name:** Elephant **Type:** Mammal **Habitat:** Grasslands/Jungle **Size (weight):** 16,000 pounds
Name: Sailfish **Type:** Fish **Habitat:** Ocean **Size (weight):** 150 pounds	**Name:** Anaconda **Type:** Reptile **Habitat:** Grasslands/Jungle **Size (weight):** 500 pounds	**Name:** Dolphin **Type:** Mammal **Habitat:** Ocean **Size (weight):** 600 pounds
Name: Leatherback Turtle **Type:** Reptile **Habitat:** Ocean **Size (weight):** 1,400 pounds	**Name:** Komodo Dragon **Type:** Reptile **Habitat:** Grasslands/Jungle **Size (weight):** 200 pounds	**Name:** Great Barracuda **Type:** Fish **Habitat:** Ocean **Size (weight):** 90 pounds
Name: Manta Ray **Type:** Fish **Habitat:** Ocean **Size (weight):** 300 pounds	**Name:** Hippopotamus **Type:** Mammal **Habitat:** Grasslands/Jungle **Size (weight):** 8,250 pounds	**Name:** *Tyrannosaurus rex* **Type:** Reptile **Habitat:** Grasslands/Jungle **Size (weight):** 12,000 pounds

Practice
Sort the animal cards from the previous page according to the questions below. Then answer each question.

1. How many types of animals are represented by the cards? List them.
   3 types: mammals, fish, and reptiles

2. Organize the animals on the cards according to their habitat. How many of the animals live in the ocean? List them.
   8 animals: saltwater crocodile, blue whale, whale shark, sailfish, dolphin, leatherback turtle, great barracuda, manta ray

3. How many of the animals can be found in the grasslands or the jungle? List them.
   7 animals: giraffe, cheetah, elephant, anaconda, Komodo dragon, hippopotamus, *Tyrannosaurus rex*

4. How many more animals live in the ocean than live in the grasslands or jungle?
   1 animal

5. Name the animal that weighs the most.
   blue whale

6. Organize the animals on the cards according to their types. How many of them are reptiles? List them.
   5 animals: saltwater crocodile, anaconda, leatherback turtle, Komodo dragon, *Tyrannosaurus rex*

Reflect
Analyzing data often leads to new questions. What are some new questions you could ask about this data set?
Sample answer: Of the three heaviest animals, how many live in the ocean?

3 Reflect   10 ▶

Extended Response   REASONING
Review students' answers to the Reflect prompt at the bottom of student page 7.

Discuss aloud students' suggestions for how to further classify the animals.
- **How else can we classify these animals?**

Suggest certain ideas such as the colors of the animals or what the animals eat.

Real-World Application   APPLYING
Instruct students to think of a situation other than the animal cards or the **Orchestra Activity** in which items are classified. Give them the following example to begin their brainstorming. What different plants grow around the school? Have students think of categories such as flowers, shrubs, trees, and weeds; and list plants and flowers in each category.

4 Assess

Informal Assessment
Use the Student Assessment Record, **Assessment,** page 100, to record informal observations.

UNDERSTANDING

Orchestra Activity
Did the student
- ☐ make important observations?
- ☐ extend or generalize learning?
- ☐ provide insightful answers?
- ☐ pose insightful questions?

Lesson 4

Objective

Students ask appropriate questions to analyze data and draw conclusions.

Materials

Additional Materials

Magazines and newspapers

Access Vocabulary

survey A method of collecting information by asking people the same question

Creating Context

Ask students what sports and games they play or watch. Make a list of the features of each game. The following questions will help start students thinking about what information is and is not useful. What is the goal of the game? How many points are scored? Is this a team sport? How do you decide who goes first?

1 Warm Up 5

Concept Building REASONING

Arrange students into groups of three or four. Have students think of a group of people to be surveyed. Then have students come up with interesting questions they would ask the group. Have each group share with the class their survey and possible questions.

2 Engage 30

Skill Building

Students should remain in the same groups during the Skill-Building activity.

Good Question UNDERSTANDING

For each question generated in the Concept-Building activity, have the students discuss the following:

- **Who might ask this question and why?**
- **What conclusion might be drawn from the survey?**
- **When a conclusion has been drawn from the survey, how might this lead to more questions and surveys?** For example, suppose it is determined that most fourth, fifth, and sixth graders prefer a certain brand of cereal. The makers of this cereal might conclude that their advertising should focus on this age group. Then they might want to ask questions to determine the most effective medium for advertising the cereal to this age group.

Monitoring Student Progress

If . . . students are struggling to come up with good ideas for questions,

Then . . . have them think of the commercials they have seen on television or in advertisements in magazines. Have them discuss what questions may have been asked and when those advertisements were planned.

MathTools Use the Spreadsheet tool to demonstrate and explore organizing data.

Using Student Pages

Have students complete **Workbook,** pages 8–9, on their own.

Week 1 — Examining Data

Lesson 4

Key Idea
In order to analyze data to reach conclusions, the appropriate questions need to be asked.

Try This
Use the data below to answer each question.

Wilderness Camp Field Trip Favorite Activities			
	Third Graders	Fourth Graders	Fifth Graders
Swimming	9	12	6
Fishing	8	5	10
Canoeing	5	8	7
Hiking	2	4	3

1. How many students in each grade went on the field trip?
24 third graders, 29 fourth graders, 26 fifth graders

2. What was the most popular activity among fourth graders?
swimming

3. How many more fifth graders said that canoeing was their favorite activity than said hiking?
4 students

4. Of the students surveyed, how many said that swimming was their favorite activity?
27 students

5. What was the least favorite activity among all the students?
hiking

Practice
Use the data below to answer the following questions.

How Do You Get to School?	
Transportation	Students
Bus	88
Walk	19
Car	35
Bicycle	22

6. How many students walk or ride their bicycles to school each day?
Forty-one students walk or ride their bicycles to school.

7. Who might ask Question 6? What might be the reason for asking this question? Explain.
Sample answer: School or city officials might be interested in knowing how many students walk or ride their bicycles to school because they are concerned about safety.

8. What conclusion might be drawn from the data about the number of students who walk or ride their bicycles to school each day?
A large number of students walk or ride their bicycles to school each day. Therefore, the school might take extra precautions to ensure their safety.

Reflect
Sometimes interesting results come from studying data. These results can lead to new questions. Suppose the results of a cafeteria study show that the most popular type of sandwich among students is peanut butter and jelly.

If the cafeteria does not offer peanut butter and jelly sandwiches, what questions might this result cause them to ask?
Sample answer: The cafeteria might consider how much it costs to offer peanut butter and jelly sandwiches and how this compares to the costs of other items already offered by the cafeteria.

---

3 Reflect   10

Extended Response REASONING

Review students' answers to the Reflect prompt at the bottom of student page 9.

■ **Is there only one right answer to this problem?** no

Explain that there are not necessarily any incorrect answers, but some questions might lead to more useful conclusions. Invite students to share their examples with the class.

Real-World Application APPLYING

Have students look through magazines, newspapers, and online articles to find real-world examples of actual surveys. They should look for topics of interest to elementary school students, such as favorite roller coasters and favorite television programs. Discuss with them the following questions:

■ **Who asked the survey question?**
■ **What conclusions have been drawn?**
■ **What are some further questions that could be asked?**

---

4 Assess

Informal Assessment

Use the Student Assessment Record, *Assessment,* page 100, to record informal observations.

UNDERSTANDING

Good Question
Did the student
☐ make important observations?
☐ extend or generalize learning?
☐ provide insightful answers?
☐ pose insightful questions?

Examining Data • Lesson 4   8–9

Lesson 5 Review

Objective

Students review skills learned this week and complete the weekly assessment.

Materials

Review materials will be selected from those used in previous activities.

Creating Context

Have students create a brief survey regarding what languages are spoken by students in class.

1 Warm Up 5

Concept Building UNDERSTANDING

■ **Name a fact that is numerical data.** Sample fact: A tortoise can live up to 150 years.

■ **Fourth grade is the best grade in the school. Is this statement fact or opinion?** opinion

2 Engage 20

Skill Building

Free-Choice Activity ENGAGING

For the last day of the week, allow students to choose an activity from the previous lessons. Some activities they may choose include the following:

• **Conduct a Survey**
• **Fact or Fiction?**
• **Orchestra Activity**
• **Good Question**

Make a note of the activities students select. Do they prefer easy or challenging activities? If you believe your students would benefit from extra practice on specific skills, choose an activity for them.

Monitoring Student Progress

If . . . students are accurately identifying facts and opinions on a consistent basis,

Then . . . invite them to research topics at home and bring in a statement they think will "stump" the class.

Using Student Pages

Have students complete *Workbook,* pages 10–11, on their own.

3 Reflect 10

Extended Response REASONING

Review students' answers to the Reflect prompts at the bottom of student pages 10–11.

Discuss the answers with the group to reinforce Week 1 concepts.

Lesson 5 Review

This week you learned how to decide what was asked in a survey, why the survey might have been conducted, and how to sort, analyze, and draw conclusions about data.

Lesson 1 Use the data displays below to answer questions in Lessons 1, 3, and 4.

After-School Activities		
Activity	**Tally**	**Numbers**
Sports	||| ||	7
Video games	|||| ||||	9
Computer	||||		6
Bicycle riding	||||	4

Favorite Zoo Exhibits	
Exhibit	**Students**
Reptile	7
Shark	14
Monkey	9
Elephant	11

1. What question was likely asked in the After-School Activities survey?
   What is your favorite after-school activity?

2. What question was likely asked in the Favorite Zoo Exhibits survey?
   Which zoo exhibit do you like best?

Lesson 2

3. Describe the results of the After-School Activities Survey as facts or opinions and categorical or numerical.
   opinions, numerical

4. Describe the results of the Favorite Zoo Exhibits Survey as facts or opinions and categorical or numerical.
   opinions, numerical

Reflect

Who might have asked the question about the zoo? Explain.
Sample answer: The teacher who took the students on the field trip might have asked the question to find the exhibit with the most student interest.

10   Data Analysis and Applications • Week 1

Lesson 3

5. What is one way the after-school activities in the survey can be classified?
   Sample answer: indoor and outdoor activities

6. What is one way zoo exhibits can be categorized?
   Sample answer: by the type of animal: reptile, fish, or mammal

Lesson 4

7. What conclusion can you draw about the popularity of the shark tank as compared to the reptile area?
   The shark tank was twice as popular as the reptile area.

8. What conclusion can you draw about how students spend their time after school?
   Students like to play video games more than the other activities in the survey.

Reflect

Use the categories named in Problems 5 and 6 to write a statement about each display that tells a conclusion made about the data.
Sample answer: The indoor activities of video games and computers combined were more popular than the outdoor activities of sports and bicycle riding combined.
Sample answer: The two mammal exhibits combined were more popular than the fish exhibit.

Examining Data • Lesson 5 Review   11

 Assess   10

A Gather Evidence

Formal Assessment

Have students complete the Weekly test on **Assessment**, pages 64–65. Record progress on the Student Assessment Record, **Assessment**, page 100.

B Summarize Findings

Determine whether students have Minimal, Basic, or Secure understanding of the concepts presented in Week 1.

C Differentiate Instruction

Based on your observations, use these teaching strategies next week to follow up.

Assessment, pp. 64–65

Minimal Understanding

- Repeat the Warm-Up and Engage activities to develop Week 1 concepts.
- Use **eMathTools** Spreadsheet to develop and reinforce this week's concepts.

Basic Understanding

- Repeat Engage activities in subsequent weeks.
- Use **eMathTools** Spreadsheet to develop and reinforce this week's concepts.

Secure Understanding

- Use **eMathTools** Spreadsheet to develop and reinforce this week's concepts.
- Use variations of the weekly Warm-Up and Engage activities using higher numbers and multiple steps.

Examining Data • Lesson 5 Review   10–11

Week 2

Data and Bar Graphs

Week at a Glance

This week, students continue with **Number Worlds,** Level F, Data Analysis and Applications, by investigating how to create and interpret bar graphs and double bar graphs. Students will continue to analyze data and to draw conclusions from data sets presented in data displays.

Skills Focus

- Create bar graphs and double bar graphs from a set of data
- Interpret bar graphs and double bar graphs, and use them to draw conclusions

Math Background

In today's rapidly changing information age, students are constantly being called on to answer questions that are based on an analysis of data. The results of surveys, studies, experiments, and polls are commonplace in newspapers and magazines, on the nightly news, and on the Internet.

—Chapin, Suzanne, Alice Koziol, Jennifer MacPherson, and Carol Rezba. 2003. *Navigating Through Data Analysis and Probability in Grades* 3–5. National Council of Teachers of Mathematics. Page 11.

How Students Learn

Students need meaningful ways to practice interpreting and analyzing data. It is only after students have developed such skills that they will be able to determine the accuracy of the presented data. Teachers need to offer students many experiences in answering questions about a data display and posing questions of their own. It is through this exposure that students will learn to collect, organize, and display data that address issues of interest to them.

Teaching for Understanding

Observe closely while evaluating the assigned tasks this week to see whether students can demonstrate the following understandings.

Benchmark after Lesson 2: Students can create bar graphs from a set of data and to interpret the graphs.

Benchmark after Lesson 3: Students can create double bar graphs from a set of data and to interpret the graphs.

Benchmark after Lesson 4: Students can identify the types of graphs appropriate for particular situations.

Math at Home

Give one copy of the Letter to Home, page A22, to each student. Encourage students to share and complete the activity with their caregivers.

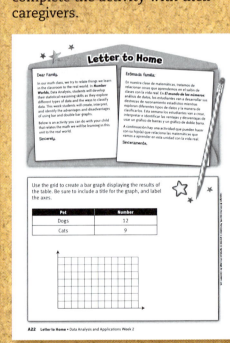

Letter to Home, Teacher Edition, p. A22

PACING	LESSON	LEARNING GOALS	NCTM	MATERIALS	TECHNOLOGY
DAY 1	pages 12–13	Students create bar graphs from a set of categorical data and use the graphs to interpret the data.	• Data Analysis and Applications • Communication • Connections • Representation	• Graph paper • Poster board or butcher paper	**e MathTools** Graphing Tool
DAY 2	pages 14–15	Students create bar graphs from a set of numerical data and use the graphs to interpret the data.	• Data Analysis and Applications • Communication • Connections • Representation	Graph paper	**e MathTools** Graphing Tool
DAY 3	pages 16–17	Students create double bar graphs from a set of data and use the graphs to interpret the data.	• Data Analysis and Applications • Communication • Connections • Representation	• Graph paper • Colored pencils • Sample Bar Graphs, p. B24	**e MathTools** Graphing Tool
DAY 4	pages 18–19	Students identify the types of graphs appropriate for particular situations.	• Data Analysis and Applications • Problem Solving • Communication	No additional materials needed	**e MathTools** Graphing Tool
DAY 5	**Review and Assess** pages 20–21	Students review skills learned this week and complete the weekly assessment.	• Data Analysis and Applications • Problem Solving • Communication • Connections • Representation	Materials will be selected from Lessons 1–4.	**e MathTools** Review previous activities

Math Vocabulary

bar graph A data display that uses bars to compare different categories of data

double bar graph A data display that uses two differently colored or shaded bars to compare data from two different sets

English Learners

SPANISH COGNATES

English	Spanish
graphs	gráficas
bar graphs	gráfica de barra

ALTERNATE VOCABULARY

interpret the data To look at the information on the graph, and draw conclusions about what it means

Lesson 1

Objective

Students create bar graphs from a set of categorical data and use the graphs to interpret the data.

Materials

Additional materials
- Graph paper
- Poster board or butcher paper

Access Vocabulary

bar graph A data display that uses bars to compare different categories of data

data table A graphic organizer that displays data to help make comparisons and draw conclusions

Creating Context

An excellent strategy to aid English Learners in acquiring the academic language of mathematics and understanding complex concepts is to use graphics and visuals. In this lesson, English Learners will see examples of how this skill is used and can examine the characteristics of the real graphs used in authentic situations.

1 Warm Up   5

Concept Building COMPUTING

Display a sample bar graph on the board or overhead projector.

- **Bar graphs use either horizontal or vertical bars to represent the counts for different categories. The categories can be numerical or categorical.**

Point out the axes, categories, scale, and title. Discuss the importance of having an accurate title, and labeling the axes correctly.

2 Engage   30

Skill Building

Present the Eye Color Data table on the board or overhead projector.

Eye Color Data	
Blue eyes	8 students
Green eyes	6 students
Brown eyes	10 students
Hazel eyes	12 students

Eye Color UNDERSTANDING

Provide students with graph paper to create a bar graph using data displayed on the board. As a group discuss the following parts of a bar graph:

- Labeled axes
- Appropriate scale for the vertical axis
- Bars for each data set
- Title of the bar graph

As each topic is discussed, add that part to the bar graph you are creating on the board.

Students should create a bar graph on their papers that matches what you have on the board. When students are finished, ask the following questions:

- **Which eye color did the most students have?** hazel
- **How many students were included in the results?** 36
- **Which eye color did the least number of students have?** green

Monitoring Student Progress

If . . . students do not use an appropriate scale for the vertical axis,

Then . . . ask them what the largest number of students is for any one color. The scale will need to be at least this number.

 MathTools For additional practice creating bar graphs, use *eMathTools* Graphing Tool.

Using Student Pages

Have students complete *Workbook,* pages 12−13, on their own.

Lesson 1

Key Idea
Bar graphs use bars to compare different categories of data. The graphs can be used to display both categorical and numerical data.

Try This
The table shows the results of a survey taken by fourth graders about their favorite team sport. Follow the steps to make a bar graph of the data.

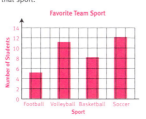

Favorite Team Sport Among Fourth Graders											
V	F	S	B	S	S	V	S	F	B	S	V
B	S	V	V	S	V	F	B	V	S	B	S
S	V	V	B	V	B	F	F	V	S	S	B

Key: F = Football, V = Volleyball, B = Basketball, S = Soccer

Step 1 How many categories of sports are there? Create a tally chart to show how many students voted for each sport. Which sport received the least votes? Which sport received the most votes? How many students voted for each sport?

4 categories; least votes:
football; most votes: soccer;
F: 5, V: 11, B: 8, S: 12

Step 2 A bar graph has two axes. Label the horizontal axis with the categories of sports. Label the vertical axis with a scale for the number of students. What increments should you use? Give the graph a title.

a scale from 0 to 14 with marks
at every even number

Step 3 Draw a bar for each category. The height of each bar should match the number of students who voted for that sport.

Favorite Team Sport

(bar graph: vertical axis labeled "Number of Students" from 0 to 14; horizontal axis labeled "Sport" with categories Football, Volleyball, Basketball, Soccer)

Practice
Use the bar graph that you created about favorite team sports to answer each question.

1. Which sport in the survey was the most popular? How many students voted for this sport?
soccer; 12 students

2. Which sport in the survey was the least popular? How many students voted for this sport?
football; 5 students

3. How many more students in the survey voted for soccer than voted for basketball?
Four more students voted for soccer than voted for basketball.

4. How many students voted for volleyball or basketball?
19 students

5. How many fourth graders were surveyed altogether?
36 students

Reflect
Which data display makes it easier to analyze the results of the survey—the table, the tally chart, or the bar graph you created? Explain and give an example of one way in which one data display is better than the other.
Sample answer: The bar graph makes it easier to analyze the results of the survey because its bars can be used to easily compare the categories. For example, someone can look at the bar graph and know immediately which team sport was most popular, which was least popular, how much more popular a certain sport was compared to another, and so on. With the table, each letter must be counted before these conclusions can be drawn. With the tally chart, you can easily see which has the most votes and the least votes, but you still need to count to know the actual number of votes for each sport.

---

3 Reflect
10

Extended Response REASONING
Review students' answers to the Reflect prompt at the bottom of student page 13.

- **In what ways is the tally sheet better than the data table?** The tally sheet separates data by category, so you can see which category is more popular.

- **In what ways is the bar graph better than the tally sheet?** The bars make it easier to see the relative sizes of the data points.

Real-World Application APPLYING
Bar graphs are used in newspapers, television commercials, infomercials, magazines, and on the Internet. Share examples of bar graphs in the real world. Include at least one example of a graph with horizontal bars and one example of a graph with vertical bars. Ask students to identify key characteristics of the graphs. Explain that using horizontal bars does not change the meaning of a bar graph.

4 Assess

Informal Assessment
Use the Student Assessment Record, *Assessment,* page 100, to record informal observations.

UNDERSTANDING

Eye Color
Did the student
❏ make important observations?
❏ extend or generalize learning?
❏ provide insightful answers?
❏ pose insightful questions?

Lesson 2

Objective

Students create bar graphs from a set of numerical data and use the graphs to interpret the data.

Materials

Additional Materials

Graph paper

Access Vocabulary

horizontal axis A straight line across a page from left to right, like a horizon

vertical axis A straight line up and down on a page

increment The amount of increase marked along the axes of a graph

Creating Context

Not all students have access to the Internet or magazines in the home. Provide other resources for graphs and data, including the school library or daily newspaper.

1 Warm Up 5

Concept Building UNDERSTANDING

Bar graphs can also be used to display numerical data. Explain that the labels on the horizontal axis are usually ranges of numbers. Discuss with students different sets of data that can be displayed in a bar graph, such as points scored in a basketball game, hours spent studying, and other ideas students suggest. For each data set, discuss ranges of numbers that could be used on the horizontal axis of the bar graph and the scale that should be used on the vertical axis.

2 Engage 30

Strategy Building

Discuss with students how much time they typically spend doing homework each night. Draw a table on the board to record responses for each student in the class, and label the table "Homework Time."

More Homework! UNDERSTANDING

Have each student write on the board the number of minutes he or she spends doing homework each night. Discuss with students the decisions that need to be made before they can begin to create a bar graph to display this data.

- **What ranges of times should be used for the horizontal axis?** Answers will vary.
- **What scale should be used for the vertical axis?** Answers will vary.

Arrange students into groups of three or four to create a bar graph of the data they collected. Allow time for them to complete the task. Then ask them questions about their graphs.

- **How many students do at least thirty minutes of homework each night?** Answers will vary.
- **How many students do more than one hour of homework each night?** Answers will vary.

Monitoring Student Progress

If . . . students are struggling with what ranges of numbers to use on the horizontal axis,	**Then . . .** ask them if 15- or 30-minute increments would be better. Ask them how many ranges of numbers would be created using each increment.

eMathTools For additional practice creating bar graphs, students should complete **eMathTools** Graphing Tool.

Using Student Pages

Have students complete **Workbook,** pages 14–15, on their own.

Lesson 2

Key Idea

Bar graphs can also be used to present numerical data. In this type of bar graph, each bar represents a number or range of numbers instead of a category.

Try This

Jeremy surveyed several students at his school. He asked how many hours each spends watching television during the week. The results are shown in the table. Follow the steps to make a bar graph of the data.

Hours of Television (each week)

2	4	6	8	9	10	2	12	0	8
4	15	4	0	3	8	3	9	4	5
7	2	4	6	5	4	1	5	9	12

Step 1 Use the ranges given below to create a tally chart of the data. Determine how many students are in each range.

Ranges: 0–3 hours   4–6 hours
7–9 hours   10 or more hours

0–3: 8 students; 4–6: 11 students;
7–9: 7 students; 10+: 4 students

Step 2 Label the horizontal axis with the ranges named above. Label the vertical axis with a scale for the number of students. What range and increment should you use on the vertical axis? Give the graph a title.

a scale of 0 to 12

Step 3 Draw a bar for each range of hours. The height of the bar should match the number of students that fall in the range. If you draw the bars without spaces between them, you have created a histogram. A histogram is a special type of bar graph.

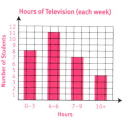
Hours of Television (each week)

Practice

Use the bar graph that you created about watching television to answer each question.

1. In which range of hours do most students watch television? How many students watch this number of hours of television each week?
4–6 hours; 11 students

2. In which range of hours do the fewest students watch television? How many students are there who watch this number of hours of television each week?
10+ hours; 4 students

3. How many of the students surveyed watch at least 4 hours of television each week?
22 students

4. How many students did Jeremy survey altogether?
30 students

5. Draw a conclusion about the number of students who fall in the 0–3 hour range as compared to the number who fall in the 10+ hour range.
Sample answer: Twice as many students watch between 0 and 3 hours of television each week than watch 10 or more hours each week.

Reflect

Suppose Jeremy wants to know how many students watch exactly twelve hours of television each week. Is the bar graph or the table better for answering this question? Explain.

The table is better for determining how many students watch a specific number of hours of television because the bar graph uses ranges of hours. This information cannot be determined from the bar graph. However, the bar graph is better for comparing the numbers of students in each range.

3 Reflect   10

Extended Response   REASONING

Review students' answers to the Reflect prompt at the bottom of student page 15.

- **In what ways is the data table better than the bar graph?** Answers will vary.
- **In what ways is the bar graph better than the data table?** Answers will vary.

Guide students to understand that when they use a bar graph to display numerical data, they will not be able to see the individual values. A bar graph allows them to compare how many times certain ranges of numbers occur in a numerical data set.

Real-World Application   APPLYING

As students learned in Week 1, bar graphs are used often to display real-world data and surveys. Ask students to look for a bar graph in magazines and newspapers or on the Internet. Students should write questions about their graphs similar to the questions they answered on the student pages of this lesson.

4 Assess

Informal Assessment

Use the Student Assessment Record, **Assessment,** page 100, to record informal observations.

UNDERSTANDING

More Homework!
Did the student
❏ make important observations?
❏ extend or generalize learning?
❏ provide insightful answers?
❏ pose insightful questions?

Data and Bar Graphs • Lesson 2   14–15

Lesson 3

Objective

Students create double bar graphs from a set of data and use the graphs to interpret the data.

Materials

Program Materials
Sample Bar Graphs, p. B24

Additional Materials
- Graph paper
- Colored pencils

Access Vocabulary

double bar graph A data display that uses two differently colored or shaded bars to compare data from two different sets

Creating Context

In the Skill-Building activity of this lesson, students make a double bar graph comparing different recess games. Some games may have different names in other countries. Review with English Learners each of the games to be sure they understand the nature of the game.

1 Warm Up 5

Concept Building UNDERSTANDING

Present 4th Graders and 5th Graders graphs from Sample Bar Graphs, page B24, on the board or overhead protector.

- **Would it be easier to make comparisons if the data for fourth and fifth graders were on the same graph?** yes

2 Engage 30

Skill Building

Present the Favorite Recess activity on the board or overhead projector. Use the steps below to instruct students how to make a double bar graph. Students will need graph paper and colored pencils.

Favorite Recess Activity			
	Four Square	Tetherball	Tag
Fourth Graders	7	5	4
Fifth Graders	5	6	3

Recess Activity APPLYING

- **Step 1** Draw the axes on the graph paper. Label the horizontal axis with the recess activities. Label the vertical axis with a scale for the number of students. Give the graph a title.
- **Step 2** Make the bar for the fourth graders who like Four Square best. Choose a color to shade the bar. To the right of that bar, make a bar for the fifth graders who like Four Square best. Color this bar a different color from the first bar.
- **Step 3** Repeat Step 2 for the other two activities.

Monitoring Student Progress

If . . . a student is having difficulty getting started,

Then . . . have the student begin by setting up a suitable set of axes with appropriate labels. Explain that each activity category will have two bars associated with it, one for the fourth graders and one for the fifth graders.

eMathTools For additional practice making double bar graphs, students should complete *eMathTools* Graphing Tool.

Using Student Pages

Have students complete *Workbook,* pages 16–17, on their own.

Lesson 3

Key Idea
Bar graphs are useful for comparing categorical and numerical data for two or more groups. A *double bar graph* uses two different-colored or shaded bars to compare data from two different groups.

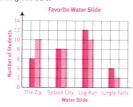

Try This
Follow the steps to make a double bar graph of the data in the table.

Favorite Water Slide		
Slide	Fourth Graders	Fifth Graders
The Zip	6	10
Splash City	8	8
Log Run	12	10
Jungle Falls	4	2

Step 1 Label the horizontal axis with the names of the water slides. Label the vertical axis with a scale for the number of students. Give the graph a title.

Step 2 Draw a bar to represent the number of fourth graders who voted for the first slide named on the horizontal axis. Color or shade that bar. Then draw another bar beside it using a different color or shading to represent the number of fifth graders who voted for that slide.

Step 3 Draw a fourth-grade bar and a fifth-grade bar for each of the other slides. The color or shading of all fourth-grade bars needs to be the same. The color or shading of all fifth-grade bars needs to be the same but different from the fourth-grade bars.

Favorite Water Slide

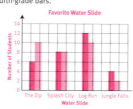

Practice
Use the double bar graph that you created about water slides to answer each question.

1 What was the most popular water slide in each class?
For the fourth graders it was Log Run; for the fifth graders it was a tie between The Zip and Log Run.

2 What was the least popular water slide in each class?
Jungle Falls was the least popular water slide in both classes.

3 Which water slide had the same number of fourth and fifth graders vote for it? How many students from each grade voted for this ride as their favorite?
Splash City; 8 students from each grade

4 How many more fifth graders voted for The Zip than did fourth graders?
4 more students

5 How many fourth graders voted for either Splash City or Jungle Falls as their favorite water slide?
12 students

6 How many fifth graders voted for either The Zip or Log Run as their favorite water slide?
20 students

Reflect
Which water slide would you expect to have the longest line on a hot summer afternoon? Which slide would you expect to have the shortest line? Who might be interested in knowing the results of this survey? Explain.
Sample answer: Log Run is the most popular water slide, and Jungle Falls is the least popular. So Log Run would have the longest line, and Jungle Falls would have the shortest line. The water park owners might want to know this information so that they can use their more popular rides in advertising. Also, they might want to allow more space for lines on the more popular rides.

3 Reflect   10

Extended Response REASONING
Review students' answers to the Reflect prompt at the bottom of student page 17.

- **How is the popularity of a particular water slide related to the length of the line at the ride?** The more popular the ride is, the longer the line will be.

- **Can you think of any other situations like this?** Answers will vary.

Real-World Application APPLYING
Double bar graphs are often used to compare data from two different time periods. For example, a U.S. Census worker might compare the populations of several cities in Pennsylvania in 1990 and in 2000.

- **Display the Census Bar Graph on the board or overhead projector.**

- **What are the categories on the horizontal axis?** the cities

- **What are the labels on the bars for each category?** 1990 and 2000

4 Assess

Informal Assessment
Use the Student Assessment Record, *Assessment*, page 100, to record informal observations.

APPLYING

Recess Activity
Did the student
- ❏ apply learning to new situations?
- ❏ contribute concepts?
- ❏ contribute answers?
- ❏ connect mathematics to the real world?

Lesson 4

Objective

Students identify the advantages and disadvantages of a bar graph and a double bar graph.

Materials

No additional materials needed

Access Vocabulary

bar graph A data display that uses bars to compare different categories of data

double bar graph A data display that uses two differently colored or shaded bars to compare data from two different sets

circle graph A data display that divides a circle into sections to compare parts of a group to the whole

Creating Context

A circle graph is often called a pie graph. Ask English Learners why they think it has that name. Remind them that a circle graph is best used when comparing a part to the whole.

1 Warm Up 5

Concept Building UNDERSTANDING

- Look back over the many bar graphs that you have created so far this week.
- Is it harder to read a double bar graph than a single bar graph? no
- Can you think of a situation where you wouldn't want to use a double bar graph? to display one data set, or to display the combined results of two data sets

2 Engage 30

Strategy Building

Organize students into three groups.

One Bar or Two? REASONING

Describe each situation listed below. Ask the groups to discuss which type of graph they would use to display the data. After each group has discussed the situation, have each group present its choice and the reasons for that choice. If groups disagree, give students a chance to present their ideas again.

- Hours that boys spend playing video games and hours that girls spend playing video games
- Combined hours that boys and girls spend playing video games
- The number of times a family eats breakfast together each week and the number of times a family eats dinner together each week

Extend the lesson to discuss probability.

- **Are these events more likely, less likely, or equally likely to occur?**

Monitoring Student Progress

| **If . . .** students need more practice creating double bar graphs, | **Then . . .** have them choose single bar graphs made earlier in the week and make up a second set of data to graph on each graph. |

 MathTools For additional practice using bar graphs and circle graphs, students should complete *eMathTools* Graphing Tool.

Using Student Pages

Have students complete **Workbook,** pages 18–19, on their own.

Lesson 4

Key Idea
The bars in a bar graph can be either vertical or horizontal.

Try This
These bar graphs display the results of the same survey given to fourth-grade and fifth-grade students.

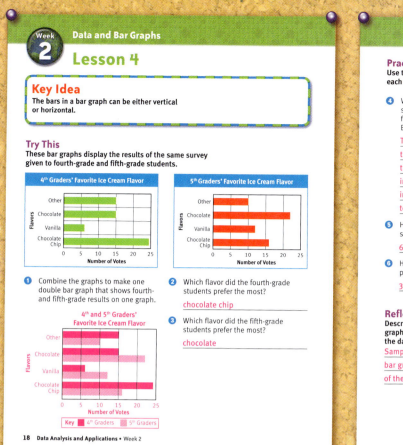

4th Graders' Favorite Ice Cream Flavor

5th Graders' Favorite Ice Cream Flavor

❶ Combine the graphs to make one double bar graph that shows fourth- and fifth-grade results on one graph.

4th and 5th Graders' Favorite Ice Cream Flavor

Key ■ 4th Graders ▨ 5th Graders

❷ Which flavor did the fourth-grade students prefer the most?
chocolate chip

❸ Which flavor did the fifth-grade students prefer the most?
chocolate

Practice
Use the bar graphs about ice cream flavors to answer each question.

❹ Which data display makes it easier to see that vanilla is the least popular flavor of ice cream of both grades? Explain.
The double bar graph shows the results of both grades, and the vanilla bars are the shorter in each color. If you look at the individual bar graphs, you have to look at both graphs.

❺ How many students altogether were surveyed?
60 students

❻ How many of those surveyed preferred chocolate ice cream?
37 students

❼ Create a bar graph other than a double bar graph that shows the results of both grades combined.

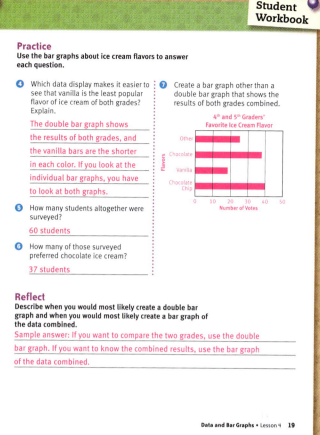

4th and 5th Graders' Favorite Ice Cream Flavor

Reflect
Describe when you would most likely create a double bar graph and when you would most likely create a bar graph of the data combined.
Sample answer: If you want to compare the two grades, use the double bar graph. If you want to know the combined results, use the bar graph of the data combined.

3 Reflect   10

Extended Response REASONING
Review students' answers to the Reflect prompt at the bottom of student page 19.

Select several students to share their answers with the rest of the class. Discuss with the class whether they agree or disagree and why.

■ **Which graph would be easier to read?** The answer depends upon what information you are looking for.

Real-World Application APPLYING
Suppose a cereal maker wants to show that 35 percent of elementary school students eat its brand of cereal every morning. Should the cereal maker use a bar graph or a circle graph? Explain.
The cereal maker should use a circle graph because this type of display compares a part to the whole.

4 Assess

Informal Assessment
Use the Student Assessment Record, *Assessment,* page 100, to record informal observations.

REASONING

One Bar or Two?
Did the student
❑ provide a clear explanation?
❑ communicate reasons and strategies?
❑ choose appropriate strategies?
❑ argue logically?

Data and Bar Graphs • Lesson 4   18–19

Lesson 5 Review

Objective

Students review skills learned this week and complete the weekly assessment.

Materials

Review materials will be selected from those used in previous activities.

Creating Context

An excellent strategy to use with English Learners is to provide writing templates for them to complete. In this way, they can learn grammar, structure, and writing conventions as well as the key concepts.

2 Engage 20

Strategy Building

Free-Choice Activity ENGAGING

For the last day of the week, allow students to choose an activity from the previous lessons. Some activities they may choose include the following:

- **Eye Color**
- **More Homework**
- **Recess Activity**
- **One Bar or Two?**

Make a note of the activities students select. Do they prefer easy or challenging activities? If you believe your students would benefit from extra practice on specific skills, choose an activity for them.

Monitoring Student Progress

If . . . students want to make a bar graph such as those found in newspapers and magazines,

Then . . . have them use plain paper and a straightedge. *USA Today* newspapers include many bar graphs, especially on the lower corner of the front page.

Using Student Pages

Have students complete **Workbook,** pages 20–21, on their own.

1 Warm Up 5

Concept Building UNDERSTANDING

- **In your opinion, which type of data graph presents information most clearly?** Answers will vary.
- **Which type of data graph do you most often see in newspapers and magazines?** Answers will vary.

3 Reflect 10

Extended Response REASONING

Review students' answers to the Reflect prompt at the bottom of student page 21.

Discuss the answers with the group to reinforce Week 2 concepts.

Week 2 — Data and Bar Graphs

Lesson 5 Review

This week you studied bar graphs. You learned how to create a bar graph and a double bar graph and how to read data from the graphs.

Lesson 1 Make a bar graph of the data in the table.

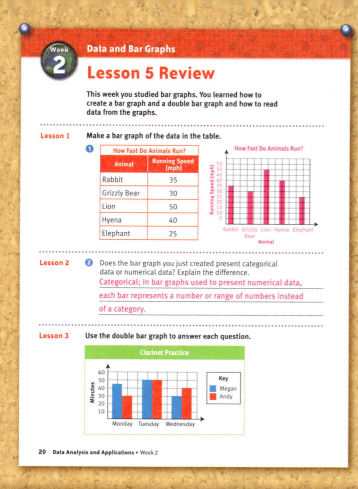

❶

How Fast Do Animals Run?	
Animal	Running Speed (mph)
Rabbit	35
Grizzly Bear	30
Lion	50
Hyena	40
Elephant	25

Lesson 2 ❷ Does the bar graph you just created present categorical data or numerical data? Explain the difference.

Categorical; In bar graphs used to present numerical data, each bar represents a number or range of numbers instead of a category.

Lesson 3 Use the double bar graph to answer each question.

❸ Who spent more time practicing the clarinet on Monday? How many more minutes did this student practice than the other?

Megan practiced 15 minutes more than Andy on Monday.

❹ How much time altogether did Megan spend practicing the clarinet during the three-day period?

2 hours 5 minutes

❺ How much time altogether did Andy spend practicing the clarinet during the three-day period?

2 hours

Lesson 4 Use the double bar graph to answer each question.

❻ On which day did Megan and Andy practice the same number of minutes? Explain.

Tuesday; Their bars are the same height.

❼ On which day did Megan and Andy practice the least number of minutes combined? Explain.

Wednesday; Their combined practice time totals 70 minutes, the least amount of practice during the week.

Reflect

Why do you think the clarinet practice data was presented on a double bar graph instead of two individual bar graphs?

It is easier to compare each student's data when it is side-by-side.

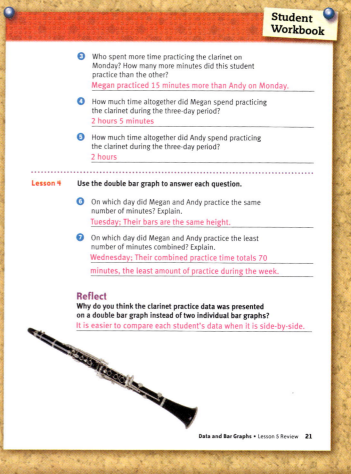

---

4 Assess   10

A Gather Evidence

Formal Assessment

Have students complete the Weekly test, **Assessment,** pages 66–67. Record progress on the Student Assessment Record, **Assessment,** page 100.

B Summarize Findings

Determine whether students have Minimal, Basic, or Secure understanding of the concepts presented in Week 2.

C Differentiate Instruction

Based on your observations, use these teaching strategies next week to follow up.

Minimal Understanding
- Repeat the Warm-Up and Engage activities to develop Week 2 concepts.
- Use *eMathTools* Graphing Tool to develop and reinforce this week's concepts.

Basic Understanding
- Repeat Engage activities in subsequent weeks.
- Use *eMathTools* Graphing Tool to reinforce this week's concepts.

Secure Understanding
- Use *eMathTools* Graphing Tool to reinforce this week's concepts.
- Use variations of the weekly Warm-Up and Engage activities using higher numbers and multiple steps.

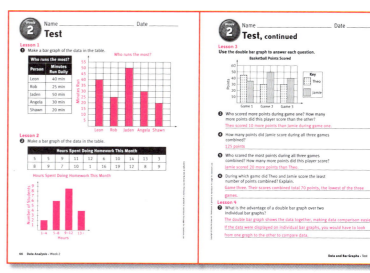

Assessment, pp. 66–67

Showing Data in Line Plots

Week at a Glance

This week, students continue with *Number Worlds,* Level F, Data Analysis and Applications, by investigating how to create and interpret line plots. Students will continue to analyze data and to draw conclusions from data sets presented in data displays.

Math Background

Simplistic views can lead to the use of recipe approaches to data analysis and to the treatment of data as numbers only, stripped of context and practical importance. Conversely, staying grounded in the data and attentive to what they have to say keeps the tools of data analysis—collecting, graphing, and analyzing—in their appropriate subservient roles.

—Kilpatrick, Jeremy, W. Gary Martin, and Deborah Schifter. 2003. *A Research Companion to Principles and Standards for School Mathematics.* National Council of Teachers of Mathematics. Page 194–195.

How Students Learn

It is important to help students see data as a means of learning something. Data are not just numbers to be picked up and manipulated. Students need to see data as numbers in context that are worthy of having questions posed about them and interesting enough that people want to seek the answers to the questions.

Teaching for Understanding

Observe closely while evaluating the assigned tasks this week to see whether students can demonstrate the following understandings.

Benchmark after Lesson 2: Students can identify the range, mode, outliers, and clusters of a data set presented in a line plot.

Benchmark after Lesson 3: Students can design an investigation and collect data to study that investigation.

Benchmark after Lesson 4: Students can construct and interpret a line plot to analyze the results of an investigation.

Skills Focus

- Construct and interpret line plots
- Identify the range, mode, outliers, and clusters of data presented in a line plot
- Conduct an investigation and use a line plot to analyze the results

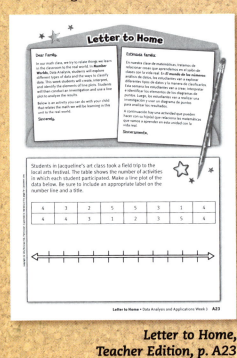

Math at Home

Give one copy of the Letter to Home, page A23, to each student. Encourage students to share and complete the activity with their caregivers.

Letter to Home, Teacher Edition, p. A23

Week 3 Planner   Showing Data in Line Plots

PACING	LESSON	LEARNING GOALS	NCTM	MATERIALS	TECHNOLOGY
DAY 1	pages 22–23	Students construct line plots from a set of data.	• Data Analysis and Applications • Communication • Representation	• Straightedge • Graph paper	**MathTools** Graphing Tool
DAY 2	pages 24–25	Students analyze line plots and identify the range, mode, outliers, and clusters.	• Data Analysis and Applications • Communication • Representation	• Straightedge • Graph paper	**MathTools** Graphing Tool
DAY 3	pages 26–27	Students design an investigation, make predictions, and collect data.	• Data Analysis and Applications • Communication • Connections	• Straightedge • Graph paper	**MathTools** Graphing Tool
DAY 4	pages 28–29	Students construct a line plot and use it to analyze the results of their investigation.	• Data Analysis and Applications • Problem Solving • Communication • Representation	• Straightedge • Graph paper • Results from Lesson 3 activity	**MathTools** Graphing Tool
DAY 5	Review and Assess pages 30–31	Students review skills learned this week and complete the weekly assessment.	• Data Analysis and Applications • Problem Solving • Communication • Connections • Representation	Materials will be selected from Lessons 1–4.	**MathTools** Review previous activities

Math Vocabulary

line plot A data display that uses a number line and Xs to show a set of numerical data

outlier A value that is much higher or much lower than the rest of the numbers in the set

cluster A group of data values that are close to one another

English Learners

SPANISH COGNATES

English	Spanish
data analysis	análisis de datos
probability	probabilidad
mode	modo
range	rango
investigation	investigación
results	resultados

ALTERNATE VOCABULARY

range The difference between the greatest and the least values of a data set

mode The value(s) that occur(s) most often in a data set

Lesson 1

Objective

Students construct line plots from a set of data.

Materials

Additional Materials
- Straightedge
- Graph paper

Access Vocabulary

line plot A data display that uses a number line and *Xs* to show a set of numerical data

range The difference between the greatest and the least number in a data set

Creating Context

English Learners may benefit from a discussion of the various ways to describe the extremes of the set of data points. For example, the data points can be described as least to greatest, or smallest to largest.

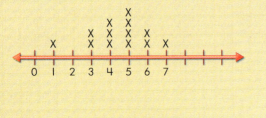

① Warm Up   5

Concept Building   UNDERSTANDING

Present the sample line plot shown above on the board or overhead projector.

- ■ A line plot, which is another way of displaying data, is built on a number line. Each *X* above a number shows how many times that number is in the data set.
- ■ The range is the difference between the greatest number and the least number of the data. **What is the range of this set?** 6

② Engage   30

Skill Building

Present the following data table on the board or overhead projector. Use the steps below to instruct the students in making a line plot. Students will need a straightedge and graph paper.

Number of Passengers in Cars That Passed the School in 1 Hour				
2	1	2	1	3
1	1	3	2	1
2	2	4	3	1

How Many Passengers?   APPLYING

- ■ We need to label the number line. Look at the data. **What is the least number in the data?** 1
- ■ **What is the greatest number in the data?** 4
- ■ Label the tick marks from 1 to 4 on the number line. **How many numbers are in the data set?** 15
- ■ So, there needs to be 15 *Xs* above the number line.
- ■ How many cars passed by that had only 1 person in the car? How many *Xs* should you place above the 1? 6; 6

Guide students to place the appropriate number of *Xs* above each number. Invite a volunteer to draw the line plot on the board for others to compare with his or her own.

Monitoring Student Progress

If . . . students do not include all the values on the number line between the least and greatest values in a data set,

Then . . . ask them to discuss the effect of leaving out these values.

 MathTools For additional practice creating bar graphs, students should complete *eMathTools* Graphing Tool.

Using Student Pages

Have students complete **Workbook**, pages 22–23, on their own.

Lesson 1

Key Idea

Line plots are data displays that are useful for plotting numerical data on a number line. A number line is plotted using numbers that range from at least the lowest data value to the highest data value. Then an *X* is placed over each number every time it occurs in the data set.

Try This

The table shows the number of pets that the students in Miss Nelson's homeroom have. Follow the steps to create a line plot of the data.

Number of Pets Miss Nelson's Homeroom Students								
2	1	4	3	2	1	7	1	0	2
0	2	1	1	3	0	2	3	4	1

Step 1 What is the least number in the data? What is the greatest number in the data? Draw a number line that will include all the values in the data set.

_____0; 7_____

Step 2 Place an *X* above the number line every time a number appears in the table.

Step 3 Label the number line, and give the graph a title.

Students should follow the steps above to create a line plot with 3 *X*s over 0, 6 *X*s over 1, 5 *X*s over 2, 3 *X*s over 3, 2 *X*s over 4, 0 *X*s over 5, 0 *X*s over 6, and 1 *X* over 7.

Practice

Make a line plot of each set of data. Be sure to include an appropriate label on the number line and a title.

1 Students in Natalie's science class took a walk in the woods to see how many different kinds of leaves they could find. The table shows the number of different leaves collected by each student.

4	6	5	5	7	5	3	4
6	3	5	4	2	4	5	5
5	6	3	2	3	4	5	4

Students should create a line plot with 0 *X*s over 1, 2 *X*s over 2, 4 *X*s over 3, 6 *X*s over 4, 8 *X*s over 5, 3 *X*s over 6, 1 *X* over 7, and 0 *X*s over 8.

2 Coach Hernandez timed how long it took students in his gym class to run one mile. The times (in minutes) are shown in the table.

8	9	8	10	7	11	12
10	6	9	8	9	9	7
7	8	10	9	9	12	6

Students should create a line plot with 0 *X*s over 5, 2 *X*s over 6, 3 *X*s over 7, 4 *X*s over 8, 6 *X*s over 9, 3 *X*s over 10, 1 *X* over 11, 2 *X*s over 12, and 0 *X*s over 13.

Reflect

Why might it be helpful to write all of the values in a data set in order from least to greatest before you begin to make a line plot? Explain.

Writing all the values in order from least to greatest can help you avoid missing any values on the number line. It also makes it easier to see the range of numbers that needs to be included on the number line.

---

3 Reflect   10

Extended Response   REASONING

Review students' answers to the Reflect prompt at the bottom of student page 23.

Discuss with students how writing all the data points from a set in order from least to greatest can help them determine what numbers to include on the number line. It can also help them be sure that they don't miss any data points when making the plot.

■ **How would you explain this to someone who has never done this before?**

Real-World Application   APPLYING

Teachers can use a line plot to display the results of their students' scores on a math quiz. Suppose a quiz consists of ten questions that students can answer correctly or incorrectly.

Sketch a sample line plot for the grades of a class with 18 students.

4 Assess

Informal Assessment

Use the Student Assessment Record, *Assessment,* page 100, to record informal observations.

APPLYING

How Many Passengers?
Did the student
❏ apply learning to new situations?
❏ contribute concepts?
❏ contribute answers?
❏ connect mathematics to the real world?

Lesson 2

Objective

Students analyze line plots and identify the range, mode, outliers, and clusters.

Materials

Additional Materials
- Straightedge
- Graph paper

Access Vocabulary

range The difference between the greatest and the least values of a data set

mode The value(s) that occur(s) most often in a data set

outlier A value that is much higher or much lower than the rest of numbers in the set

cluster A group of data values that are close to each other

Creating Context

Some of the key terms in this lesson are words used in other contexts. Remind English Learners to ask what a word means if they are uncertain.

1 | Warm Up   5

Concept Building   UNDERSTANDING

Line plots are useful for seeing how data are distributed.

Review with students the vocabulary for this lesson.

Present the line plot created from the previous lesson on the board or overhead projector.

- **Which number has the most Xs above it? This is the mode.** 1
- **Between which numbers is most data clustered, or grouped together?** 1 and 2

2 | Engage   30

Strategy Building

Present the following data on the board or overhead projector.

Number of Pencils in Students' Desks				
2	3	5	1	2
1	2	1	2	1
2	1	3	1	1

School Supplies   COMPUTING

Have students use the data about the number of pencils in each student's desk to make a line plot. Students will need straightedges and graph paper.

- **What is the range in the number of pencils?** 4
- **What is the mode of the data set?** 1
- **Are there any clusters of data on the line plot?** Data are clustered around 1 and 2.
- **Are there any outliers in the data set?** Yes; 5

Monitoring Student Progress

If . . . students are having difficulty identifying clusters of data,

Then . . . remind them that clusters of data are groups of values that are close to each other.

MathTools For additional practice creating bar graphs, students should complete **eMathTools** Graphing Tool.

Using Student Pages

Have students complete **Workbook,** pages 24–25, on their own.

Showing Data in Line Plots

Lesson 2

Key Idea

Line plots can help you describe a set of data.

- The **range** of a set of data is the difference between the greatest and least value.
- The **mode** of a set of data is the number that occurs most often.
- An **outlier** is a value in the set that is a lot greater than or a lot less than the rest of the values in the set.
- A **cluster** is a large group of data values that are close to each other.

Try This

Use the line plot to answer each question.

Daily High Temperature

Temperature (°F)

1 What is shown by the line plot? What question was asked?
Daily high temperatures for fourteen days; sample question: "What was the high temperature each day over the past two weeks?"

2 What is the highest temperature shown in the line plot? What is the lowest temperature? What is the range of the data?
80°F; 65°F; 15

3 What is the mode of the data?
78°F

4 On how many days was the daily high temperature at least 70°F?
13 days

5 Do any data values appear to be outliers in the data set? Explain.
Yes; the temperature of 65°F is much lower than the other temperatures in the line plot.

Practice
Use the line plot to answer each question.

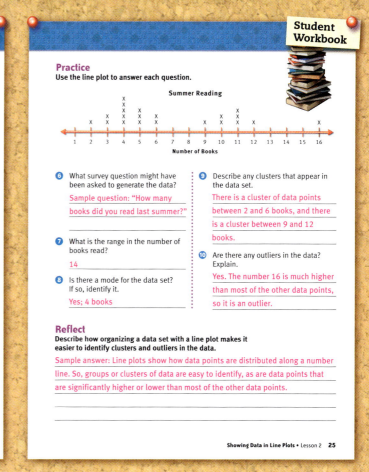

Summer Reading

Number of Books

6 What survey question might have been asked to generate the data?
Sample question: "How many books did you read last summer?"

7 What is the range in the number of books read?
14

8 Is there a mode for the data set? If so, identify it.
Yes; 4 books

9 Describe any clusters that appear in the data set.
There is a cluster of data points between 2 and 6 books, and there is a cluster between 9 and 12 books.

10 Are there any outliers in the data? Explain.
Yes. The number 16 is much higher than most of the other data points, so it is an outlier.

Reflect
Describe how organizing a data set with a line plot makes it easier to identify clusters and outliers in the data.
Sample answer: Line plots show how data points are distributed along a number line. So, groups or clusters of data are easy to identify, as are data points that are significantly higher or lower than most of the other data points.

Reflect   10

Extended Response   REASONING

Review students' answers to the Reflect prompt at the bottom of student page 25.

■ **Are there other strategies that could be used to answer this question?**
Have students share with the class some of the examples of clusters and outliers they have drawn on their line plots.

Real-World Application   APPLYING

Outliers can occur in real-world data sets. For example, suppose points scored by the players on a basketball team for one game are 7, 4, 5, 6, 4, 23, 5, and 7. One player had an exceptionally good game and scored many more points than the other players. Have students suggest other real-world examples in which an outlier might occur.

Assess

Informal Assessment

Use the Student Assessment Record, *Assessment,* page 100, to record informal observations.

COMPUTING

School Supplies
Did the student
❏ respond accurately?
❏ respond quickly?
❏ respond with confidence?
❏ self-correct?

Lesson 3

Objective

Students design an investigation, make predictions, and collect data.

Materials

Additional Materials
- Straightedge
- Graph paper

Access Vocabulary

prediction A statement about what you expect the outcome of an investigation to be

investigation A study or direct research to answer a question

Creating Context

A weather forecast uses special vocabulary and often is presented in the future verb tense. Suggest that English Learners listen to a weather broadcast in English on TV or radio to capture some of the basic weather features of your area. Provide a template for talking about the coming weather conditions using the future tense.

1 Warm Up   5

Concept Building   UNDERSTANDING

Discuss the reasons one might make a prediction about the outcome of an investigation. Making predictions about one outcome can help you think about other possible outcomes. This will help you ask better questions during the investigation.

2 Engage   30

Strategy Building

Invite a student volunteer to the board to record the data that is gathered in a tally chart. Students will need straightedges and graph paper.

Make a Prediction   UNDERSTANDING

■ **How many pets do you have at home?**
Before collecting any data, have each student make a prediction about what he or she thinks the mode of the data will be. Students should then draw a prediction line plot on their papers.

Skill Building

Gather Data   APPLYING

Ask students one-by-one to say how many pets they have at home. The student volunteer will record the data on the board. Save this data for the Skill-Building activity in Lesson 4. Students will compare the collected data with their predictions in the next lesson and analyze the results.

Monitoring Student Progress

If . . . students struggle to sketch a prediction line plot,

Then . . . ask them what they think the most common number of pets (the mode) will be. Instruct them that they should place the greatest number of *X*s above this number. Emphasize that there is no correct or incorrect prediction line plot. The prediction plot is an opinion of the expected results.

 MathTools For additional practice creating line plots, students should complete *eMathTools* Graphing Tool.

Using Student Pages

Have students complete **Workbook,** pages 26–27, on their own.

Lesson 3

Key Idea

When you make a *prediction*, you make a statement about the expected outcome of an investigation.

Try This

Follow the steps below to design an investigation.

Step 1 Decide on an investigation question that you find interesting. The investigation should involve numerical data. Here are some sample investigation questions that you can use, or you can create your own.

- How many basketball free throws can you make in two minutes?
- What is your height in inches?
- How many hours of television do you watch each week?
- How many books do you read each month?
- How far (in feet or yards) can you kick a soccer ball?

When choosing your investigation question, keep in mind what group of people you will be surveying when you conduct the investigation.

❶ What is your investigation question?

Answers will vary.

Step 2 Find a sample group of people to survey or investigate. This group can be family members, students in your school, teachers, children in your neighborhood, or others. Choose a group so you will be able to survey between ten and twenty people.

❷ Describe the group you will survey.

Descriptions will vary.

Practice

Continue following the steps below to make a prediction about your investigation and to gather data.

Step 3 Make a prediction about what you expect to discover with your investigation.

❸ Express this prediction as a statement. Then draw a prediction line plot to display what you think the data will look like.

Answers will vary.

Step 4 Collect data by asking your group of people the survey question.

❹ Record the results in a table or a tally chart in the space provided. You will use the results of this investigation in Lesson 4.

Survey results will vary.

Reflect

Does the group of people you survey depend on the investigation question you want to ask? Explain and give an example.

Yes. Some groups of people would not be appropriate for certain survey questions. For example, if you want to investigate how much homework students do each night, it would not make sense to survey teachers in your school.

---

3 Reflect 10

Extended Response REASONING

Review students' answers to the Reflect prompt at the bottom of student page 27. Discuss with students how asking a survey question of the wrong group can lead to inaccurate results.

- **Why would the results be inaccurate?** Invite individuals to share with the rest of the class their thoughts of how this could happen.
- **Can you think of any other examples of this happening?**

Real-World Application APPLYING

Weather forecasters make predictions about what the weather will be during the next several days.

- **Do you think weather forecasts are more accurate for tomorrow or for five days from now? Explain.** The forecasts for tomorrow are generally more accurate. The further into the future the forecast is, the more time there is for conditions to change.

4 Assess

Informal Assessment

Use the Student Assessment Record, **Assessment,** page 100, to record informal observations.

UNDERSTANDING	APPLYING
Make a Prediction	**Gather Data**
Did the student	Did the student
❏ make important observations?	❏ apply learning to new situations?
❏ extend or generalize learning?	❏ contribute concepts?
❏ provide insightful answers?	❏ contribute answers?
❏ pose insightful questions?	❏ connect mathematics to the real world?

Lesson 4

Objective

Students construct a line plot and use it to analyze the results of their investigation.

Materials

Additional Materials
- Straightedge
- Graph paper
- Results from Lesson 3 activity

Access Vocabulary

line plot A data display that uses a number line and *X*s to show a set of numerical data

Creating Context

Graphic organizers are an excellent way for English Learners to extend their vocabulary knowledge in a particular academic area. In this lesson, students will construct an investigation including a survey. Help students develop a topic and brainstorm categories using a word web to capture brainstorming ideas.

1 Warm Up   5

Concept Building UNDERSTANDING

Have students discuss how they can compare the actual results of an investigation with their original predictions. Suggestions should include examining the ranges and modes of the two sets of data, as well as identifying clusters and outliers.

2 Engage   30

Skill Building

Remind students of the data they collected about how many pets their classmates have at home and of the prediction line plots they made in Lesson 3. Present the table of data on the board.

Analyze the Results APPLYING

Have each student make a line plot using the data collected about the number of pets. Invite a volunteer to construct the line plot on the board. Students should compare their line plots with the line plot completed on the board. Then, compare the actual results with the prediction line plots by asking students such questions as the following:

- How does your predicted range compare with the actual range of the data?
- How does your predicted mode compare with the actual mode of the data?
- Is the appearance of the actual line plot similar to your prediction line plot?

Monitoring Student Progress

If . . . students are not sure what information to look at to analyze the results of the investigation and to compare to a prediction line plot,

Then . . . guide them to identify the mode and range of the data. Then compare these directly with the mode and the range of their prediction line plots.

 MathTools For additional practice creating line plots, students should complete *eMathTools* Graphing Tool.

Using Student Pages

Have students complete *Workbook,* pages 28–29, on their own.

Left workbook page (28):

Week 3 — **Showing Data in Line Plots**

Lesson 4

Key Idea
You can analyze data and compare results using line plots.

Try This
Use the data that you collected in your investigation in Lesson 3 to make a line plot in the space below.

Plots will vary.

Practice
Use the results of your investigation to answer the following questions.

1. Do the results of your investigation agree with what you predicted would happen? Explain.

   Answers will vary.

2. Are there any results that are surprising to you? Explain and use examples from the data of the line plot to support your response.

   Answers will vary.

3. What is the range of the data that you collected?

   Answers will vary.

28 Data Analysis and Applications • Week 3

Right workbook page (29):

4. Is there a mode in your data? If so, identify it.

   Answers will vary.

5. Are there any clusters of data in your line plot? If so, describe them.

   Answers will vary.

6. Are there any outliers in your line plot? If so, identify them.

   Answers will vary.

7. What conclusions, if any, can you draw about your investigation question?

   Answers will vary.

8. After analyzing your data, are there any other follow-up questions that you think would be interesting to study?

   Answers will vary.

Reflect
Suppose Rafael and Deena each conducted an investigation to study the favorite fruit drink flavor of elementary school students. Rafael surveyed sixty elementary students at the mall on a Saturday. Deena asked twelve fourth graders in her homeroom. Whose investigation do you think gave more accurate results? Explain.

Rafael's survey had more accurate results. He asked a larger group of students, and he did not restrict himself to only fourth graders.

Showing Data in Line Plots • Lesson 4   29

---

3 Reflect   10

Extended Response   REASONING
Review students' answers to the Reflect prompt at the bottom of student page 29.

- **How would using a larger sample help give more accurate results?**
- **How would it help eliminate outliers?**

Real-World Application   APPLYING
Suppose the most commonly served meals at Erin's school cafeteria are grilled cheese sandwiches and hamburgers. A survey of students as they wait in line shows that the most popular meal is pizza. How might the cafeteria workers use the results of this survey to plan their meals? Explain. Sample answer: They might start to serve pizza more often. This will make the students happier, and the cafeteria will likely serve more meals.

4 Assess

Informal Assessment
Use the Student Assessment Record, **Assessment,** page 100, to record informal observations.

APPLYING

Analyze the Results
Did the student
❏ apply learning to new situations?
❏ contribute concepts?
❏ contribute answers?
❏ connect mathematics to the real world?

Showing Data in Line Plots • Lesson 4   28–29

Lesson 5 Review

Skill Building

Free-Choice Activity ENGAGING

For the last day of the week, allow students to choose an activity from the previous lessons. Some activities they may choose include the following:

- **How Many Passengers?**
- **School Supplies**
- **Gather Data**
- **Analyze the Results**

Make a note of the activities students select. Do they prefer easy or challenging activities? If you believe your students would benefit from extra practice on specific facts, choose an activity for them.

Monitoring Student Progress

If . . . students like to play one game in particular,	**Then . . .** encourage them to record the rules and make copies of any game boards needed so that they can play at home with a caregiver and friends.

Using Student Pages

Have students complete **Workbook,** pages 30–31, on their own.

Objective

Students review skills learned this week and complete the weekly assessment.

Materials

Review materials will be selected from those used in previous activities.

Creating Context

Discuss with English Learners that they may receive phone calls from people saying that they are taking a survey. Caution students not to give any personal information about themselves or their family over the phone to someone they don't know.

3 Reflect 10

Extended Response REASONING

Review students' answers to the Reflect prompts at the bottom of student pages 30–31.

Discuss the answers with the group to reinforce Week 3 concepts.

1 Warm Up 5

Concept Building COMPUTING

Review the different types of data: fact or opinion and numerical or categorical. Review bar graphs and line plots.

- **If a data set is categorical opinions, which type of data display should you use?** bar graph
- **If a data set is numerical opinions, which type of data display can you use?** either type of graph

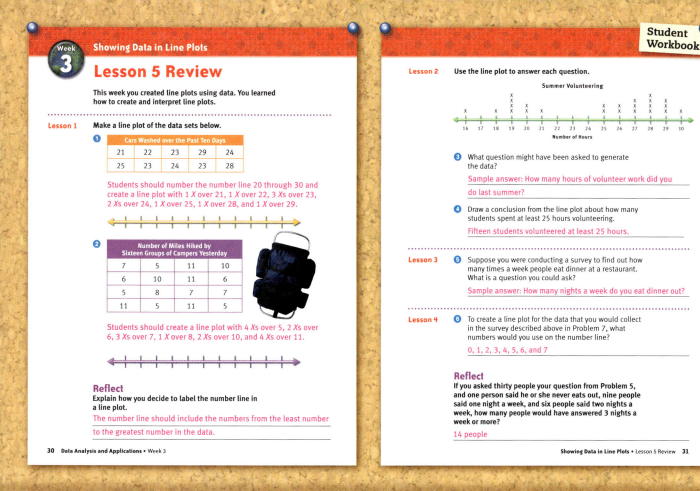

Lesson 5 Review

This week you created line plots using data. You learned how to create and interpret line plots.

Lesson 1   Make a line plot of the data sets below.

1

Cars Washed over the Past Ten Days				
21	22	23	29	24
25	23	24	23	28

Students should number the number line 20 through 30 and create a line plot with 1 *X* over 21, 1 *X* over 22, 3 *X*s over 23, 2 *X*s over 24, 1 *X* over 25, 1 *X* over 28, and 1 *X* over 29.

2

Number of Miles Hiked by Sixteen Groups of Campers Yesterday			
7	5	11	10
6	10	11	6
5	8	7	7
11	5	11	5

Students should create a line plot with 4 *X*s over 5, 2 *X*s over 6, 3 *X*s over 7, 1 *X* over 8, 2 *X*s over 10, and 4 *X*s over 11.

Reflect

Explain how you decide to label the number line in a line plot.

The number line should include the numbers from the least number to the greatest number in the data.

Lesson 2   Use the line plot to answer each question.

Summer Volunteering
Number of Hours

3 What question might have been asked to generate the data?

Sample answer: How many hours of volunteer work did you do last summer?

4 Draw a conclusion from the line plot about how many students spent at least 25 hours volunteering.

Fifteen students volunteered at least 25 hours.

Lesson 3   **5** Suppose you were conducting a survey to find out how many times a week people eat dinner at a restaurant. What is a question you could ask?

Sample answer: How many nights a week do you eat dinner out?

Lesson 4   **6** To create a line plot for the data that you would collect in the survey described above in Problem 7, what numbers would you use on the number line?

0, 1, 2, 3, 4, 5, 6, and 7

Reflect

If you asked thirty people your question from Problem 5, and one person said he or she never eats out, nine people said one night a week, and six people said two nights a week, how many people would have answered 3 nights a week or more?

14 people

 Assess   10

Gather Evidence

Formal Assessment

Have students complete the Weekly test, *Assessment,* pages 68–69. Record progress on the Student Assessment Record, *Assessment,* page 100.

B Summarize Findings

Determine whether students have Minimal, Basic, or Secure understanding of the concepts presented in Week 3.

C Differentiate Instruction

Based on your observations, use these teaching strategies next week to follow up.

Minimal Understanding

- Repeat the Warm-Up and Engage activities to develop Week 3 concepts.
- Use *eMathTools* Graphing Tool to develop and reinforce this week's concepts.

Basic Understanding

- Repeat Engage activities in subsequent weeks.
- Use *eMathTools* Graphing Tool to reinforce this week's concepts.

Secure Understanding

- Use *eMathTools* Graphing Tool to reinforce this week's concepts.
- Use variations of the weekly Warm-Up and Engage activities using higher numbers and multiple steps.

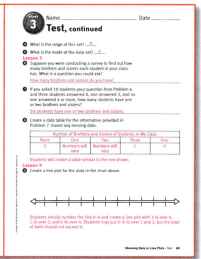

Assessment, pp. 68–69

Showing Data in Line Plots • Lesson 5 Review   30–31

Week 4 — Data Shown in Line Graphs

Week at a Glance

This week, students continue with **Number Worlds,** Level F, Data Analysis and Applications, by using line graphs as a way of recording, organizing, and analyzing data that represent a change over time.

Math Background

As adults, students will use their statistical literacy to analyze commercials, political programs, and medical claims. Technology plays an increasingly important role in analyzing data and graphical displays of data. Students at all levels use computers to generate raw data and to create all sorts of graphs.

—Chapin, Suzanne, and Art Johnson. 2000. *Math Matters: Grades K–6, Understanding the Math You Teach.* Math Solutions Publications. Page 213.

How Students Learn

Viewing data analysis in this cyclical process makes sense because it is the way researchers address an investigation. In actuality, scientists look ahead to imagine the analysis and conclusion stages before they begin to collect the data. Students, especially those who learn best by seeing the "big picture," will benefit from looking ahead to imagine analyzing data that they collect and drawing conclusions about it before they actually begin to collect their data.

Teaching for Understanding

Observe closely while evaluating the assigned tasks this week to see whether students can demonstrate the following understandings.

Benchmark after Lesson 2: Students can create a line graph to represent a set of data and to use the graph to analyze the data.

Benchmark after Lesson 3: Students can use double line graphs to analyze and compare two different sets of data.

Benchmark after Lesson 4: Students can collect and organize the data in a line graph to answer a statistical question.

Skills Focus

- Construct and interpret line graphs
- Interpret double line graphs
- Conduct an investigation and use a line graph to analyze the results

Math at Home

Give one copy of the Letter to Home, page A24, to each student. Encourage students to share and complete the activity with their caregivers.

Letter to Home, Teacher Edition, p. A24

PACING	LESSON	LEARNING GOALS	NCTM	MATERIALS	TECHNOLOGY
DAY 1	pages 32–33	Students use line graphs to analyze data.	• Data Analysis and Applications • Problem Solving • Communication	• Straightedge • Graph paper	MathTools Graphing Tool
DAY 2	pages 34–35	Students create line graphs from a set of data and use the graphs to analyze the data.	• Data Analysis and Applications • Problem Solving • Communication • Representation	• Straightedge • Graph paper	MathTools Graphing Tool
DAY 3	pages 36–37	Students use double line graphs to analyze and compare two different sets of data.	• Data Analysis and Applications • Problem Solving • Communication • Representation	• Straightedge • Graph paper	MathTools Graphing Tool
DAY 4	pages 38–39	Students design an investigation and collect data to answer a statistical question.	• Data Analysis and Applications • Problem Solving • Communication • Connections	• Straightedge • Graph paper • Clock or watch with a second hand	MathTools Graphing Tool
DAY 5	Review and Assess pages 40–41	Students review skills learned this week and complete the weekly assessment.	• Data Analysis and Applications • Problem Solving • Communication • Connections • Representation	Materials will be selected from Lessons 1–4.	MathTools Review previous activities

Math Vocabulary

line graph A data display that is used to show how data change over a period of time

coordinate grid A system used to identify a location by describing the distances from two axes that are perpendicular lines

double line graph A line graph that compares two sets of data on the same coordinate grid

English Learners

SPANISH COGNATES

English	Spanish
graph	gráfica
results	resultados
investigation	investigación

ALTERNATE VOCABULARY

weather patterns Predictable temperatures and precipitation (rain, snowfall) in an area over time

hike To walk a long distance for fun and recreation, usually in the woods or mountains

online Accessing the Internet via the computer

Lesson 1

Objective
Students use line graphs to analyze data.

Materials
Additional Materials
- Straightedge
- Graph paper

Access Vocabulary
line graph A data display that is used to show how data change over a period of time

Creating Context
Students in different parts of the country may be more familiar with different types of weather. Ask students to list different weather conditions and post them on a two-column chart. One column heading is *Local Weather,* and the other heading is *Weather Elsewhere.*

1 Warm Up 5

Concept Building UNDERSTANDING
Sketch a coordinate grid on the board similar to the one on **Workbook,** page 32. Number each axis 1–6. Have students volunteer to come to the board to plot the data points (5, 1); (3, 3); and (4, 6). Explain that the skill of being able to plot data points is needed to create a line graph.

2 Engage 30

Strategy Building
- **Does anyone know the predicted high temperature for today?** Allow students to answer, and allow students to use thermometers to measure the temperature.
- **How does the average high temperature for our area change throughout the year?** Allow students to answer.

Present the St. Louis temperature data on the board or overhead projector. Ask students to describe any patterns that they notice.

St. Louis, Missouri, Average Temperature			
Jan	29°	July	85°
Feb	34°	Aug	88°
Mar	45°	Sept	72°
Apr	57°	Oct	68°
May	66°	Nov	42°
June	75°	Dec	35°

Climate Study APPLYING
On the board, create a line graph of the temperature data. Students should re-create on their papers what is displayed on the board.
- **Which month has the warmest average temperature?** August
- **How many degrees does the average temperature increase between March and April?** 12°
- **How much warmer would you expect a day in February to be than a day in January?** 5° warmer

Monitoring Student Progress
If . . . students are taking a long time to create their line graphs, : **Then . . .** you can provide grid paper that already has the axes drawn.

MathTools For additional practice creating line graphs, students should complete *eMathTools* Graphing Tool.

Using Student Pages
Have students complete **Workbook,** pages 32–33, on their own.

Lesson 1

Key Idea

A *line graph* is a data display that is used to show how data change over a period of time.

Line graphs consist of data points that are connected by line segments and are often used to show trends.

Try This

Use the line graph below to answer each question.

Bus Stop—Tuesday Morning

1 What investigation is shown in the line graph?

how the number of people at the bus stop changes during the morning hours

2 Describe the overall trend in the data shown. Describe the story being told by the graph.

The number of people at the bus stop increases between 7:00 and 8:00, and then the number of people decreases. The greatest number of people waiting at the bus stop occurs at 8:00 in the morning. This is probably the time most people leave for work.

Practice

Use the line graph below to answer each question.

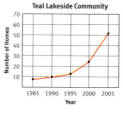

Teal Lakeside Community

3 What investigation is shown in the line graph?

how the number of homes in the Teal Lakeside Community has changed over the past two decades

4 Describe the overall trend in the data shown. Describe the story being told by the graph.

The number of homes in the Teal Lakeside Community was fairly low between 1985 and 1995. However, in the past ten years, the number of homes has increased sharply. More people are building homes in the Teal Lakeside Community.

5 Predict the number of homes in the community by the year 2010.

Sample answer: about 70–75 homes by 2010

Reflect

Line graphs are used to shows trends in data. What does it mean to "show a trend"?

A line graph shows how data change over time. The overall picture of whether the data are increasing or decreasing is easily visible.

3 Reflect   10 ▶

Extended Response   REASONING

Review students' answers to the Reflect prompt at the bottom of student page 33.

- **Can trends in data be as easily visible in a bar graph? Explain.** Answers will vary.
- **Is there one type that works best for each situation? Explain.** yes; Explanations will vary.

Real-World Application   APPLYING

Line graphs are useful for tracking different kinds of weather patterns. Help students research climate-related data, other than temperature, for the local area. An idea for climate-related data is the average amount of rainfall or snowfall. Display the data in a line graph.

4 Assess

Informal Assessment

Use the Student Assessment Record, **Assessment,** page 100, to record informal observations.

APPLYING

Climate Study

Did the student
- ❏ apply learning to new situations?
- ❏ contribute concepts?
- ❏ contribute answers?
- ❏ connect mathematics to the real world?

Lesson 2

Objective

Students create line graphs from a set of data and use the graphs to analyze the data.

Materials

Additional Materials
- Straightedge
- Graph paper

Access Vocabulary

coordinate grid A system used to identify a location by describing the distances from two axes that are perpendicular lines

Creating Context

Graphic organizers are excellent tools to use with English Learners because they do not require extensive language skill but do allow us to exercise critical thinking.

1 Warm Up 5

Concept Building COMPUTING

Sketch a coordinate grid on the board similar to the one you created in Lesson 1. Have students other than Day 1 volunteers come to the board to plot the data points (3, 2); (1, 4); and (5, 6). Explain that the skill of being able to plot data points is needed to create a line graph.

2 Engage 30

Strategy Building

A student needs graph paper, a pencil, and a straightedge to create a line graph. Present Marisa's growth data on the board or overhead projector.

Growth Chart for Marisa, Born August 8, 2005			
Birth	20 in.	18 weeks	25 in.
3 weeks	21 in.	24 weeks	26 in.
6 weeks	23 in.	30 weeks	27 in.
12 weeks	24 in.	52 weeks	31 in.

Infant Growth APPLYING

Have each student sketch a line graph of the data on their papers. Then have students answer the following questions.

- **How many inches did Marisa grow from birth to 6 weeks?** 3 inches
- **How long was Marisa at 24 weeks?** 26 inches
- **What length do you think Marisa was at 27 weeks?** about 26.5 inches

Monitoring Student Progress

If . . . students are interested in comparing their own growth charts,

Then . . . have them get that information from their caregiver(s) so that they can graph their data.

e MathTools For additional practice creating line graphs, students should complete **eMathTools** Graphing Tool.

Using Student Pages

Have students complete **Workbook,** pages 34–35, on their own.

Data Shown in Line Graphs
Lesson 2

Key Idea
You can create line graphs when given data.

Try This
Nina recorded in the table the amount of money she had each day last week. Follow the steps below to create a line graph from the data set.

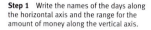

Amount of Money	
Day	**Money**
Monday	$1.25
Tuesday	$2.25
Wednesday	$2.25
Thursday	$1.50
Friday	$1.75

Step 1 Write the names of the days along the horizontal axis and the range for the amount of money along the vertical axis.

Step 2 Plot a point for each day in the table. Then connect the data points with line segments.

Step 3 Label the axes of the line graph, and give the graph a title.

Amount of Money Nina Had

Practice
Use your line graph from the previous page to answer each question.

1. On which two days did Nina have the same amount of money?
   Tuesday and Wednesday

2. On which day did Nina have the least amount of money?
   Monday

3. How much money did Nina have on Friday?
   $1.75

4. How much more money did Nina have on Friday than on Monday?
   50 cents

5. Did Nina have more money on Thursday or on Friday?
   Friday

6. How much more money did Nina have on Friday than on Thursday?
   25 cents

Reflect
Would you use a bar graph, a line plot, or a line graph to represent each data set? Each type of graph should be used only once.

- the outside temperature over the past eight hours ____line graph____

- the favorite color of students in Sharon's class ____bar graph____

- the quiz grades of students in Miss Wilson's class ____line plot____

3 Reflect 10

Extended Response REASONING

Review students' answers to the Reflect prompt at the bottom of student page 35.

Have students give examples of data sets that should be displayed using line plots, sets that should be displayed using bar graphs, and sets that should be displayed using line graphs.

- **Could the data be displayed using any type of graph?** yes
- **Is there one type that works best for each situation?** usually

Discuss the differences in the types of data shown in each graph.

Real-World Application APPLYING

Pediatricians and other health-care providers use growth charts to evaluate a child's growth for the first twelve years of life. Each time a child is taken in for a checkup, the weight and height are recorded. The doctor can then compare the growth rate of the child with a line graph of average growth rates.

4 Assess

Informal Assessment

Use the Student Assessment Record, **Assessment,** page 100, to record informal observations.

APPLYING

Infant Growth
Did the student
- ☐ apply learning to new situations?
- ☐ contribute concepts?
- ☐ contribute answers?
- ☐ connect mathematics to the real world?

Lesson 3

Objective

Students use double line graphs to analyze and compare two different sets of data.

Materials

Additional Materials
- Straightedge
- Graph paper

Access Vocabulary

double line graph A line graph that compares two sets of data on the same coordinate grid

Creating Context

Ask students if their family has a place where they mark the height of growing children in the family. Explain that doctors collect data of the changing height of their patients to determine whether the child is healthy and developing on schedule.

1 Warm Up   5

Concept Building REASONING

- **Do you think siblings grow at the same rate?**
  Answers will vary. Allow students time to discuss their answers and reasons.
- **Can anyone think of a way we could compare the growth rates of siblings?** Allow time to discuss.
- **What type of graph could we use to display this data?** Allow time to discuss.

2 Engage   30

Skill Building

Display the Jones Children's data on the board or overhead projector. On the board, create a line graph for Casey. Then on the same graph, plot the data for Terrance. Be sure to use a different color or line style to distinguish the lines. Students should re-create on their papers what is displayed on the board.

Growth of Jones Children		
Week	Casey	Terrance
2	19 in.	21 in.
6	20 in.	22 in.
10	23 in.	24 in.
36	29 in.	30 in.

Sibling Rivalry UNDERSTANDING

Have students answer the following questions about the double line graph you have created on the board.

- **Which child grew the most between week 6 and week 18?** Casey
- **How many inches did Terrance grow from week 2 to week 36?** 9 inches
- **Of those weeks shown, during which weeks did Casey's and Terrance's height differ only by 1 inch?** week 18 and week 36

Monitoring Student Progress

If . . . students cannot remember which line represents which set of data,

Then . . . encourage them to highlight the legend or to write a reminder tip next to each line on the graph.

MathTools For additional practice creating line graphs, students should complete *eMathTools* Graphing Tool.

Using Student Pages

Have students complete *Workbook,* pages 36–37, on their own.

Data Shown in Line Graphs
Lesson 3

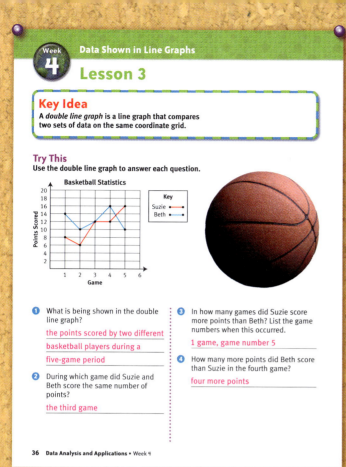

Key Idea

A *double line graph* is a line graph that compares two sets of data on the same coordinate grid.

Try This

Use the double line graph to answer each question.

Basketball Statistics

Key
Suzie
Beth

1 What is being shown in the double line graph?
the points scored by two different basketball players during a five-game period

2 During which game did Suzie and Beth score the same number of points?
the third game

3 In how many games did Suzie score more points than Beth? List the game numbers when this occurred.
1 game, game number 5

4 How many more points did Beth score than Suzie in the fourth game?
four more points

Practice
Use the double line graph to answer each question.

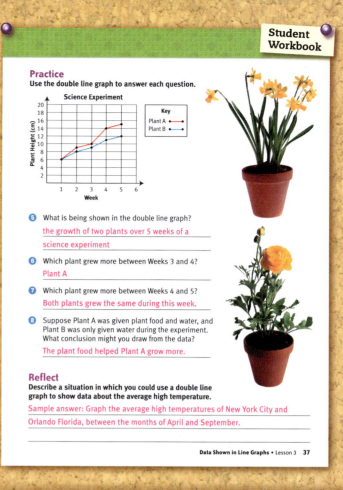

Science Experiment

Key
Plant A
Plant B

5 What is being shown in the double line graph?
the growth of two plants over 5 weeks of a science experiment

6 Which plant grew more between Weeks 3 and 4?
Plant A

7 Which plant grew more between Weeks 4 and 5?
Both plants grew the same during this week.

8 Suppose Plant A was given plant food and water, and Plant B was only given water during the experiment. What conclusion might you draw from the data?
The plant food helped Plant A grow more.

Reflect
Describe a situation in which you could use a double line graph to show data about the average high temperature.
Sample answer: Graph the average high temperatures of New York City and Orlando Florida, between the months of April and September.

Reflect   10

Extended Response   REASONING

Review students' answers to the Reflect prompt at the bottom of student page 37.

Have students share with the class their examples of data sets that can be displayed using a double line graph. Students should re-create their graphs on the board while the graph is explained.

■ **Could the data be displayed using another type of graph?** yes

■ **Is this type of graph the best for some situations?** yes

Discuss the strengths and weaknesses of the various types of graphs.

Real-World Application   APPLYING

In a triathlon, athletes compete by swimming, bicycling, and running three consecutive legs of a race. A double line graph can be used to compare the progress of two competitors at various points along the race. What would the labels for each axis of the graph be? vertical axis—distance, horizontal axis—time

Assess

Informal Assessment

Use the Student Assessment Record, *Assessment,* page 100, to record informal observations.

UNDERSTANDING

Sibling Rivalry
Did the student
❏ make important observations?
❏ extend or generalize learning?
❏ provide insightful answers?
❏ pose insightful questions?

Lesson 4

Objective

Students design an investigation and collect data to answer a statistical question.

Materials

Additional Materials
- Straightedge
- Graph paper
- Clock or watch with a second hand

Access Vocabulary

line graph A data display that is used to show how data change over a period of time

Creating Context

English Learners may benefit from clarification of some common phrases and words that proficient English speakers probably know. Occasionally words have more than one meaning or are used in potentially puzzling idiomatic expressions. Before or during the lesson, be sure to clarify the words that may be confusing to these students.

1 Warm Up 5

Have students discuss how they can compare their predictions about the outcome of an investigation with the actual results. The focus of the discussion should be on analyzing data that change over time.

2 Engage 30

Skill Building

You will need a clock or watch with a second hand. Students will see how long they can perform a certain activity. For example, students can see how long they can stand on one foot or how long they can hold their arms outstretched to the side.

How Many Are Left? COMPUTING

Discuss with students how long they think they can remain in the given position. If time permits, have them create a prediction line graph of how many students will be in the given position after increments of 10 seconds, 20 seconds, and so on up to 1 minute.

- **Make a prediction about how many students will be left after each minute.**

Have students begin in the given position and record how many students remain after increments of 10 seconds. When all students are finished, organize the results in a line graph.

- **How did your prediction compare to the actual results?** Answers will vary.

Monitoring Student Progress

| If . . . students are unable to participate in the activity, | Then . . . assign the tasks of being official timekeeper and data recorder. |

MathTools For additional practice creating line graphs, students should complete *eMathTools* Graphing Tool.

Using Student Pages

Have students complete *Workbook,* pages 38–39, on their own.

Lesson 4

Key Idea
You can make a prediction, design an investigation, collect data, and use the results to answer questions and check your prediction.

Try This
Follow the steps below to design an experiment and make a prediction.

Step 1 Decide on an investigation question that you find interesting. The investigation should involve numerical data that change over time. Here are some sample investigation questions that you can use, or you can create your own.

- How many students are on the playground at different times during the school day?
- How much rain has your city had during the past several months?
- What was the high or low temperature over the past several days?

❶ What is your investigation question?

Questions will vary.

Step 2 Make a prediction about what you expect to find in your investigation. Draw a prediction line graph of what you expect the data to look like.

❷ What is your prediction?

Predictions will vary.

Graphs will vary.

Practice
Continue following the steps below to collect your investigation data and analyze the results.

Step 3 Collect the data for your investigation.

❸ Record your data in a table.

Tables will vary.

Step 4 Display your data.

❹ Create a line graph.

Graphs will vary.

Step 5 Analyze the results of your investigation.

❻ Do the results agree or disagree with your predictions? What conclusions can you draw from the investigation? Explain.

Answers will vary.

Reflect
Suppose you collected data about the water temperature of Lake Erie between the months of September and February. Would you expect the line segments to show an increase or a decrease in the temperatures? Explain.

The temperature of the water should be getting colder during these months.

The segments in the line graph should be decreasing.

Reflect   10

Extended Response   REASONING

Review students' answers to the Reflect prompt at the bottom of student page 39.

- **What are some other examples of data sets that would probably be only increasing or only decreasing?** Sample answer: The amount of water in a bathtub after the drain plug is pulled would be only decreasing.

- **If you graphed the water temperature of Lake Erie for an entire year, what would the graph show?** increase and decrease

Real-World Application   APPLYING

Parents and doctors often keep a record of children's heights as they grow. If students have access to this data, have them display the past several years in a line graph. If the data are not available, have students make a line graph speculating how tall they think they were in second grade, in third grade, and so on.

Assess

Informal Assessment

Use the Student Assessment Record, *Assessment,* page 100, to record informal observations.

COMPUTING

How Many Are Left?

Did the student
- ❏ respond accurately?
- ❏ respond quickly?
- ❏ respond with confidence?
- ❏ self-correct?

Data Shown in Line Graphs • Lesson 4   38–39

Lesson 5 Review

Objective

Students review skills learned this week and complete the weekly assessment.

Materials

Review materials will be selected from those used in previous activities.

Creating Context

Review with English Learners the types of data displays we have studied. Ask questions that allow English Learners of early proficiency levels to answer by pointing to their choices. For English Learners who are at an intermediate level of proficiency or higher, pose comparing questions.

 Unit Assessment

Students should complete the Data Analysis Test found on **Assessment,** pages 82–83. Using the key on **Assessment,** page 99, identify incorrect responses. Reteach and review the Warm-Up and Engage activities to reinforce concept understanding.

1 Warm Up 5

Concept Building REASONING

■ Would it be appropriate to graph data on a line graph that shows the calories you consume and burn throughout the day? yes

■ Would it be appropriate to graph data on a line graph that shows the time it took for Ethan to read four different books? no

Allow time for students to discuss and explain their answers.

2 Engage 20

Skill Building

Free-Choice Activity ENGAGING

For the last day of the week, allow students to choose an activity from the previous lessons. Some activities they may choose include the following:

• **Climate Study**
• **Infant Growth**
• **Sibling Rivalry**
• **How Many Are Left?**

Make a note of the activity students select. Do they prefer easy or challenging activities? If you believe your students would benefit from extra practice on specific facts, choose an activity for them.

Monitoring Student Progress

If . . . students need more practice creating line graphs,	Then . . . they should choose **Infant Growth** for their **Free-Choice** activity. They should gather new data to use.

Using Student Pages

Have students complete **Workbook,** pages 40–41, on their own.

3 Reflect 10

Extended Response REASONING

Review students' answers to the Reflect prompts at the bottom of student pages 40–41.

Discuss the answers with the group to reinforce Week 4 concepts.

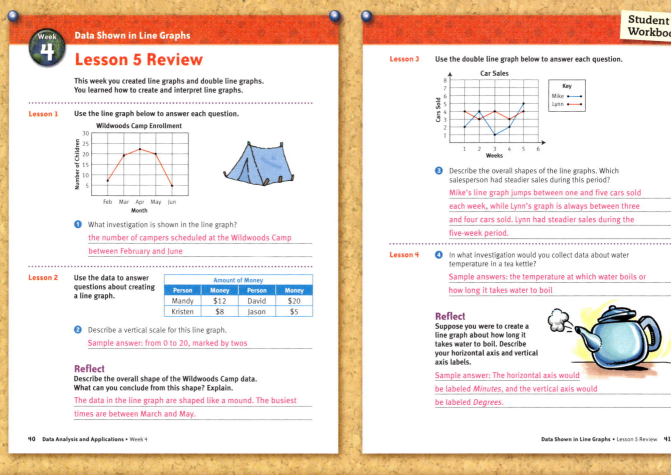

Lesson 5 Review

This week you created line graphs and double line graphs. You learned how to create and interpret line graphs.

Lesson 1 Use the line graph below to answer each question.

Wildwoods Camp Enrollment

❶ What investigation is shown in the line graph?

the number of campers scheduled at the Wildwoods Camp between February and June

Lesson 2 Use the data to answer questions about creating a line graph.

Amount of Money

Person	Money	Person	Money
Mandy	$12	David	$20
Kristen	$8	Jason	$5

❷ Describe a vertical scale for this line graph.

Sample answer: from 0 to 20, marked by twos

Reflect

Describe the overall shape of the Wildwoods Camp data. What can you conclude from this shape? Explain.

The data in the line graph are shaped like a mound. The busiest times are between March and May.

Student Workbook

Lesson 3 Use the double line graph below to answer each question.

Car Sales

Key
Mike
Lynn

❸ Describe the overall shapes of the line graphs. Which salesperson had steadier sales during this period?

Mike's line graph jumps between one and five cars sold each week, while Lynn's graph is always between three and four cars sold. Lynn had steadier sales during the five-week period.

Lesson 4 ❹ In what investigation would you collect data about water temperature in a tea kettle?

Sample answers: the temperature at which water boils or how long it takes water to boil

Reflect

Suppose you were to create a line graph about how long it takes water to boil. Describe your horizontal axis and vertical axis labels.

Sample answer: The horizontal axis would be labeled *Minutes*, and the vertical axis would be labeled *Degrees*.

---

4 Assess 10

A Gather Evidence

Formal Assessment

Have students complete the Weekly test, *Assessment,* pages 70–71. Record progress on the Student Assessment Record, *Assessment,* page 100.

B Summarize Findings

Determine whether students have Minimal, Basic, or Secure understanding of the concepts presented in Week 4.

C Differentiate Instruction

Based on your observations, use these teaching strategies next week to follow up.

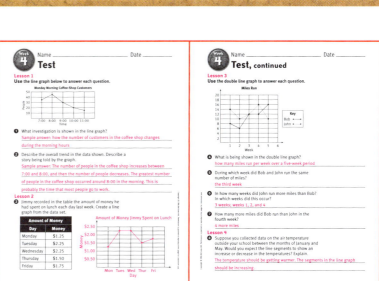

Assessment, pp. 70–71

Minimal Understanding

- Repeat the Warm-Up and Engage activities to develop Week 4 concepts.
- Use *eMathTools* Graphing Tool to develop and reinforce this week's concepts.

Basic Understanding

- Repeat Engage activities in subsequent weeks.
- Use *eMathTools* Graphing Tool to reinforce this week's concepts.

Secure Understanding

- Use *eMathTools* Graphing Tool to reinforce this week's concepts.
- Use variations of the weekly Warm-Up and Engage activities using higher numbers and multiple steps.

Data Shown in Line Graphs • Lesson 5 Review 40–41

Week 1 — Examining Data

Practice

Use the data displays below to answer the following questions.

Students' Family Pets		
Animal	Tally	Numbers
Fish	IIII	5
Dog	IIII IIII	9
Hamster	IIII I	6
Cat	IIII	5

Students' Favorite Color	
Color	Students
Red	12
Blue	16
Orange	8
Green	5

1 What question was likely asked in the Family Pets survey?
Sample answer: What kind of pet do you have at home?

2 What question was likely asked in the Favorite Color survey?
Sample answer: What is your favorite color?

3 What conclusion can you draw about the number of students with fish compared to the number of students with cats?
The same number of students have fish as students who have cats.

4 What conclusion can you draw about the color blue as compared to the color orange?
The color blue is twice as popular as the color orange.

Tell whether each question asks for *facts* or *opinions*. Then state whether the data gathered will be *categorical* or *numerical*.

5 What is your favorite type of music?
opinions; categorical

6 How many students went to camp over the summer?
facts; numerical

7 How long would you like to stay at the park?
opinions; numerical

Week 2 — Data and Bar Graphs

Practice

Make a bar graph of the data in the list.

Number of CDs You Own									
25	29	24	15	23	10	5	11	24	20
24	17	25	21	28	12	19	30	16	30

1

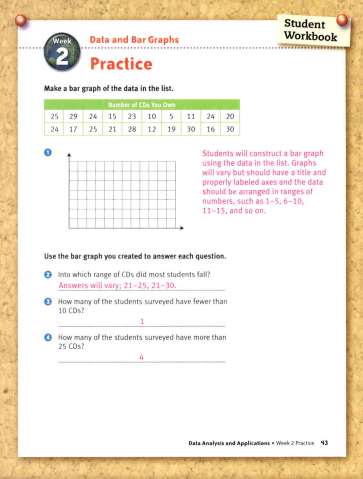

Students will construct a bar graph using the data in the list. Graphs will vary but should have a title and properly labeled axes and the data should be arranged in ranges of numbers, such as 1–5, 6–10, 11–15, and so on.

Use the bar graph you created to answer each question.

2 Into which range of CDs did most students fall?
Answers will vary; 21–25, 21–30.

3 How many of the students surveyed have fewer than 10 CDs?
1

4 How many of the students surveyed have more than 25 CDs?
4

Unit 6
Data Analysis and Applications Practice Pages

Weeks 1 and 2

1 Assign the Practice pages, *Workbook,* pages 42–43, at the end of Weeks 1 and 2. Students should complete these pages independently.

2 Check student answers using the annotated pages above.

3 If students have difficulty with a Practice page, review the activities completed throughout the week, and have students complete the weekly practice again before they complete the weekly test.

Week 3 — Showing Data in Line Plots

Practice

Make a line plot of the data below.

1 Number of home runs hit this season by 16 players.

11	14	19	16	15	18	15	13
15	19	15	13	14	16	14	10

Students should place 1 *X* over 10, 1 *X* over 11, 2 *X*s over 13, 3 *X*s over 14, 4 *X*s over 15, 2 *X*s over 16, 1 *X* over 18, and 2 *X*s over 19.

2 What statistical question might have been asked to generate the data?
Sample answer: "How many home runs did you hit this season?"

3 What is the range in the number of home runs hit this season?
9 home runs

4 Does the data set have any modes? If so, identify them.
yes; 15 home runs

5 Describe any clusters that appear in the data set.
A cluster of data points appears between 13 and 16 home runs.

6 Draw a conclusion from the line plot about how many players hit at least 16 home runs.
Five players hit 16 or more home runs.

Week 4 — Data Shown in Line Graphs

Practice

Use the data to answer questions about creating a line graph.

Daily Earnings				
Monday	Tuesday	Wednesday	Thursday	Friday
$ 5	$8	$9	$4	$3

1

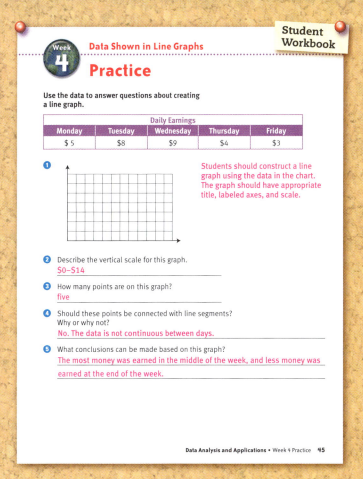

Students should construct a line graph using the data in the chart. The graph should have appropriate title, labeled axes, and scale.

2 Describe the vertical scale for this graph.
$0–$14

3 How many points are on this graph?
five

4 Should these points be connected with line segments? Why or why not?
No. The data is not continuous between days.

5 What conclusions can be made based on this graph?
The most money was earned in the middle of the week, and less money was earned at the end of the week.

Unit 6
Data Analysis and Applications Practice Pages

Weeks 3 and 4

1 Assign the Practice pages, *Workbook,* pages 44–45, at the end of Weeks 3 and 4. Students should complete these pages independently.

2 Check student answers using the annotated pages above.

3 If students have difficulty with a Practice page, review the activities completed throughout the week, and have students complete the weekly practice again before they complete the weekly test.

Appendix

Letter to Home

Dear Family,

In our math class, we try to relate things we learn in the classroom to the real world. This week students will explore different ways to visualize and represent equal quantities. In **Number Worlds,** Number Sense, your child will recognize equivalent representations for the same number and generate them by composing and decomposing quantities. He or she will separate and regroup double-digit and triple-digit numbers into hundreds, tens, and ones and gain experience with regrouping.

Below is an activity you can do with your child that relates the math we will be learning in this unit to the real world.

Sincerely,

Estimada familia:

En nuestra clase de matemáticas, tratamos de relacionar cosas que aprendemos en el salón de clases con la vida real. Esta semana los estudiantes explorarán las formas de visualizar y representar cantidades iguales. En **El mundo de los números:** Sentido numérico, su hijo(a) reconocerá las representaciones equivalentes para el mismo número y las generará al componer y descomponer cantidades. Su hijo(a) separará y reagrupará números de dos y tres dígitos en centenas, decenas y unidades, y ganará más experiencia al reagrupar.

Abajo, hay una actividad que puede hacer con su hijo que relaciona las matemáticas que estaremos aprendiendo en esta unidad con la vida real.

Sinceramente,

Cut construction paper into 15 or more rectangles, and label them so you have 13 or more "one-dollar" bills and 2 or more "ten-dollar" bills. Play money can also be used. Show the same amount of money using various combinations of bills. Practice trading 10 one-dollar bills for 1 ten-dollar bill. Later in the week, add a one-hundred-dollar bill along with more one- and ten-dollar bills.

Letter to Home

Dear Family,

In our math class, we try to relate things we learn in the classroom to the real world. This week students will investigate the meaning of place value. In **Number Worlds,** Number Sense, students will understand the place-value structure of the base-ten number system for numbers through the thousands place. They will identify the value for each digit in any four-digit number, and they will use expanded notation to represent numbers.

Below is an activity you can do with your child that relates the math we will be learning in this unit to the real world.

Sincerely,

Estimada familia:

En nuestra clase de matemáticas, tratamos de relacionar las cosas que aprendemos en el salón de clases con la vida real. Esta semana los estudiantes investigarán el significado del valor posicional. En **El mundo de los números:** Sentido numérico, su hijo(a) entenderá la estructura del valor posicional en un sistema numérico de base diez para los números hasta la posición de los millares. Ellos identificarán el valor de cada dígito en cualquier número de cuatro dígitos y usarán la notación desarrollada para representar números.

Abajo, hay una actividad que usted puede hacer con su hijo que relaciona las matemáticas que estaremos aprendiendo en esta unidad con la vida real.

Sinceramente,

Label ten index cards with the numerals 0 through 9, and mix them up. Draw three cards, and place one in each column on the Number Construction Mat. Say the number aloud. Repeat this activity with three new cards.

Hundreds	Tens	Ones

Letter to Home

Dear Family,

In our math class, we try to relate things we learn in the classroom to the real world. In **Number Worlds,** Number Sense, your child will gain experience with numbers 1–100 in a ten-by-ten arrangement. He or she should understand that in a ten-by-ten arrangement, horizontal movement affects the digit in the ones place and vertical movement affects the digit in the tens place. Your child will also practice successive addition mental computations.

Below is an activity you can do with your child that relates the math we will be learning in this unit to the real world.

Sincerely,

Estimada familia:

En nuestra clase de matemáticas, tratamos de relacionar las cosas que aprendemos en el salón de clases con la vida real. En **El mundo de los números:** Sentido numérico, su hijo(a) obtendrá experiencia con números del 1 al 100 en patrones de diez en diez. Su hijo(a) debe entender que en un patrón de diez en diez, el movimiento horizontal afecta el dígito en la posición de las unidades y el movimiento vertical afecta el dígito en la posición de las decenas. Su hijo(a) también practicará el cálculo mental de suma consecutiva.

Abajo, hay una actividad que usted puede hacer con su hijo que relaciona las matemáticas que estaremos aprendiendo en esta unidad con la vida real.

Sinceramente,

Place your finger on any number (e.g., 12). Choose a single-digit number (e.g., 5), and add it to the first number by counting and moving your finger along the row until you reach the sum of the two numbers (17). Say, "Twelve plus five equals seventeen." Do this several times with other numbers. Then skip count by tens and other numbers.

91	92	93	94	95	96	97	98	99	100
81	82	83	84	85	86	87	88	89	90
71	72	73	74	75	76	77	78	79	80
61	62	63	64	65	66	67	68	69	70
51	52	53	54	55	56	57	58	59	60
41	42	43	44	45	46	47	48	49	50
31	32	33	34	35	36	37	38	39	40
21	22	23	24	25	26	27	28	29	30
11	12	13	14	15	16	17	18	19	20
1	2	3	4	5	6	7	8	9	10

Letter to Home

Dear Family,

In our math class, we try to relate things we learn in the classroom to the real world. In **Number Worlds,** Number Sense, your child will compare and order whole numbers to 9,999, compare and order money amounts containing combinations of coins and bills, and develop an informal understanding of the fractions ½ and ¼. Your child will also correctly use the terms *greater than, less than,* and *equal to* when relating two numbers.

Below is an activity you can do with your child that relates the math we will be learning in this unit to the real world.

Sincerely,

Estimada familia:

En nuestra clase de matemáticas, tratamos de relacionar las cosas que aprendemos en el salón de clases con la vida real. En **El mundo de los números:** Sentido numérico, su hijo(a) comparará y ordenará números enteros hasta 9,999, comparará y ordenará cantidades de dinero combinando monedas y billetes, y desarrollará un razonamiento informal de las fracciones ½ y de ¼. Su hijo(a) también usará correctamente los términos *mayor que, menor que* e *igual a* cuando se relacionan dos números.

Abajo, hay una actividad que usted puede hacer con su hijo que relaciona las matemáticas que estaremos aprendiendo en esta unidad con la vida real.

Sinceramente,

Use coins to model various amounts of money. Say aloud each amount in dollars and cents. Practice until you are comfortable with this, and then show two different amounts at the same time. Say aloud the amounts, and use the terms *greater than, less than,* or *equal to* to compare the amounts.

Letter to Home

Dear Family,

In our math class, we try to relate things we learn in the classroom to the real world. Patterns are all around us. Patterns help us notice regularity, variety, and the ways topics interconnect. In **Number Worlds,** Number Patterns and Relationships, students will identify and build growing patterns. This week students will identify missing terms from patterns and create growing patterns using Pattern Blocks.

Below is an activity you can do with your child that relates the math we will be learning in this unit to the real world.

Sincerely,

Estimada familia:

En nuestra clase de matemáticas, tratamos de relacionar cosas que aprendemos en el salón de clases con la vida real. Hay patrones en todas partes. Nos ayudan a observar la regularidad, la variedad y la manera como se relacionan los temas. En **El mundo de los números:** patrones y relaciones numéricos, los estudiantes van a identificar y formar patrones de crecimiento. Esta semana los estudiantes van a identificar términos que faltan en los patrones y crear patrones de crecimiento usando figuras geométricas.

A continuación hay una actividad que pueden hacer con su hijo(a) que relaciona las matemáticas que vamos a aprender en esta unidad con la vida real.

Sinceramente,

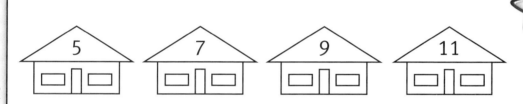

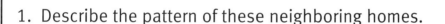

1. Describe the pattern of these neighboring homes.

2. What house number would be right before house number 5?

3. What house number would be right after house number 11?

4. What house number would be right after house number 15?

Letter to Home

Dear Family,

In our math class, we try to relate things we learn in the classroom to the real world. Patterns are all around us. Patterns help us notice regularity, variety, and the ways topics interconnect. In **Number Worlds,** Number Patterns and Relationships, students will continue to learn about patterns. Students will use tables to model real-world patterns and to help them make decisions.

Below is an activity you can do with your child that relates the math we will be learning in this unit to the real world.

Sincerely,

Estimada familia:

En nuestra clase de matemáticas, tratamos de relacionar cosas que aprendemos en el salón de clases con la vida real. Hay patrones en todas partes. Nos ayudan a observar la regularidad, la variedad y la manera como se relacionan los temas. En **El mundo de los números:** patrones y relaciones numéricos, los estudiantes seguirán aprendiendo sobre los patrones y usarán tablas para representar patrones de la vida real y tomar decisiones.

A continuación hay una actividad que pueden hacer con su hijo(a) que relaciona las matemáticas que vamos a aprender en esta unidad con la vida real.

Sinceramente,

Use the circle pattern to determine the relationship between sets and number of circles. Draw the missing set and fill in the missing values on the table.

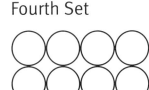

| First Set | Second Set | Third Set | Fourth Set |

Set	Number of Circles
1	
2	
3	
4	
5	
6	
10	

Letter to Home

Dear Family,

In our math class, we try to relate things we learn in the classroom to the real world. Patterns are all around us. Patterns help us notice regularity, variety, and the ways topics interconnect. In *Number Worlds,* Number Patterns and Relationships, students will construct and interpret graphs showing patterns from tables. Students will compare line graphs of two related patterns and use graphs to create stories.

Below is an activity you can do with your child that relates the math we will be learning in this unit to the real world.

Sincerely,

Estimada familia:

En nuestra clase de matemáticas, tratamos de relacionar cosas que aprendemos en el salón de clases con la vida real. Hay patrones en todas partes. Nos ayudan a observar la regularidad, la variedad y la manera como se relacionan los temas. En *El mundo de los números:* patrones y relaciones numéricos, los estudiantes van a construir e interpretar gráficos que muestran patrones a partir de tablas. Los estudiantes van a comparar gráficos lineales de dos patrones relacionados y usar gráficos para crear historias.

A continuación hay una actividad que pueden hacer con su hijo(a) que relaciona las matemáticas que vamos a aprender en esta unidad con la vida real.

Sinceramente,

The table below shows the amount of money Leslie puts into her piggy bank each week. Complete the table and create a coordinate graph of the total from the table below.

Week	Deposit	Total
1	$.50	
2	$.20	
3	$.10	
4	$.75	
5	$.50	
6	$.40	
7	$0	
8	$.15	
9	$.25	
10	$.30	

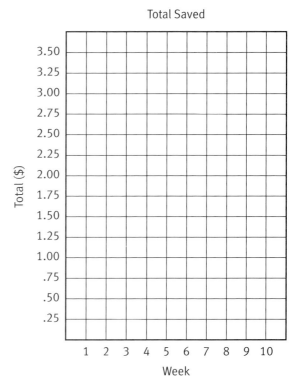

Total Saved

Letter to Home

Dear Family,

In our math class, we try to relate things we learn in the classroom to the real world. Patterns are all around us. Patterns help us notice regularity, variety, and the ways topics interconnect. In **Number Worlds,** Number Patterns and Relationships, students will explore variables and equations. Students will find unknown values in equations and use the concept of equality to help them solve problems involving weights.

Below is an activity you can do with your child that relates the math we will be learning in this unit to the real world.

Sincerely,

Estimada familia:

En nuestra clase de matemáticas, tratamos de relacionar cosas que aprendemos en el salón de clases con la vida real. Hay patrones en todas partes. Nos ayudan a observar la regularidad, la variedad y la manera como se relacionan los temas. **En El mundo de los números:** patrones y relaciones numéricos, los estudiantes van a explorar variables y ecuaciones, hallar valores desconocidos en ecuaciones y usar los conceptos de igualdad para resolver problemas que tienen que ver con el peso.

A continuación hay una actividad que pueden hacer con su hijo(a) que relaciona las matemáticas que vamos a aprender en esta unidad con la vida real.

Sinceramente,

On Monday, Leslie had $10.50 in her piggy bank. She spent some of her piggy bank money on Tuesday. How much money did Leslie spend if she now has $7.45 in her piggy bank? Use the equation below to help you solve this problem.

Equation: $10.50 − _____ = $7.45

After spending $4.30 at the student store, Mike now has $8 in his piggy bank. How much money was in Mike's piggy bank before he spent anything at the student store? Use the equation below to help you solve this problem.

Equation: _____ − $4.30 = $8.00

Letter to Home

Dear Family,

In our math class, we try to relate things we learn in the classroom to the real world. This week students will develop and extend their skills with addition and subtraction by solving word problems. In *Number Worlds,* Addition and Subtraction, your child will decide the operation needed, represent the problem with a number sentence, and solve the problem.

Below is an activity you can do with your child that relates the math we will be learning in this unit to the real world.

Sincerely,

Estimada familia:

En nuestra clase de matemáticas, tratamos de relacionar cosas que aprendemos en el salón de clases con la vida real. Esta semana, los estudiantes van a resolver problemas para desarrollar y ampliar sus destrezas con la suma y la resta. En *El mundo de los números:* la suma y la resta, su hijo(a) va a decidir qué operación se necesita, cómo representar el problema con un enunciado numérico y cómo resolverlo.

A continuación hay una actividad que pueden hacer con su hijo(a) que relaciona las matemáticas que vamos a aprender en esta unidad con la vida real.

Sinceramente,

Solve the following word problems, and then create and solve a word problem of your own using real situations and numbers from your life.

Janet read 16 books over the summer. Veronica read 5 more books than Janet. How many books did Veronica read? _____

Tony went to Andrew's house to watch two movies. The boys watched one movie for 148 minutes and a second movie for 119 minutes. How many minutes did the boys spend watching movies? _____

Bryan's metal building set had 2,675 pieces when it was new. Now Bryan has 1,278 pieces in his set. How many pieces of the metal building set has Bryan lost? _____

Letter to Home

Dear Family,

In our math class, we try to relate things we learn in the classroom to the real world. This week students will practice and more fully develop their computation skills. In *Number Worlds,* Addition and Subtraction, your child will also explore which numbers in a trio have the greatest and least sum and the greatest and least difference.

Below is an activity you can do with your child that relates the math we will be learning in this unit to the real world.

Sincerely,

Estimada familia:

En nuestra clase de matemáticas, tratamos de relacionar cosas que aprendemos en el salón de clases con la vida real. Esta semana, los estudiantes van a practicar y desarrollar más sus destrezas de cálculo. En *El mundo de los números:* la suma y la resta, su hijo(a) también va a explorar cuáles tres números: al sumarlos tienen la mayor y menor suma, y al restarlos tienen la mayor y menor diferencia.

A continuación hay una actividad que pueden hacer con su hijo(a) que relaciona las matemáticas que vamos a aprender en esta unidad con la vida real.

Sinceramente,

Discuss a situation such as going to a grocery store with $10 to buy four items. Talk about the need to estimate the prices of the four items to be sure that the sum of the prices is less than $10. In the space below, write the four items and an estimate of what each will cost. Add the estimates, and decide if you would have enough money for the purchase.

Letter to Home

Dear Family,

In our math class, we try to relate things we learn in the classroom to the real world. This week students will explore when and why to estimate sums and differences. In **Number Worlds,** Addition and Subtraction, your child will discover that estimation serves as an important companion to computation. It provides a tool for judging the reasonableness of calculator, mental, and paper-and-pencil computations.

Below is an activity you can do with your child that relates the math we will be learning in this unit to the real world.

Sincerely,

Estimada familia:

En nuestra clase de matemáticas, tratamos de relacionar cosas que aprendemos en el salón de clases con cosas de la vida real. Esta semana, los estudiantes van a explorar cuándo y por qué deben estimar sumas y diferencias. En **El mundo de los números:** la suma y la resta, su hijo(a) va a descubrir que la estimación es un elemento importante de la computación. Ésta provee una herramienta para determinar si los cálculos hechos con la calculadora, los mentales y los hechos con papel y lápiz son razonables.

A continuación hay una actividad que pueden hacer con su hijo(a), la cual relaciona las matemáticas que vamos a aprender en esta unidad con la vida real.

Sinceramente,

Choose five items from around the house, and list them below. Estimate what each item would cost if purchased today, and write that number next to the item. Find the actual cost of each item using sales papers, catalogs, or the Internet. Write down the actual cost of the item, and decide if your estimate was over, under, or close, and calculate the difference.

Item	Estimate	Actual Cost	Over/Under	Difference
Box of Cereal	$2	$2.49	under	49 cents

Letter to Home

Dear Family,

In our math class, we try to relate things we learn in the classroom to the real world. This week students experience addition and subtraction in the context of real-world situations using measurements. In **Number Worlds,** Addition and Subtraction, the focus is not on teaching measurement but on using measurement in context for practicing addition and subtraction.

Below is an activity you can do with your child that relates the math we will be learning in this unit to the real world.

Sincerely,

Estimada familia:

En nuestra clase de matemáticas, tratamos de relacionar cosas que aprendemos en el salón de clases con la vida real. Esta semana los estudiantes van a trabajar con la suma y la resta de medidas en situaciones de la vida real. En **El mundo de los números:** la suma y la resta, el objetivo no es enseñar a medir sino usar las medidas para practicar la suma y la resta.

A continuación hay una actividad que pueden hacer con su hijo(a) que relaciona las matemáticas que vamos a aprender en esta unidad con la vida real.

Sinceramente,

Copy a recipe in the space below. Examine the recipe together. Use empty measuring cups and measuring spoons to pretend to mix the recipe. At each step, be sure to choose the correct measure. For more of a challenge, double the recipe.

Letter to Home

Dear Family,

In our math class, we try to relate things we learn in the classroom to the real world. In **Number Worlds,** Multiplication, students will develop and deepen their conceptual knowledge of multiplication through the use of models. This week students use "groups" as models for multiplication. Students represent these groups using pictures.

Below is an activity you can do with your child that relates the math we will be learning in this unit to the real world.

Sincerely,

Estimada familia:

En nuestra clase de matemáticas, tratamos de relacionar cosas que aprendemos en el salón de clases con la vida real. En *El mundo de los números:* la multiplicación, los estudiantes van a desarrollar y profundizar su conocimiento conceptual de la multiplicación mediante el uso de modelos. Esta semana los estudiantes usarán "grupos" como modelos para la multiplicación. Ellos representan estos grupos usando dibujos.

A continuación hay una actividad que pueden hacer con su hijo(a) que relaciona las matemáticas que vamos a aprender en esta unidad con la vida real.

Sinceramente,

Look around the house, and make a list of items that belong in groups. Enter the items and their group number into the chart. Decide if the group number is usually the same or if the item commonly appears in other groups as well.

Item	Number in Group	Name of Group	Is This Number Consistent?
eggs	12	carton/dozen	yes
sneakers	2	pair	yes
cans of soda	6	six-pack	no; 12, 24, 2-liter, etc.

Letter to Home

Dear Family,

In our math class, we try to relate things we learn in the classroom to the real world. In *Number Worlds,* Multiplication, students will develop and deepen their conceptual knowledge of multiplication through the use of models. This week students continue to use "groups" as models for multiplication and turn those models into number sentences.

Below is an activity you can do with your child that relates the math we will be learning in this unit to the real world.

Sincerely,

Estimada familia:

En nuestra clase de matemáticas, tratamos de relacionar cosas que aprendemos en el salón de clases con la vida real. En *El mundo de los números:* la multiplicación, los estudiantes van a desarrollar y profundizar su conocimiento conceptual de la multiplicación mediante el uso de modelos. Esta semana los estudiantes usan "grupos" como modelos para multiplicar y los convierten en enunciados numéricos.

A continuación hay una actividad que pueden hacer con su hijo(a) que relaciona las matemáticas que vamos a aprender en esta unidad con la vida real.

Sinceramente,

Look around the house, and make a list of items that belong in groups, or use the same list from last week. Enter the items and their group numbers into the chart. Create and solve a number sentence using the information on the chart.

Number of Groups (multiplier)	Item	Number in Group	Number Sentence
2	dozen eggs	12	$2 \times 12 = 24$
3	pairs of sneakers	2	$3 \times 2 = 6$
4			
5			
6			
2			
3			
4			
5			

Letter to Home

Dear Family,

In our math class, we try to relate things we learn in the classroom to the real world. In **Number Worlds,** Multiplication, students will develop and deepen their conceptual knowledge of multiplication through the use of models. This week students continue to use "groups" as models for multiplication and turn those models into number sentences.

Below is an activity you can do with your child that relates the math we will be learning in this unit to the real world.

Sincerely,

Estimada familia:

En nuestra clase de matemáticas, tratamos de relacionar cosas que aprendemos en el salón de clases con la vida real. En **El mundo de los números:** la multiplicación, los estudiantes van a desarrollar y profundizar su conocimiento conceptual de la multiplicación mediante el uso de modelos. Esta semana los estudiantes seguirán usando "grupos" como modelos para multiplicar y los convertirán en enunciados numéricos.

A continuación hay una actividad que pueden hacer con su hijo(a) que relaciona las matemáticas que vamos a aprender en esta unidad con la vida real.

Sinceramente,

Make a list of items that belong in specific groups (3–9). Be creative, as some of these may be difficult. Write the items on the chart. Choose any number as the multiplier, and create a number sentence using the information on the chart.

Item	Number in Group	Number of Groups	Number Sentence
pair of sneakers	2	3	$2 \times 3 = 6$
	3		
	4		
	5		
	6		
	7		
	8		
	9		

Letter to Home

Dear Family,

In our math class, we try to relate things we learn in the classroom to the real world. In **Number Worlds,** Multiplication, students will develop and deepen their conceptual knowledge of multiplication through the use of models. This week students learn and develop thinking strategies for multiplication.

Below is an activity you can do with your child that relates the math we will be learning in this unit to the real world.

Sincerely,

Estimada familia:

En nuestra clase de matemáticas, tratamos de relacionar cosas que aprendemos en el salón de clases con la vida real. En **El mundo de los números:** la multiplicación, los estudiantes van a desarrollar y profundizar su conocimiento conceptual de la multiplicación mediante el uso de modelos. Esta semana los estudiantes aprenderán y desarrollarán estrategias de razonamiento para la multiplicación.

A continuación hay una actividad que pueden hacer con su hijo(a) que relaciona las matemáticas que vamos a aprender en esta unidad con la vida real.

Sinceramente,

Many times in the real world, items need to be multiplied by various factors to arrive at a final number. For example, socks come in pairs, and pairs of socks usually come in packages of six. A parent may buy two packages for two children. Below is a number sentence showing the total number of socks mentioned above.

$2 \times 6 \times 2 = 24$ socks

Describe two additional situations in which the original number is multiplied by several factors before arriving at a final number. Show a number sentence for each.

Letter to Home

Dear Family,

In our math class, we try to relate things we learn in the classroom to the real world. In **Number Worlds,** Geometry and Measurement, students will develop their visualization and thinking skills as they build, draw, and measure geometric shapes. This week students will recognize, construct and describe the characteristics of shapes—particularly quadrilaterals—and create geometric designs based on shape descriptions.

Below is an activity you can do with your child that relates the math we will be learning in this unit to the real world.

Sincerely,

Estimada familia:

En nuestra clase de matemáticas, tratamos de relacionar cosas que aprendemos en el salón de clases con la vida real. En **El mundo de los números:** geometría y medidas, los estudiantes van a desarrollar sus destrezas de visualización y razonamiento a medida que forman, trazan y miden figuras geométricas. Esta semana los estudiantes van a reconocer, construir y describir las características de las figuras, en especial las de los cuadriláteros, y crear diseños geométricos basados en descripciones de figuras.

A continuación hay una actividad que pueden hacer con su hijo(a) que relaciona las matemáticas que vamos a aprender en esta unidad con la vida real.

Sinceramente,

Look around the house, and find three items that are made of triangles or are roughly triangular in shape. Draw a picture of each item.

Letter to Home

Dear Family,

In our math class, we try to relate things we learn in the classroom to the real world. In **Number Worlds,** Geometry and Measurement, students will develop their visualization and thinking skills as they have hands-on experiences in sorting, building, drawing, manipulating, tracing, and measuring geometric shapes. This week students will identify congruent shapes and perform simple and complex transformations.

Below is an activity you can do with your child that relates the math we will be learning in this unit to the real world.

Sincerely,

Estimada familia:

En nuestra clase de matemáticas, tratamos de relacionar cosas que aprendemos en el salón de clases con la vida real. En **El mundo de los números:** geometría y medidas, los estudiantes van a desarrollar sus destrezas de visualización y razonamiento a medida que practican cómo clasificar, construir, dibujar, manipular, trazar y medir figuras geométricas. Esta semana los estudiantes van a identificar figuras congruentes y hacer transformaciones sencillas y complejas.

A continuación hay una actividad que pueden hacer con su hijo(a) que relaciona las matemáticas que vamos a aprender en esta unidad con la vida real.

Sinceramente,

Find three items around the house that are made of congruent shapes. Draw the items here.

Letter to Home

Dear Family,

In our math class, we try to relate things we learn in the classroom to the real world. In **Number Worlds,** Geometry and Measurement, students continue to manipulate and measure geometric shapes. This week students will define area and the units of measurement for area and calculate the area of rectangles and shapes consisting of multiple rectangles.

Below is an activity you can do with your child that relates the math we will be learning in this unit to the real world.

Sincerely,

Estimada familia:

En nuestra clase de matemáticas, tratamos de relacionar cosas que aprendemos en el salón de clases con la vida real. En **El mundo de los números:** geometría y medidas, los estudiantes seguirán manipulando y midiendo figuras geométricas. Esta semana los estudiantes van a definir el área y las unidades de medida para el área y calcular el área de rectángulos y figuras compuestas de varios rectángulos.

A continuación hay una actividad que pueden hacer con su hijo(a) que relaciona las matemáticas que vamos a aprender en esta unidad con la vida real.

Sinceramente,

Find five items around the house that can be modeled with a two-dimensional shape. List the items from those having the smallest area to those having the largest area.

Example: A table can be modeled by a rectangle.

Letter to Home

Dear Family,

In our math class, we try to relate things we learn in the classroom to the real world. In **Number Worlds,** Geometry and Measurement, students will develop their visualization and thinking skills as they have hands-on experiences in sorting, building, drawing, manipulating, tracing, and measuring geometric shapes. This week students investigate the various types of conversions and convert measurements for length, weight, volume, and time.

Below is an activity you can do with your child that relates the math we will be learning in this unit to the real world.

Sincerely,

Estimada familia:

En nuestra clase de matemáticas, tratamos de relacionar cosas que aprendemos en el salón de clases con la vida real. En **El mundo de los números:** geometría y medidas, los estudiantes van a desarrollar sus destrezas de visualización y razonamiento mientras practican cómo clasificar, construir, dibujar, manipular, trazar y medir figuras geométricas. Esta semana los estudiantes van a investigar varios tipos de conversiones y las medidas de conversión de longitud, peso, volumen y tiempo.

A continuación hay una actividad que pueden hacer con su hijo(a) que relaciona las matemáticas que vamos a aprender en esta unidad con la vida real.

Sinceramente,

Find ten large items around the house. Measure and record the length of each item in inches and feet. Which unit of measurement (inches or feet) is better for measuring each item? Explain.

Item	Inches	Feet	Explanation
Table	48	4	Four feet is a simpler measurement.

Letter to Home

Dear Family,

In our math class, we try to relate things we learn in the classroom to the real world. In **Number Worlds,** Data Analysis, students will explore different types of data and the ways to classify data. This week students will ask appropriate questions about a set of data and draw conclusions, categorize data as being fact or opinion, and sort elements in a data set.

Below is an activity you can do with your child that relates the math we will be learning in this unit to the real world.

Sincerely,

Estimada familia:

En nuestra clase de matemáticas, tratamos de relacionar cosas que aprendemos en el salón de clases con la vida real. En **El mundo de los números:** análisis de datos, los estudiantes van a explorar diferentes tipos de datos y la manera de clasificarlos. Esta semana los estudiantes van a hacer preguntas sobre un conjunto de datos y sacar conclusiones, categorizar datos como hecho u opinión, y clasificar elementos en un conjunto de datos.

A continuación hay una actividad que pueden hacer con su hijo(a) que relaciona las matemáticas que vamos a aprender en esta unidad con la vida real.

Sinceramente,

Ms. Wilbecca asked her students to identify their favorite type of pet. The following table shows the results of Ms. Wilbecca's survey.

Pet	Number
Dogs	12
Cats	9
Hamsters	3
Birds	2
Fish	2

What conclusions can be drawn?

_____

How many students are in Ms. Wilbecca's class? How did you know?

_____

Letter to Home

Dear Family,

In our math class, we try to relate things we learn in the classroom to the real world. In **Number Worlds,** Data Analysis, students will develop their statistical reasoning skills as they explore different types of data and the ways to classify data. This week students will create, interpret, and identify the advantages and disadvantages of using bar and double bar graphs.

Below is an activity you can do with your child that relates the math we will be learning in this unit to the real world.

Sincerely,

Estimada familia:

En nuestra clase de matemáticas, tratamos de relacionar cosas que aprendemos en el salón de clases con la vida real. En **El mundo de los números:** análisis de datos, los estudiantes van a desarrollar sus destrezas de razonamiento estadístico mientras exploran diferentes tipos de datos y la manera de clasificarlos. Esta semana los estudiantes van a crear, interpretar e identificar las ventajas y desventajas de usar un gráfico de barras y un gráfico de doble barra.

A continuación hay una actividad que pueden hacer con su hijo(a) que relaciona las matemáticas que vamos a aprender en esta unidad con la vida real.

Sinceramente,

Use the grid to create a bar graph displaying the results of the table. Be sure to include a title for the graph, and label the axes.

Pet	Number
Dogs	12
Cats	9

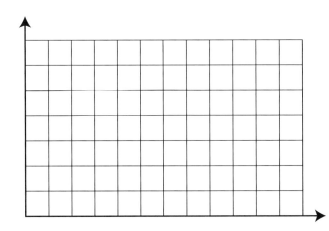

Letter to Home

Dear Family,

In our math class, we try to relate things we learn in the classroom to the real world. In **Number Worlds,** Data Analysis, students will explore different types of data and the ways to classify data. This week students will create, interpret, and identify the elements of line plots. Students will then conduct an investigation and use a line plot to analyze the results.

Below is an activity you can do with your child that relates the math we will be learning in this unit to the real world.

Sincerely,

Estimada familia:

En nuestra clase de matemáticas, tratamos de relacionar cosas que aprendemos en el salón de clases con la vida real. En **El mundo de los números:** análisis de datos, los estudiantes van a explorar diferentes tipos de datos y la manera de clasificarlos. Esta semana los estudiantes van a crear, interpretar e identificar los elementos de los diagramas de puntos. Luego, los estudiantes van a realizar una investigación y usar un diagrama de puntos para analizar los resultados.

A continuación hay una actividad que pueden hacer con su hijo(a) que relaciona las matemáticas que vamos a aprender en esta unidad con la vida real.

Sinceramente,

Students in Jacqueline's art class took a field trip to the local arts festival. The table shows the number of activities in which each student participated. Make a line plot of the data below. Be sure to include an appropriate label on the number line and a title.

4	3	2	5	5	3	1	4
4	4	3	1	2	3	5	4

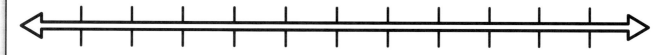

Copyright © SRA/McGraw-Hill. Permission is granted to reproduce this page for classroom use.

Letter to Home

Dear Family,

In our math class, we try to relate things we learn in the classroom to the real world. In **Number Worlds**, Data Analysis, students will explore different types of data and the ways to classify data. This week students will construct and interpret both line graphs and double line graphs and use a line graph to analyze the results of a conducted investigation.

Below is an activity you can do with your child that relates the math we will be learning in this unit to the real world.

Sincerely,

Estimada familia:

En nuestra clase de matemáticas, tratamos de relacionar cosas que aprendemos en el salón de clases con la vida real. En **El mundo de los números:** análisis de datos, los estudiantes van a explorar diferentes tipos de datos y la manera de clasificarlos. Esta semana los estudiantes van a construir e interpretar gráficos lineales y gráficos de doble línea y usar un gráfico lineal para analizar los resultados de una investigación dirigida.

A continuación hay una actividad que pueden hacer con su hijo(a) que relaciona las matemáticas que vamos a aprender en esta unidad con la vida real.

Sinceramente,

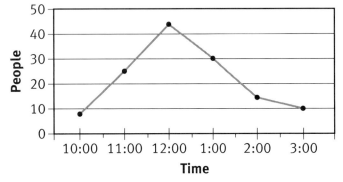

Number of Customers in Restaurant

Describe the story being told by the graph.

_____

_____

_____

_____

Number Construction Mat

Hundreds	Tens	Ones

Place Value Mat

Tens Waiting Area	Ones Waiting Area

Base-Ten Money Table

Amount	Combination of Coins

3-Digit Window

Hundreds	Tens	Ones

Hundreds	Tens	Ones

4-Digit Window

Thousands	Hundreds	Tens	Ones

Thousands	Hundreds	Tens	Ones

Blank Check Form

Date _____ **101**

Pay to the
Order of _____ | $ _____

_____ Dollars

Bank of **NUMBER WORLDS**™

_____   _____
Memo   Signature

Date _____ **102**

Pay to the
Order of _____ | $ _____

_____ Dollars

Bank of **NUMBER WORLDS**™

_____   _____
Memo   Signature

Date _____ **103**

Pay to the
Order of _____ | $ _____

_____ Dollars

Bank of **NUMBER WORLDS**™

_____   _____
Memo   Signature

Monster Cards

The Room Service Hotel

Materials
- Hotel Game Board
- Order Cards (slips of paper with room numbers)
- Pawns, one for each player
- Playing Chips to be used as "tokens"
- Blank Delivery Slips, several for each student

Remind the children how to move around on the Hotel Game Board.

- Have the students find the kitchen (door 0) and the stairs (going up the left side of the hotel).
- When a player receives an order for a ground floor room, he or she will need to exit the kitchen on the right and move down the hallway one door at a time to the destination room.
- When a player receives an order for a room above the ground floor, he or she will need to go up the correct number of flights of stairs (behind and above the kitchen) and through the hallway to the destination room.
- The only way to get from one level to another is to use the stairs; there is one flight of stairs between each floor, and each flight has ten steps.

Explain the rules of the game.

- Students will take turns picking an Order Card and moving their pawns to the appropriate room.
- Each time a student makes a delivery, he or she will earn one token for each flight of stairs climbed, to be collected at the end of the game.
- Because deliveries made to the ground floor are so close to the kitchen, they are considered "complimentary," and no tokens are earned for them.

- Players will use a Delivery Slip to keep track of their tokens by writing the number that shows which level they traveled to and how many tokens they earned.
- Each player will use a new Delivery Slip at the end of each turn.
- The player who earns the most tokens wins the game.

When all the players have had a turn and have filled out a Delivery Slip, explain that the student who is the closest to the kitchen will return to the kitchen first, and the student who is the farthest away will return last.

- Discuss who is the closest to the kitchen.
- During each player's turn to return to the kitchen, have the student say how he or she plans to travel back to the kitchen. For example, a child might say, "To get back to the kitchen, I will walk past five rooms and down three flights of stairs."

When each player has completed 10 deliveries or when time runs out, have the students use their Delivery Slips to figure out who earned the most tokens and wins the game.

- Allow the students to determine their totals in any way they choose.

Payment Options

Anna's Options

Anna's neighbors have hired her to pet-sit their dog for 7 days. They have offered two different options for being paid.

- **Option 1:** Anna receives $10 for the first day and $2 per day after the first day.
- **Option 2:** Anna receives $1 for the first day. Every day after the first day, she receives an additional amount that is $1 more than the previous day.

Option 1	
Day	**Amount Earned for the Day**
1	$10
2	$2
3	$2
4	$2
5	$2
6	$2
7	$2

Option 2	
Day	**Amount Earned for the Day**
1	$1
2	$2
3	$3
4	$4
5	$5
6	$6
7	$7

Jim's Options

Jim was hired to do yard work for his neighbor. The neighbor expects the work to last 5 days, but it could last 7 days. Payment options are as follows:

- **Option 1:** Jim receives $12 for the first day and $2 per day after the first day.
- **Option 2:** Jim receives $2 for the first two days. Every day after the second day, he receives an amount that is $1 more than the previous day.

Option 1		
Day	**Amount Earned for the Day**	**Total Earnings**
1	$12	$12
2	$2	$14
3	$2	$16
4	$2	$18
5	$2	$20
6	$2	$22
7	$2	$24

Option 2		
Day	**Amount Earned for the Day**	**Total Earnings**
1	$2	$2
2	$2	$4
3	$3	$7
4	$4	$11
5	$5	$16
6	$6	$22
7	$7	$29

Paddle Signs

Quick Estimate

Exact Answer

Picture Cards A

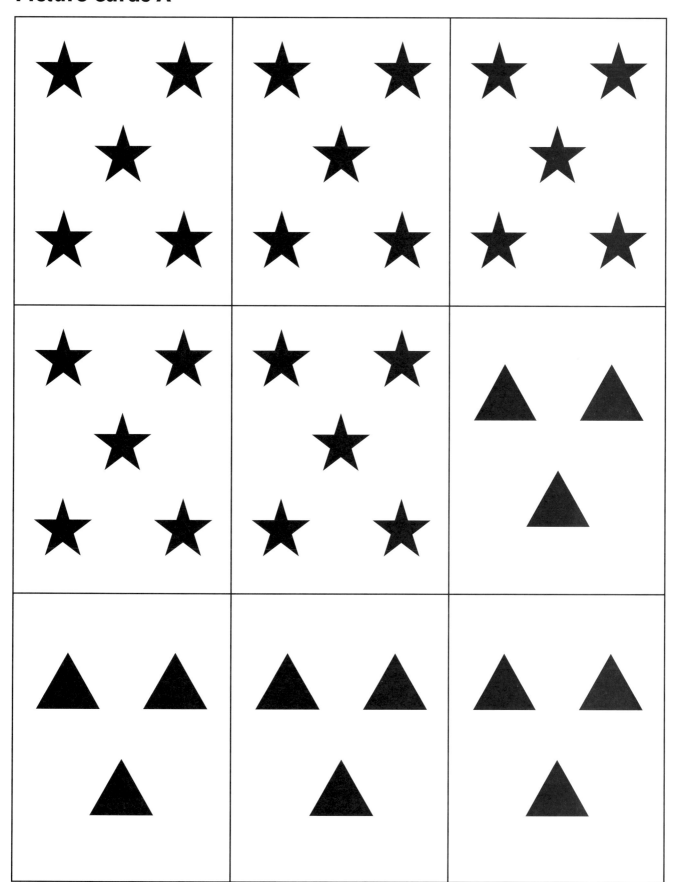

Picture Cards B

Recording Chart

Number of Groups	Number in Each Group	Multiplication Sentence

Number Chart 0–99

90	91	92	93	94	95	96	97	98	99
80	81	82	83	84	85	86	87	88	89
70	71	72	73	74	75	76	77	78	79
60	61	62	63	64	65	66	67	68	69
50	51	52	53	54	55	56	57	58	59
40	41	42	43	44	45	46	47	48	49
30	31	32	33	34	35	36	37	38	39
20	21	22	23	24	25	26	27	28	29
10	11	12	13	14	15	16	17	18	19
0	1	2	3	4	5	6	7	8	9

Multiplication Bingo

Materials
- Multiplication Bingo Cards, one for each player
- Playing Chips
- 2 Dot 1–6 Cubes
- Board or Flip Chart

Distribute a Bingo Mat and Playing Chips to each student.

- Tell the students that the object of the game is to be the first player to cover all the squares in a row or column or on the diagonal.
- Designate one student to be the Bingo Leader, to roll the dice, and to write the equations.

Explain the rules of the game.

- The Bingo Leader will roll two Dot Cubes, and the result of the multiplication of the two Dot Cubes will tell the players which square they can cover with a chip.
- For example, if the Dot Cubes show a 2 and a 3, players may cover either two rows of three dots or three rows of two dots on their Bingo Mats.
- On a board or flip chart, the Bingo Leader will write a multiplication equation using

the numbers on the Dot Cubes, such as $2 \times 3 =$, and will ask volunteers to say the answer and explain how they figured it out.

- If a combination that has already been rolled is rolled again, the players will select one of their lettered squares to cover.
- When a player has covered all the squares in a row or column or on the diagonal, he or she will call "Bingo!" and read each array in the winning line by stating the equation, such as "$2 \times 3 = 6$."
- While the player is reading the equations, the Bingo Leader and the other players should scan the list of equations on the board to verify that they were called.
- If the player makes a mistake, the Bingo Leader and other players will help him or her fix the mistake, and play will continue.
- The first person to call "Bingo!" correctly is the winner.

Multiplication Bingo Cards

Building Facts Chart

Array	_____ group(s) of _____	Multiplication Problem	Product

Multiplication Table

2-facts	3-facts	4-facts	5-facts	6-facts	7-facts	8-facts	9-facts

Multiplication Memory Game Cards A

1 × 2	1 × 3	1 × 4	1 × 5	1 × 6	1 × 7	1 × 8	1 × 9
2 × 2	2 × 3	2 × 4	2 × 5	2 × 6	2 × 7	2 × 8	2 × 9
3 × 2	3 × 3	3 × 4	3 × 5	3 × 6	3 × 7	3 × 8	3 × 9
4 × 2	4 × 3	4 × 4	4 × 5	4 × 6	4 × 7	4 × 8	4 × 9
5 × 2	5 × 3	5 × 4	5 × 5	5 × 6	5 × 7	5 × 8	5 × 9
6 × 2	6 × 3	6 × 4	6 × 5	6 × 6	6 × 7	6 × 8	6 × 9
7 × 2	7 × 3	7 × 4	7 × 5	7 × 6	7 × 7	7 × 8	7 × 9
8 × 2	8 × 3	8 × 4	8 × 5	8 × 6	8 × 7	8 × 8	8 × 9
9 × 2	9 × 3	9 × 4	9 × 5	9 × 6	9 × 7	9 × 8	9 × 9
10 × 2	10 × 3	10 × 4	10 × 5	10 × 6	10 × 7	10 × 8	10 × 9

Multiplication Memory Game Cards B

2	3	4	5	6	7	8	9
4	6	8	10	12	14	16	18
6	9	12	15	18	21	24	27
8	12	16	20	24	28	32	36
10	15	20	25	30	35	40	45
12	18	24	30	36	42	48	54
14	21	28	35	42	49	56	63
16	24	32	40	48	56	64	72
18	27	36	45	54	63	72	81
20	30	40	50	60	70	80	90

Product Bingo Cards A

B	I	N	G	O
18	4	5	7	18
20	24	6	14	27
3	16	10	21	36
12	40	12	28	40
2	8	60	35	90

B	I	N	G	O
2	8	6	35	9
6	16	24	42	16
9	24	15	56	18
27	32	12	63	36
30	40	45	70	56

B	I	N	G	O
4	12	45	7	8
12	20	15	21	16
14	28	20	35	27
24	40	35	49	32
9	4	50	63	48

B	I	N	G	O
3	36	54	14	9
12	28	18	28	18
18	20	24	42	24
21	40	42	56	45
20	12	60	70	80

Product Bingo Cards B

B	I	N	G	O
2	4	5	14	24
12	12	10	28	27
15	16	20	56	32
27	20	40	63	36
16	28	15	70	40

B	I	N	G	O
4	8	6	14	45
3	40	12	28	54
21	24	24	35	63
2	16	48	56	90
30	12	18	7	8

B	I	N	G	O
30	4	25	21	16
24	12	60	49	36
16	20	50	70	40
4	28	12	56	45
6	36	36	7	56

B	I	N	G	O
2	8	30	21	24
6	16	45	35	18
9	24	10	49	45
10	32	12	63	56
16	40	36	70	80

Cover-Up Game Board

Sample Bar Graphs

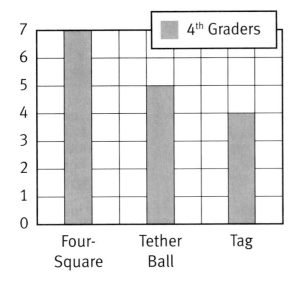

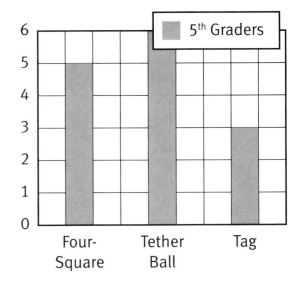

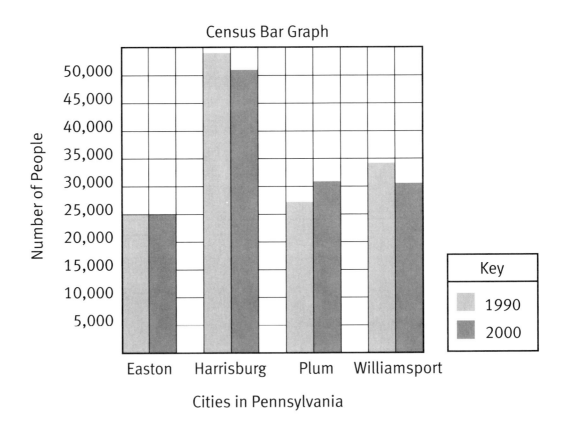

Census Bar Graph

Right Isosceles Triangles

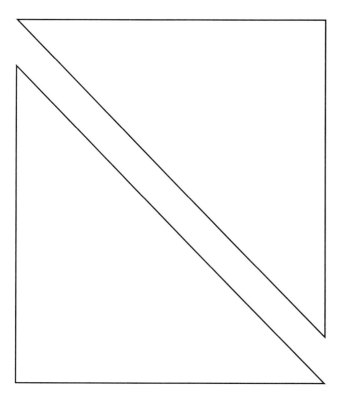

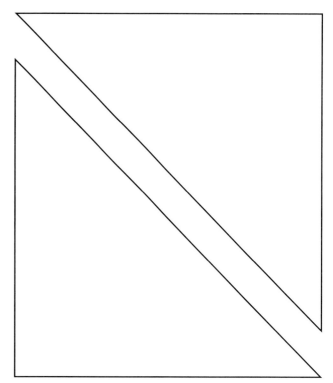

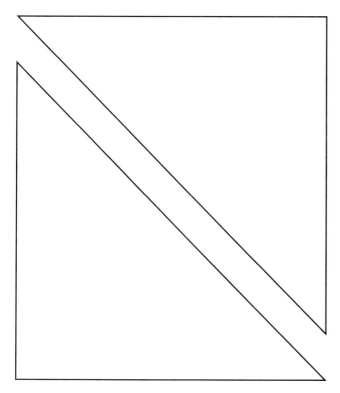

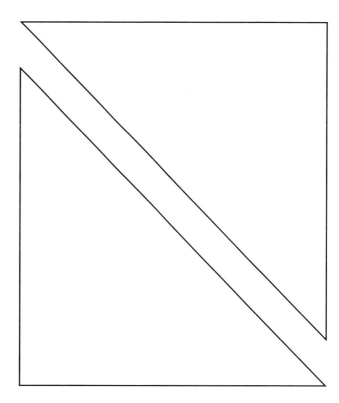

Tetronimos and Transformation Arrows

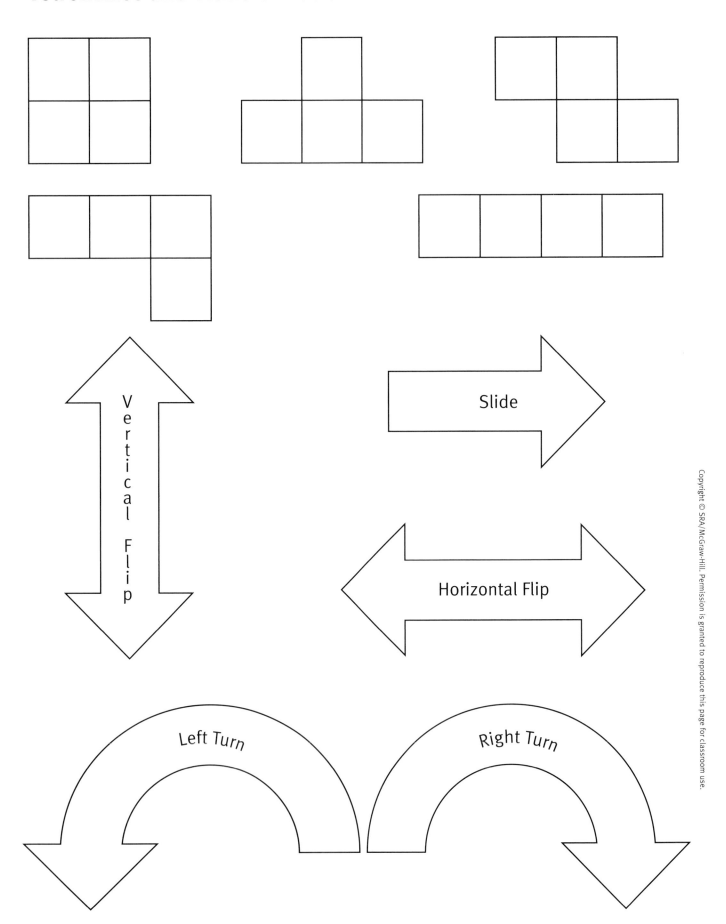

About Mathematics Intervention

What is an intervention?

An intervention is any instructional or practice activity designed to help students who are not making adequate progress.

- For struggling students, this requires an acceleration of development over a sufficient period of time.
- If the problem is small, the intervention might be brief.
- If the problem or lag in development is large, the intervention may last for weeks, months, or all year.

How is *Number Worlds* mathematics intervention different?

Number Worlds instruction is:
- More explicit
- More systematic
- More intense
- More supportive

What is the three-tier approach?

Layers of intervention are organized in tiers to enable schools to respond to student needs.

Tier 1
Core Instruction: high quality, comprehensive instruction for all students. It reduces the number of students who later become at-risk for academic problems.

Tier 2
Supplemental Intervention: Addresses essential content for students who are not making adequate progress in the core program. It reduces the need for more intensive intervention.

Tier 3
Intensive Intervention: Increases in intensity and duration for students with low-level skills and a sustained lack of adequate progress within Tiers 1 and 2.

Number Worlds is designed for both Tier 2 and Tier 3 students. Tier 2 students may spend a brief time in the program learning a key concept and then be quickly reintegrated back into the core instructional program. Levels A–C are also appropriate for Tier 1 students.

Students in Tier 3 will most likely need to complete the entire *Number Worlds* curriculum at their learning levels. Tier 3 requires intensive intervention for students with low skills and a sustained lack of adequate progress within Tiers 1 and 2. Teaching at this level is more intensive and includes more explicit instruction that is designed to meet the individual needs of struggling students. Group size is smaller, and the duration of daily instruction is longer.

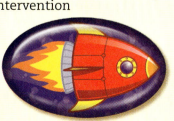

Building Number Sense with *Number Worlds:*

A Mathematics Program for Young Children

Sharon Griffin

What is number sense? We all know number sense when we see it but, if asked to define what it is and what it consists of, most of us, including the teachers among us, would have a much more difficult time. Yet this is precisely what we need to know to teach number sense effectively. Consider the answers three kindergarten children provide when asked the following question from the Number Knowledge Test (Griffin & Case, 1997): Which is bigger: seven or nine?"

> Brie responds quickly, saying "Nine." When asked how she figured it out, she says, "Well, you go, 'seven' (pause) 'eight', 'nine' (putting up two fingers while saying the last two numbers). That means nine has two more than seven. So it's bigger."
>
> Leah says, hesitantly, "Nine?" When asked how she figured it out, she says, "Because nine's a big number."
>
> Caitlin looks genuinely perplexed, as if the question was not a sensible thing to ask, and says, "I don't know."

Kindergarten teachers will immediately recognize that Brie's answer provides evidence of a well-developed number sense for this age level and Leah's answer, a more fragile and less-developed number sense. The knowledge that lies behind this "sense" may be much less apparent. What knowledge does Brie have that enables her to come up with the answer in the first place and to demonstrate good number sense in the process?

1 **Knowledge that underlies number sense**
Research conducted with the Number Knowledge Test and several other cognitive developmental measures (see Griffin, 2002; Griffin & Case, 1997 for a summary of this research) suggests that the following understandings lie at the heart of the number sense that 5-year-olds like Brie are able to demonstrate on this problem. They know (a) that numbers indicate quantity and therefore, that numbers, themselves, have magnitude; (b) that the word "bigger" or "more" is sensible in this context; (c) that the numbers 7 and 9, like every other number from 1 to 10, occupy fixed positions in the counting sequence; (d) that 7 comes before 9 when you are counting up; (e) that numbers that come later in the sequence— that are higher up— indicate larger quantities and therefore, that 9 is bigger (or more) than 7.

Brie provided evidence of an additional component of number sense in the explanation she provided for her answer. By using the Count-On strategy to show that nine comes two numbers after seven and by suggesting that this means "it has two more than seven," Brie demonstrated that she also knows (f) that each counting number up in the sequence corresponds precisely to an increase of one unit in the size of a set. This understanding, possibly more than any of the others listed above, enables children to use the counting numbers alone, without the need for real objects, to solve quantitative problems involving the joining of two sets. In so doing, it transforms mathematics from something that can only be done out there (e.g., by manipulating real objects) to something that can be done in their own heads, and under their own control.

This set of understandings, the core of *number sense,* forms a knowledge network that Case and Griffin (1990), see also Griffin and Case (1997), have called a *central conceptual structure for number*. Research conducted by these investigators has shown that this structure is central in at least two ways (see Griffin, Case, & Siegler, 1994). First, it enables children to make sense of a broad range of quantitative problems across contexts and to answer questions, for example, about two times on a clock (Which is longer?), two positions on a path (Which is farther?), and two sets of coins (Which is worth more?). Second, it provides the foundation on which children's learning of more complex number concepts, such as those involving double-digit numbers, is built. For this reason, this network of knowledge is an important set of understandings that should be taught in the preschool years, to all children who do not spontaneously acquire them.

2 How can this knowledge be taught?

Number Worlds, a mathematics program for young children (formerly called *Rightstart*), was specifically developed to teach this knowledge and to provide a test for the cognitive developmental theory (i.e., Central Conceptual Structure theory; see Case & Griffin, 1990) on which the program was based. Originally developed for kindergarten, the program (see Griffin & Case, 1995) was expanded to teach a broader range of understandings when research findings provided strong evidence that (a) children who were exposed to the program acquired the knowledge it was designed to teach (i.e., the central conceptual structure for number), and (b) the theoretical postulates on which the program was based were valid (see Griffin & Case, 1996; Griffin, Case, & Capodilupo, 1995; Griffin et al., 1994). Programs for grades one and two were developed to teach the more complex central conceptual structures that underlie base-ten understandings (see Griffin, 1997, 1998) and a program for preschool was developed (see Griffin, 2000) to teach the "precursor" understandings that lay the foundation for the development of the central conceptual structure for number.

Because the four levels of the program are based on a well-developed theory of cognitive development, they provided a finely graded sequence of activities (and associated knowledge objectives) that recapitulates the natural developmental progression for the age range of 3–9 years and that allows each child to enter the program at a point that is appropriate for his or her own development, and to progress through the program to teach 20 or more children at any one time and every effort has been made, in the construction of the **Number Worlds** program, to make it as easy as possible for teachers to accommodate the developmental needs of individual children (or groups of children) in their classroom. Five instructional principles that lie at the heart of the program are described below and are used to illustrate several features of the program that have already been mentioned and several that have not yet been introduced.

2.1 *Principle 1: Build upon children's current knowledge*

Each new idea that is presented to children must connect to their existing knowledge if it is going to make any sense at all. Children must also be allowed to use their existing knowledge to construct new knowledge that is within reach— that is one step beyond where they are now— and a set of bridging contexts and other instructional supports should be in place to enable them to do so.

In the examples of children's thinking presented earlier, three different levels of knowledge are apparent. Brie appears to have acquired the knowledge network that underlies number sense and to be ready, therefore, to move on to the next developmental level: to connect this set of understandings to the written numerals (i.e., the formal symbols) associated with each counting word. Leah appears to have some understanding of some of the components of this network (i.e., that number have magnitude) and to be ready to use this understanding as a base to acquire the remaining understandings (e.g., that a number's magnitude and its position in the counting sequence are directly related). Caitlin demonstrated little understanding of any element of this knowledge network and she might benefit, therefore, from exposure to activities that will help her acquire the "precursor" knowledge needed to build this network, namely knowledge of counting (e.g., the one-to-one correspondence rule) and knowledge of quantity (e.g., an intuitive understanding of relative amount). Although all three children are in kindergarten, each child appears to be at a different point in the developmental trajectory and to require a different set of learning opportunities; ones that will enable each child to use her existing knowledge to construct new knowledge at the next level up.

To meet these individual needs, teachers need (a) a way to assess children's current knowledge, (b) activities that are multi-leveled so children with different entering knowledge can all benefit from exposure to them, and (c) activities that are carefully sequenced and that span several developmental levels so children with different entering knowledge can be exposed to activities that are appropriate for their level of understanding. These are all available in the **Number Worlds** program and are illustrated in various sections of this paper.

Program Research

2.2 Principle 2: Follow the natural developmental progression when selecting new knowledge to be taught

Researchers who have investigated the manner in which children construct number knowledge between the ages of 3 and 9 years have identified a common progression that most, if not all, children follow (see Griffin, 2002; Griffin and Case, 1997 for a summary of this research). As suggested earlier, by the age of 4 years, most children have constructed two "precursor" knowledge networks—knowledge of counting and knowledge of quantity—that are separate in this stage and that provide the base for the next developmental stage. Sometime in kindergarten, children become able to integrate these knowledge networks—to connect the world of counting numbers to the world of quantity—and to construct the central conceptual understandings that were described earlier. Around the age of 6 or 7 years, children connect this integrated knowledge network to the world of formal symbols and, by the age of 8 or 9 years, most children become capable of expanding this knowledge network to deal with double-digit numbers and the base-ten system. A mathematics program that provides opportunities for children to use their current knowledge to construct new knowledge that is a natural next step, and that fits their spontaneous development, will have the best chance of helping children make maximum progress in their mathematics learning and development.

Because there are limits in development on the complexity of information children can handle at any particular age/stage (see Case, 1992), it makes no sense to attempt to speed up the developmental process by accelerating children through the curriculum. However, for children who are at an age when they should have acquired the developmental milestones but for some reason haven't, exposure to a curriculum that will give them ample opportunities to do so makes tremendous sense. It will enable them to catch up to their peers and thus, to benefit from the formal mathematics instruction that is provided in school. Children who are developing normally also benefit from opportunities to broaden and deepen the knowledge networks they are constructing, to strengthen these understandings, and to use them in a variety of contexts.

2.3 Principle 3: Teach computational fluency as well as conceptual understanding

Because computational fluency and conceptual understanding have been found to go hand in hand in children's mathematical development (see Griffin, 2003; Griffin et al., 1994), opportunities to acquire computational fluency, as well as conceptual understanding, are built into every *Number Worlds* activity. This is nicely illustrated in the following activities, drawn from different levels of the program.

In The Mouse and the Cookie Jar Game (created for the preschool program and designed to give 3- to 4-year-olds an intuitive understanding of subtraction), children are given a certain number of counting chips (with each child receiving the same number but a different color) and told to pretend their chips are cookies. They are asked to count their cookies and, making sure they remember how many they have and what their color is, to deposit them in the cookie jar for safe keeping. While the children sleep, a little mouse comes along and takes one (or two) cookies from the jar. The problem that is then posed to the children is "How can we figure out whose cookie(s) the mouse took?"

Although children quickly learn that emptying the jar and counting the set of cookies that bears their own color is a useful strategy to use to solve this problem, it takes considerably longer for many children to realize that, if they now have four cookies (and originally had five), it means that they have one fewer and the mouse has probably taken one of their cookies. Children explore this problem by counting and recounting the remaining sets, comparing them to each other (e.g., by aligning them) to see who has the most or least, and ultimately coming up with a prediction. When a prediction is made, children search the mouse's hole to see whose cookie had been taken and to verify or revise their prediction. As well as providing opportunities to perfect their counting skills, this activity gives children concrete opportunities to experience simple quantity transformations and to discover how the counting numbers can be used to predict and explain differences in amount.

The *Dragon Quest Game* that was developed for the Grade 1 program teaches a much more sophisticated set of understandings. Children are introduced to Phase 1 activity by being told a story about a fire-breathing dragon that has been terrorizing the village where children live. The children playing the game are heroes who have been chosen to seek out the dragon and put out his fire. To extinguish this dragon's fire (as opposed to the other, more powerful dragons they will encounter in later phases) a hero will need at least 10 pails of water. If a hero enters into the dragon's area with less than 10 pails of water, he or she will become the dragon's prisoner and can only be rescued by one of the other players.

To play the game, children take turns rolling a die and moving their playing piece along the colored game board. If they land on a well pile (indicated by a star), they can pick a card from the face-down deck of cards, which illustrate, with images and symbols (e.g., +4) a certain number of pails of water. Children are encouraged to add up their pails of water as they receive them and they are allowed to use a variety of strategies to do so, ranging from mental math (which is encouraged) to the use of tokens to keep track of the quantity accumulated. The first child to reach the dragon's lair with at least 10 pails of water can put out the dragon's fire and free any teammates who have become prisoners.

As children play this game and talk about their progress, they have ample opportunity to connect numbers to several different quantity representations (e.g., dot patterns on the die; distance of their pawn along the path; sets of buckets illustrated on the cards; written numerals also provided on the cards) and to acquire an appreciation of numerical magnitude across these contexts. With repeated play, they also become capable of performing a series of successive addition operations *in their heads* and of expanding the well pile. When they are required to submit formal proof to the mayor of the village that they have amassed sufficient pails of water to put out the dragon's fire before they are allowed to do so, they become capable of writing a series of formal expressions to record the number of pails received and spilled over the course of the game. In contexts such as these children receive ample opportunity to use the formal symbol system in increasingly efficient ways to make sense of quantitative problems they encounter in the course of their own activity.

2.4 Principle 4: provide plenty of opportunity for hands-on exploration, problem-solving, and communication

Like the *Dragon Quest Game* that was just described, many of the activities created for the **Number Worlds** program are set in a game format that provides plenty of opportunity for hands-on exploration of number concepts, for problem-solving and for communication. Communication is explicitly encouraged in a set of question prompts that are included with each small group game (e.g., How far are you now? How many more buckets do you need to put out the dragon's fire? How do you know?) as well as in a more general set of dialogue prompts that are included in the teacher's guide. Opportunities for children to discuss what they learned during game play each day, to share their knowledge with their peers, and to make their reasoning explicit are also provided in a Wrap-Up session that is included at the end of each math lesson.

Finally, in the whole group games and activities that were developed for the Warm-Up portion of each math lesson, children are given ample opportunity to count (e.g., up from 1 and down from 10) and to solve mental math problems, in a variety of contexts. In addition to developing computational fluency, these activities expose children to the language of mathematics and give them practice using it. Although this is valuable for all children, it is especially useful for ESL children, who may know how to count in their native language but not yet in English. Allowing children to take turns in these activities and to perform individually gives teachers opportunities to assess each child's current level of functioning, important for instructional planning, and gives children opportunities to learn from each other.

2.5 Principle 5: Expose children to the major ways number is represented and talked about in developed societies

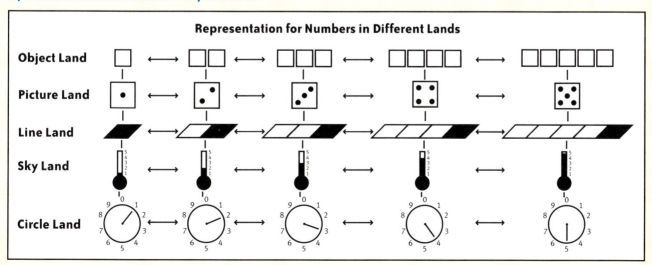

Program Research

Number is represented in our culture in five major ways: as a group of objects, a dot-set pattern, a position on a line, a position on a scale (e.g., a thermometer), and a point on a dial. In each of these contexts, number is also talked about in different ways, with a larger number (and quantity) described as "more" in the world of dot-sets, as "further along" in the world of paths and lines, as "higher up" in the world of scale measures, and as "further around" in the world of dials. Children who are familiar with these forms of representation and the language used to talk about number in these contexts have a much easier time making sense of the number problems they encounter inside and outside of school.

In the **Number Worlds** program, children are systematically exposed to these forms of representations as they explore five different "lands." Learning activities developed for each land share a particular form of number representation while they simultaneously address specific knowledge goals for each grade level. Many of the games, like *Dragon Quest*, also expose children to multiple representations of number in one activity so children can gradually come to see the ways they are equivalent.

3 Discussion

Children who have been exposed to the **Number Worlds** program do very well on number questions like the one presented in the introduction and on the Number Knowledge Test (Griffin & Case, 1997) from which this question was drawn. In several evaluation studies conducted with children from low-income communities, children who received the **Number Worlds** program made significant gains in conceptual knowledge of number and in number sense, when compared to matched-control groups who received readiness training of a different sort. These gains enabled them to start their formal schooling in grade one on an equal footing with their more advantaged peers, to perform as well as groups of children from China and Japan on a computation test administered at the end of grade one, and to keep pace with their more advantaged peers (and even outperform them on some measures) as they progressed through the first few years of formal schooling (Griffin & Case, 1997).

Teachers also report positive gains from using the **Number Worlds** program and from exposure to the instructional principles on which it is based. Although all teachers acknowledge that implementing the program and putting the principles into action is not an easy task, many claim that their teaching of all subjects has been transformed in the process. They now facilitate discussion rather than dominating it; they pay much more attention to what children say and do; and they now allow children to take more responsibility for their own learning, with positive and surprising results. Above all, they now look forward to teaching math and they and their students are eager to do more of it.

Griffin, S. "Building Number Sense with Number Worlds: A mathematics program for young children." In *Early Childhood Research Quarterly* 19 (2004) 173–180. Elsevier.

References

Case, R. (1992). *The mind's staircase: Exploring the conceptual underpinnings of children's thought and knowledge.* Hillsdale, NJ: Erlbaum.

Case, R., & Griffin, S. (1990). Child cognitive development: The role of central conceptual structures in the development of scientific and social thought. In E.A. Hauert (Ed.), *Developmental psychology: Cognitive, perceptuo-motor, and neurological perspectives* (pp. 193–230). North-Holland: Elsevier.

Griffin, S. (1997). *Number Worlds: Grade one level.* Durham, NH: Number Worlds Alliance Inc.

Griffin, S. (1998). *Number Worlds: Grade two level.* Durham, NH: Number Worlds Alliance Inc.

Griffin, S. (2000). *Number Worlds: Preschool level.* Durham, NH: Number Worlds Alliance Inc.

Griffin, S. (2002). The development of math competence in the preschool and early school years: Cognitive foundations and instructional strategies. In J. Royer (Ed.), *Mathematical cognition* (pp. 1–32). Greenwich, CT: Information Age Publishing.

Griffin, S. (2003). Laying the foundation for computational fluency in early childhood. *Teaching children mathematics,* 9, 306–309.

Griffin, S. (1997). *Number Worlds: Kindergarten level.* Durham, NH: Number Worlds Alliance Inc.

Griffin, S., & Case, R. (1996). Evaluating the breadth and depth of training effects when central conceptual structures are taught. *Society for research in child development monographs,* 59, 90–113.

Griffin, S., & Case, R. (1997). Re-thinking the primary school math curriculum: An approach based on cognitive science. *Issues in Education, 3,* 1–49.

Griffin, S., Case, R., & Siegler, R. (1994). Rightstart: Providing the central conceptual prerequisites for first formal learning of arithmetic to students at-risk for school failure. In K. McGilly (Ed.), *Classroom lessons: Integrating cognitive theory and classroom practice* (pp. 24-49). Cambridge, MA: Bradford Books MIT Press.

Griffin, S., Case, R., & Capodilupo, A. (1995). Teaching for understanding: The importance of central conceptual structures in the elementary mathematics curriculum. In A. McKeough, I. Lupert, & A. Marini (Eds.), *Teaching for transfer: Fostering generalization in learning* (pp. 121–151). Hillsdale, NJ: Erlbaum.

Mathematical proficiency has five strands. These strands are not independent; they represent different aspects of a complex whole . . . they are interwoven and interdependent in the development of proficiency in mathematics.

—Kilpatrick, J., Swafford, J., and Findell, B., eds. *Adding It Up: Helping Children Learn Mathematics.* Washington D.C.: National Research Council/National Academy Press, 2001, pp. 115–133.

Number Worlds develops all five proficiencies in each lesson, so that students build their skills, conceptual understanding, and reasoning powers as well as their abilities to apply mathematics and see it as useful.

Math Proficiencies

1 **UNDERSTANDING** (Conceptual Understanding): Comprehending mathematical concepts, operations, and relations—knowing what mathematical symbols, diagrams, and procedures mean. *Conceptual Understanding* refers to an integrated and functional grasp of mathematical ideas. Students with conceptual understanding know more than isolated facts and methods. They understand why a mathematical idea is important and the kinds of contexts in which it is useful. They have organized their knowledge into a coherent whole, which enables them to learn new ideas by connecting those ideas to what they already know. Conceptual understanding also supports retention. Because facts and methods learned with understanding are connected, they are easier to remember and use, and they can be reconstructed when forgotten. If students understand a method, they are unlikely to remember it incorrectly.

A significant indicator of conceptual understanding is being able to represent mathematical situations in different ways and knowing how different representations can be useful for different purposes.

Knowledge that has been learned with understanding provides the bases for generating new knowledge and for solving new and unfamiliar problems. When students have acquired conceptual understanding in an area of mathematics, they see the connections among concepts and procedures and can give arguments to explain why some facts are consequences of others. They gain confidence, which then provides a base from which they can move to another level of understanding.

2 **COMPUTING** (Procedural Fluency): Carrying out mathematical procedures, such as adding, subtracting, multiplying, and dividing numbers flexibly, accurately, efficiently, and appropriately. *Procedural Fluency* refers to knowledge of procedures, knowledge of when and how to use them appropriately, and skill in performing them flexibly, accurately, and efficiently. In the domain of number, procedural fluency is especially needed to support conceptual understanding of place value and the meaning of rational numbers. It also supports the analysis of similarities and differences among methods of calculating. These methods include written procedures, and mental methods for finding certain sums, differences, products, or quotients, and methods that use calculators, computers, or manipulative materials such as blocks, counters, or beads.

Students need to be efficient and accurate in performing basic computations with whole numbers without always having to refer to tables or other aids. They also need to know reasonably efficient and accurate ways to add, subtract, multiply, and divide multidigit numbers, mentally and with pencil and paper. A good conceptual understanding of place value in the base-ten system supports the development of fluency in multidigit computation. Such understanding also supports simplified but accurate mental arithmetic and more flexible ways of dealing with numbers than many students ultimately achieve.

Math Proficiencies

3 **APPLYING** (Strategic Competence): Being able to formulate problems mathematically and to devise strategies for solving them using concepts and procedures appropriately. *Strategic Competence* refers to the ability to formulate mathematical problems, represent them, and solve them. This strand is similar to what has been called problem solving and problem formulation. Although in school, students are often presented with clearly specified problems to solve, outside of school they encounter situations in which part of the difficulty is to figure out exactly what the problem is. Then they need to formulate the problem so that they can use mathematics to solve it. Consequently, they are likely to need experience and practice in problem formulating as well as in problem solving. They should know a variety of solution strategies as well as which strategies might be useful for solving a specific problem.

To represent a problem accurately, students must first understand the situation, including its key features. They then need to generate a mathematical representation of the problem that captures the core mathematical elements and ignores the irrelevant features.

Students develop procedural fluency as they use their strategic competence to choose among effective procedures. They also learn that solving challenging mathematics problems depends on the ability to carry out procedures readily, and conversely, that problem-solving experience helps them acquire new concepts and skills.

4 **REASONING** (Adaptive Reasoning): Using logic to explain and justify a solution to a problem or to extend from something known to something not yet known. *Adaptive Reasoning* refers to the capacity to think logically about the relationships among concepts and situations. Such reasoning is correct and valid, stems from careful consideration of alternatives, and includes knowledge of how to justify the conclusions. In mathematics, adaptive reasoning is the glue that holds everything together and guides learning. One uses it to navigate through the many facts, procedures, concepts, and solution methods and to see that they all fit together in some way that they make sense. In mathematics, deductive reasoning is used to settle disputes and disagreements. Answers are right because they follow from some agreed-upon assumptions through a series of logical steps. Students who disagree about a mathematical answer need not rely on checking with the teacher, collecting opinions from their classmates, or gathering data from outside the classroom. In principle, they need only check that their reasoning is valid.

Research suggests that students are able to display reasoning ability when three conditions are met: They have a sufficient knowledge base, the task is understandable and motivating, and the context is familiar and comfortable.

5 **ENGAGING** (Productive Disposition): Seeing mathematics as sensible, useful, and doable—if you work at it—and being willing to do the work. *Productive disposition* refers to the tendency to see sense in mathematics, to perceive it as both useful and worthwhile, to believe that steady effort in learning mathematics pays off, and to see oneself as an effective learner and doer of mathematics. If students are to develop conceptual understanding, procedural fluency, strategic competence, and adaptive reasoning abilities, they must believe mathematics is understandable, not arbitrary; that with diligent effort, it can be learned and used; and that they are capable of figuring it out. Developing a productive disposition requires frequent opportunities to make sense of mathematics, to recognize the benefits of perseverance, and to experience the rewards of sense making in mathematics.

Students' disposition toward mathematics is a major factor in determining their educational success. Students who have developed a productive disposition are confident in their knowledge and abilities. They see that mathematics is both reasonable and intelligible and believe that, with appropriate effort and experience, they can learn.

Content Strands of Mathematics

Number Sense and Place Value

Understanding of the significance and use of numbers in counting, measuring, comparing, and ordering.

"It is very important for teachers to provide children with opportunities to recognize the meaning of mathematical symbols, mathematical operations, and the patterns or relationships represented in the child's work with numbers. For example, the number sense that a child acquires should be based upon an understanding that inverse operations, such as addition and subtraction, undo the operations of the other. Instructionally, teachers must encourage their students to think beyond simply finding the answer and to actually have them think about the numerical relationships that are being represented or modeled by the symbols, words, or materials being used in the lesson."

Kilpatrick, J., Swafford, J. and Findell, B. eds. *Adding It Up: Helping Children Learn Mathematics.* Washington, D.C.: National Research Council/National Academy Press, 2001, p. 270–271.

Number Worlds and Number Sense

Goal: Firm understanding of the significance and use of numbers in counting, measuring, comparing and ordering. The ability to think intelligently, using numbers. This basic requirement of numeracy includes the ability to recognize given answers as absurd, without doing a precise calculation, by observing that they violate experience, common sense, elementary logic, or familiar arithmetic patterns. It also includes the use of imagination and insight in using numbers to solve problems. Children should be able to recognize when, for example, a trial-and-error method is likely to be easier to use and more manageable than a standard algorithm.

Developing number sense is a primary goal of **Number Worlds** in every grade. Numbers are presented in a variety of representations and integrated in many contexts so that students develop thorough understanding of numbers.

Algebra

Algebra is the branch of mathematics that uses symbols to represent arithmetic operations. Algebra extends arithmetic through the use of symbols and other notations such as exponents and variables. Algebraic thinking involves understanding patterns, equations, and relationships and includes concepts of functions and inverse operations. Because algebra uses symbols rather than numbers, it can produce general rules that apply to all numbers. What most people commonly think of as algebra involves the manipulation of equations and the solving of equations. Exposure to algebraic ideas can and should occur well before students first study algebra in middle school or high school. Even primary students are capable of understanding many algebraic concepts. Developing algebraic thinking in the early grades smoothes the transition to algebra in middle school and high school and ensures success in future math and science courses as well as in the workplace.

"Algebra begins with a search for patterns. Identifying patterns helps bring order, cohesion, and predictability to seemingly unorganized situations and allows one to make generalizations beyond the information directly available. The recognition and analysis of patterns are important components of the young child's intellectual development because they provide a foundation for the development of algebraic thinking."

Clements, Douglas and Sarama, J., eds. *Engaging Young Children in Mathematics: Standards for Early Childhood Mathematics Education. Mahwah,* New Jersey: Lawrence Erlbaum Associates, Publishers, 2004., p. 52.

Number Worlds and Algebra

Goal: Understanding of functional relationships between variables that represent real world phenomena that are in a constant state of change. Children should be able to draw the graphs of functions and to derive information about functions from their graphs. They should understand the special importance of linear functions and the connection between the study of functions and the solutions of equations and inequalities.

The algebra readiness instruction that begins in the Pre-K level is designed to prepare students for future work in algebra by exposing them to algebraic thinking, including looking for patterns, using variables, working with functions, using integers and exponents, and being aware that mathematics is far more than just arithmetic.

Content Strands

Arithmetic

Arithmetic, one of the oldest branches of mathematics, arises from the most fundamental of mathematical operations: counting. The arithmetic operations—addition, subtraction, multiplication, division, and place holding—form from the basis of the mathematics we use regularly. Mastery of the basic operations with whole numbers (addition, subtraction, multiplication, and division)

"Although some educators once believed that children memorized their "basic facts" as conditioned responses, research now shows that children do not move from knowing nothing about sums and differences of numbers to having the basic number combinations memorized. Instead, they move through a series of progressively more advance and abstract methods for working out the answers to simple arithmetic problems. Furthermore, as children get older, they use the procedures more and more efficiently."

Kilpatrick, J., Swafford, J., and Findell, B., eds. *Adding It Up: Helping Children Learn Mathematics*. Washington, D.C.: National Research Council/National Academy Press, 2001, pp. 182–183.

Number Worlds and Arithmetic

Goal: Mastery of the basic operations with whole numbers (addition, subtraction, multiplication, and division). Whatever other skills and understandings children acquire, they must have the ability to calculate a precise answer when necessary. This fundamental skill includes not only knowledge of the appropriate arithmetic algorithms but also mastery of the basic addition, subtraction, multiplication, and division facts and understanding of the positional notation (base ten) of the whole numbers.

Cumulative assessment occurs throughout the program to indicate when mastery of concepts and skills is expected. Once taught, arithmetic skills are also integrated into other topics such as data analysis.

Fractions, Decimals, and Percents

Understand rational numbers (fractions, decimals, and percents) and their relationships to each other, including the ability to perform calculations and to use rational numbers in measurement.

"Children need to learn that rational numbers are numbers in the same way that whole numbers are numbers. For children to use rational numbers to solve problems, they need to learn that the same rational number may be represented in different ways, as a fraction, a decimal, or a percent. Fraction concepts and representations need to be related to those of division, measurement, and ratio. Decimal and fractional representations need to be connected and understood. Building these connections takes extensive experience with rational numbers over a substantial period of time. Researchers have documented that difficulties in working with rational numbers can often be traced to weak conceptual understanding. . . . Instructional sequences in which more time is spent at the outset on developing meaning for the various representations of rational numbers and the concept of unit have been shown to promote mathematical proficiency."

Kilpatrick, J., Swafford, J., and Findell, B., eds. *Adding It Up: Helping Children Learn Mathematics*. Washington, D.C.: National Research Council/National Academy Press, 2001, pp. 415–416.

Number Worlds and Rational Numbers

Goal: Understanding of rational numbers and of the relationship of fractions to decimals. Included here are the ability to do appropriate calculations with fractions or decimals (or both, as in fractions of decimals), the use of decimals in (metric unit) measurements, the multiplication of fractions as a model for the "of" relation and as a model for areas of rectangles.

Goal: Understanding of the meaning of rates and of their relationship to the arithmetic concept of ratio. Children should be able to calculate ratios, proportions, and percentages; understand how to use them intelligently in real-life situations; understand the common units in which rates occur (such as kilometers per hour, cents per gram); understand the meaning of per; and be able to express ratios as fractions.

In **Number Worlds** understanding of rational number begins at the earliest grades with sharing activities and develops understanding of rational numbers with increasing sophistication at each grade.

Content Strands

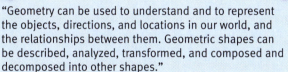
Geometry

Geometry is the branch of mathematics that deals with the properties of space. Plane geometry is the geometry of flat surfaces and solid geometry is the geometry of three-dimensional solids. Geometry has many more fields, including the study of spaces with four or more dimensions.

> "Geometry can be used to understand and to represent the objects, directions, and locations in our world, and the relationships between them. Geometric shapes can be described, analyzed, transformed, and composed and decomposed into other shapes."
>
> Clements, Douglas and Sarama, J., eds. *Engaging Young Children in Mathematics: Standards for Early Childhood Mathematics Education.* Mahwah, New Jersey: Lawrence Erlbaum Associates, Publishers, 2004., p. 39.

Number Worlds and Geometry

Goal: Understanding of an ability to use the geometric concepts of perimeter, area, volume, and congruency as applied to simple figures.

Data Analysis and Applications

Data Analysis encompasses the ability to organize information to make it easier to use and the ability to interpret data and graphs.

> "Describing data involves reading displays of data (e.g., tables, lists, graphs); that is, finding information explicitly stated in the display, recognizing graphical conventions, and making direct connections between the original data and the display. The process is essentially what has been called reading the data. . . . The process of organizing and reducing data incorporates mental actions such as ordering, grouping, and summarizing. Data reduction also includes the use of representative measures of center (often termed *measures of central tendency*) such as mean, mode, or media, and measures of spread such as range and standard deviation."
>
> Kilpatrick, J., Swafford, J., and Findell, B., eds. *Adding It Up: Helping Children Learn Mathematics.* Washington, D.C.: National Research Council/National Academy Press, 2001, pp. 289.

Number Worlds and Data Analysis

Goal: Ability to organize and arrange data for greater intelligibility. Children should develop not only the routine skills of tabulating and graphing results but also, at a higher level, the ability to detect patterns and trends in poorly organized data either before or after reorganization. In addition, children need to develop the ability to extrapolate and interpolate from data and from graphic representations. Children should also know when extrapolation or interpolation is justified and when it is not.

In **Number Worlds** students work with graphs beginning in Pre-K. In each grade the program emphasizes understanding what data show.

Problem Solving

Number Worlds and Problem Solving

Goal: Students must develop the critical thinking skills useful for solving problems. Computational skills are, of course, important, but they are not enough. Students also need an arsenal of critical-thinking skills (sometimes called problem-solving strategies and sometimes called heuristics) that they can call upon to solve particular problems. These skills should not be taught in isolation—students should learn to use them in different contexts. By doing so students are more likely to recognize in which situations a particular skill will be useful and when it is not likely to be useful. We can group critical-thinking skills into two categories—those that are useful in virtually all situations and those that are useful in specific contexts.

In **Number Worlds** students solve problems throughout the daily lessons in all levels of the program.

Technology

Technology has changed the world of mathematics. Technological tools have eliminated the need for tedious calculations and have enabled significant advances in applications of mathematics. Technology can also help to make teaching more effective and efficient. Well-designed math software activities have proven effective in advancing children's math achievements. Technology can also help teachers organize planning and instruction and manage record keeping.

Number Worlds integrates two key software programs throughout the lessons to help students develop richer mathematical skills and conceptual understanding: *eMathTools* and *Building Blocks,* which are explained on the following pages.

Building Blocks Software Activities

Building Blocks software provides computer math activities that address specific developmental levels of the math learning trajectories. **Building Blocks** software is critical to **Number Worlds**. The engaging research-based activities provide motivating development and support of concepts.

Some **Building Blocks** activities have different levels of difficulty indicated by ranges in the Activity Names below. The list provides an overview of all of the **Building Blocks** activities along with the domains, descriptions, and appropriate age ranges.

Domain: Trajectory	Activity Name	Description	Age Range (in years)
Geometry: Composition/Decomposition	Create a Scene	Students explore shapes by moving and manipulating them to make pictures.	4–12
Geometry: Composition/Decomposition	Piece Puzzler 1–5, Piece Puzzler Free Explore, and Super Shape 1–7	Students complete pictures using pattern or tangram shapes.	4–12
Geometry: Imagery	Geometry Snapshots 1–8	Students match configurations of a variety of shapes (e.g., line segments in different arrangements, 3-6 tiled shapes, embedded shapes) to corresponding configurations, given only a brief view of the goal shapes.	5–12
Geometry: Shapes (Identifying)	Memory Geometry 1–5	Students match familiar geometric shapes (shapes in same or similar sizes, same orientation) within the framework of a "Concentration" card game.	3–5
Geometry: Shapes (Matching)	Mystery Pictures 1–4 and Mystery Pictures Free Explore	Students construct predefined pictures by selecting shapes that match a series of target shapes.	3–8
Geometry: Shapes (Parts)	Shape Parts 1–7	Students build or fix some real-world object, exploring shape and properties of shapes.	5–12
Geometry: Shapes (Properties)	Legends of the Lost Shape	Students identify target shapes using textual clues provided.	8–12
Geometry: Shapes (Properties)	Shape Shop 1–3	Students identify a wide range of shapes given their names, with more difficult distracters.	8–12
Measurement: Length	Comparisons	Students are shown pictures of two objects and are asked to click on the one that fits the prompt (longer, shorter, heavier, etc.).	4–8
Measurement: Length	Deep Sea Compare	Students compare the length of two objects by representing them with a third object.	5–7
Measurement: Length	Workin' on the Railroad	Students identify the length (in non-standard units) of railroad trestles they built to span a gully.	6–9
Measurement: Length	Reptile Ruler	Students learn about linear measurement by using a ruler to determine the length of various reptiles.	7–10
Multiplication/Division	Arrays in Area	Students build arrays and then determine the length of those arrays.	8–11
Multiplication/Division	Comic Book Shop	Students use skip counting to produce products that are multiples of 10s, 5s, 2s, and 3s. The task is to identify the product, given a number and bundles.	7–9
Multiplication/Division	Egg-stremely Equal	Students divide large sets of eggs into several equal parts.	4–8
Multiplication/Division	Field Trip	Students solve multidigit multiplication problems in a field trip environment (e.g., equal number of students on each bus; the number of tickets needed for all students.)	8–11
Multiplication/Division	Snack Time	Students use direct modeling to solve multiplication problems.	6–8
Multiplication/Division	Word Problems with Tools 5–6, 10	Students use number tools to solve single and multidigit multiplication and division problems.	8–11
Multiplication/Division	Clean the Plates	Students use skip counting to produce products that are multiples of 10s, 5s, 2s, and 3s.	7–9
Numbers: Adding and Subtracting	Barkley's Bones 1–10 and Barkley's Bones 1–20	Students determine the missing addend in X + ___ = Z problems to feed bone treats to a dog. (Z = 10 or less)	5–8
Number: Adding and Subtracting	Double Compare 1–10 and Double Compare 1–20	Students compare sums of cards (to 10 or 20) to determine which sum is greater.	5–8

Building Blocks

Domain: Trajectory	Activity Name	Description	Age Range (in years)
Number: Adding and Subtracting	Word Problems with Tools 1–4, 7–9, 11–12	Students use number tools to solve single and multidigit addition and subtraction problems.	8–12
Number: Adding and Subtracting and Counting	Counting Activities (Road Race Counting Game, Numeral Train Game, etc.)	Students identify numerals or dot amounts (totals to 20) and move forward a corresponding number of spaces on a game board.	3–9
Number: Adding and Subtracting and Multiplication and Division	Function Machine 1–4	Students provide inputs to a function and examine the resulting outputs to determine the definition of that function. Functions include addition, subtraction, multiplication, or division.	6–12
Number: Comparing	Ordinal Construction Company	Students learn ordinal positions (1st through 10th) by moving objects between the floors of a building.	5–7
Number: Comparing	Rocket Blast 1–3	Given a number line with only initial and final endpoints labeled and a location on that line, students determine the number label for that location.	6–12
Number: Comparing and Counting	Party Time 1–3 and Party Time Free Explore	Students use party utensils to practice one-to-one correspondence, identify numerals that represent target amounts, and match object amounts to target numerals.	4–6
Number: Comparing and Multiplication and Division	Number Compare 1–5	Students compare two cards and choose the one with the greater value.	4–11
Number: Comparing, Counting, Adding and Subtracting	Pizza Pizzazz 1–5 and Pizza Pizzazz Free Explore	Students count items, match target amounts, and explore missing addends related to toppings on pizzas.	3–8
Number: Counting (Object)	Countdown Crazy	Students click digits in sequence to count down from 10 to 0.	5–7
Number: Counting (Object)	Memory Number 1–3	Students match displays containing both numerals and collections to matching displays within the framework of a "Concentration" card game.	4–6
Number: Counting (Object) and Adding and Subtracting	Dinosaur Shop 1–4 and Dinosaur Shop Free Explore	Students use toy dinosaurs to identify numerals representing target amounts, match object amounts to target numerals, add groups of objects, and find missing addends.	4–7
Number: Counting (Objects)	Book Stacks	Students "count on" (through at least one decade) from a given number as they load books onto a cart.	6–8
Number: Counting (Objects)	School Supply Shop	Students count school supplies, bundled in groups of ten to reach a target number up to 100.	6–8
Number: Counting (Objects)	Tire Recycling	Students use skip-counting by 2s and 5s to count tires as they are moved.	6–8
Number: Counting (Strategies)	Build Stairs 1–3 and Build Stairs Free Explore	Students practice counting, sequencing, and ordering by building staircases.	4–7
Number: Counting (Strategies)	Math-O-Scope	Students identify the numbers that surround a given number in the context of a 100s chart.	7–9
Number: Counting (Strategies)	Tidal Tally	Students identify missing addends (hidden objects) by counting forward from given addends (visible objects) to reach a numerical total.	6–9
Number: Counting (Verbal)	Count and Race	Students count up to 50 by adding cars to a racetrack one at a time.	3–6
Number: Counting (Verbal)	Before and After Math	Students identify and select numbers that come either just before or right after a target number.	4–7
Number: Counting (Verbal)	Kitchen Counter	Students click on objects one at a time while the numbers from one to ten are counted aloud.	3–6
Number: Subitizing	Number Snapshots 1–10	Students match numerals or dot collections to a corresponding numeral or collection given only a brief view of the goal collections.	3–12
Patterning	Marching Patterns 1–3	Students extend a linear pattern of marchers by one full repetition of an entire unit (AB, AAB, ABB, and ABC patterns).	5–7
Patterning	Pattern Planes 1–3	Students duplicate a linear pattern of flags based on an outline that serves as a guide (AB, AAB, ABB, and ABC patterns).	4–6
Patterning	Pattern Zoo 1–3 and Patterns Free Explore	Students identify a linear pattern of fruit that matches a target pattern to feed zoo animals (AB, AAB, ABB, and ABC patterns). Students explore patterning by creating rhythmic patterns of their own.	3–6

eMathTools

Number Worlds integrates *eMathTools* in appropriate lessons throughout the program. This component provides multimedia formats for demonstrating and exploring concepts and solving problems. The *eMathTools* are described below.

Data Organization and Display Tools

- **Spreadsheet Tool**—allows students to manage, display, sort, and calculate data. Links to the graphing tool for further data display
- **Graphing Tool**—displays data in pie charts or circle graphs, line graphs, bar graphs, or coordinate grids
- **Venn Diagram**—allows students to sort data visually

Measurement and Conversion Tools

- **Stopwatch**—measures in real time for development of counting and time concepts
- **Calendar**—an electronic calendar to develop concepts of time
- **Metric and Customary Conversion Tool**—converts metric and customary measurements in length, distance, mass and weight, time, temperature, and capacity
- **Estimating Proportion Tool**—allows visual representations of proportions to develop understanding of ratios, fractions, and decimals

Geometric Exploration Tools

- **Tessellations**—allows students to create tessellation patterns by rotating, coloring, and tiling shapes
- **Net Tool**—allows students to manipulate 2-D shapes and then print them to create 3-D shapes
- **Shape Tools**—explores and manipulates shapes to create designs
- **Geometry Sketch Tool**—allows drawing, manipulating, and measuring a wide variety of shapes
- **Pythagorean Theorem Tool**—launches right triangles to explore the Pythagorean Theorem

Calculation and Counting Tools

- **Calculator**—allows students to launch a calculator to perform mathematical operations
- **Function Machine**—an electronic version of a function machine that students use to solve missing variable problems
- **Multiplication and Division Table**—an interactive version of a table that highlights relationships between multipli-cation and division facts
- **Addition and Subtraction Table**—an interactive version of a table that highlights relationships between addition and subtraction facts
- **100s Chart**—an interactive version of a table that highlights patterns and relationships among numbers
- **Number Line**—an electronic number line that allows students to skip count and see the relationships among whole numbers, fractions, decimals, and percents
- **Number Stairs**—a tool to illustrate counting in units
- **Probability Tool**—uses number cubes, spinner, or tumble drum to test scenarios of probability
- **Set Tool**—allows students to visually represent and manipulate different sets of objects for a variety of counting activities
- **Base-Ten Blocks**—allows students to manipulate base-10 block for counting
- **Coins and Money**—uses visual representations of coins and money to represent counting
- **Fraction Tool**—represents fractional units for counting and understanding relationships
- **Array Tool**—presents arrays to represent multiplication and division patterns and relationships

English Learners

English learners enter schools at different grade levels and bring varying levels of English proficiency and academic preparation attained in their primary language. Some students have high levels of academic preparation that allow them to focus more on learning the terminology associated with the math skills they already know. These students have a great advantage as they already understand that mathematics has its own terminology and have a grasp of the basic skills and concepts needed to participate in advancing their math skills.

Other students may have had little schooling or interrupted school experiences, which means they need to catch up to their grade-level peers academically, and they must begin the task of English acquisition at the same time. Many schools use some form of primary language instruction or support to accelerate students' skill base while learning English. This approach contains the added advantage of being able to draw on the student's family as a source of support and inclusion.

Number Worlds and English Learners

Creating Context and Alternate Vocabulary

In addition to the specialized terminology of mathematics, academic language often includes the use of common English idiomatic expressions and content area words that have multiple meanings in English. Previewing these expressions with English learners and helping them keep track of their usage and meaning is excellent practice for extending knowledge of English for academic purposes.

In each *Number Worlds* lesson several words or expressions are noted along with a quick description of their meaning. Some of the words and expressions are math phrases, but many are examples of wider vocabulary used in word problems and direction lines in the lessons. Creating context and alternate vocabulary features highlight some of these key phrases and give teachers a brief explanation.

The inclusion of alternate vocabulary such as the words with double meanings that students encounter in the subject area brings to the attention of teachers those complexities that English speakers may already have mastered but that can make the difference between deep comprehension and confusion for English learners. Alternate vocabulary also highlights some potentially puzzling idiomatic expressions.

Spanish Cognates

One way to rapidly accelerate the acquisition of more sophisticated academic language is to take advantage of words with shared roots across languages. Words that are related by root or are borrowed from another language are called cognates. In English many math and science terms come from Greek, Latin, and Arabic roots. English learners who speak Romance languages such as Spanish have an advantage in these subjects because much of the important concept vocabulary is similar in both languages. By teaching students to look for some basic word parts in English, these students can take advantage of what they know in their primary language to augment their knowledge of English. Pronunciation of these words may be quite different, but the written words often look very similar and mean the same thing.

Learning Trajectories

Children follow natural developmental progressions in learning, developing mathematical ideas in their own way. Curriculum research has revealed sequences of activities that are effective in guiding children through these levels of thinking. These developmental paths are the basis for Building Blocks learning trajectories. Learning trajectories have three parts: a mathematical goal, a developmental path through which children develop to reach that goal, and a set of activities matched to each of those levels that help children develop the next level. Thus, each learning trajectory has levels of understanding, each more sophisticated than the last, with tasks that promote growth from one level to the next. The Building Blocks Learning Trajectories give simple labels, descriptions, and examples of each level. Complete learning trajectories describe the goals of learning, the thinking and learning processes of children at various levels, and the learning activities in which they might engage. This document provides only the developmental levels.

Learning Trajectories for Primary Grades Mathematics
Developmental Levels

Frequently Asked Questions (FAQ)

1. **Why use learning trajectories?** Learning trajectories allow teachers to build the *mathematics of children— the thinking of children as it develops naturally.* So, we know that all the goals and activities are within the developmental capacities of children. We know that each level provides a natural *developmental building block* to the next level. Finally, we know that the activities provide the *mathematical building blocks* for school success, because the research on which they are based typically involves higher-income children.

2. **When are children "at" a level?** Children are at a certain level when most of their behaviors reflect the thinking— ideas and skills—of that level. Often, they show a few behaviors from the next (and previous) levels as they learn.

3. **Can children work at more than one level at the same time?** Yes, although most children work mainly at one level or in transition between two levels (naturally, if they are tired or distracted, they may operate at a much lower level). Levels are not "absolute stages." They are "benchmarks" of complex growth that represent distinct ways of thinking. So, another way to think of them is as a sequence of different patterns of thinking. Children are continually learning, within levels and moving between them.

4. **Can children jump ahead?** Yes, especially if there are separate "sub-topics." For example, we have combined many counting competencies into one "Counting" sequence with sub-topics, such as verbal counting skills. Some children learn to count to 100 at age 6 after learning to count objects to 10 or more, some may learn that verbal skill earlier. The sub-topic of verbal counting skills would still be followed.

5. **How do these developmental levels support teaching and learning?** The levels help teachers, as well as curriculum developers, assess, teach, and sequence activities. *Teachers who understand learning trajectories and the developmental levels that are at their foundation are more effective and efficient.* Through planned teaching and also encouraging informal, incidental mathematics, teachers help children learn *at an appropriate and deep level.*

6. **Should I plan to help children develop just the levels that correspond to my children's ages?** No! The ages in the table are typical ages children develop these ideas. *But these are rough guides only*—children differ widely. Furthermore, the ages below are lower bounds of what children achieve without instruction. So, these are *"starting levels"* not goals. We have found that children who are provided high-quality mathematics experiences are capable of developing to levels one or more years beyond their peers.

Each column in the table below, such as "Counting," represents a main developmental progression that underlies the learning trajectory for that topic.

For some topics, there are "subtrajectories"—strands within the topic. In most cases, the names make this clear. For example, in Comparing and Ordering, some levels are about the "Comparer" levels, and others about building a "Mental Number Line." Similarly, the related subtrajectories of "Composition" and "Decomposition" are easy to distinguish. Sometimes, for clarification, subtrajectories are indicated with a note in italics after the title. For example, in Shapes, Parts and Representing are subtrajectories within the Shapes trajectory.

Clements, D. H., Sarama, J., & DiBiase, A.-M. (Eds.). (2004). *Engaging Young Children in Mathematics: Standards for Early Childhood Mathematics Education.* Mahwah, NJ: Lawrence Erlbaum Associates.

Clements, D. H., & Sarama, J. (in press). "Early Childhood Mathematics Learning." In F. K. Lester, Jr. (Ed.), *Second Handbook of Research on Mathematics Teaching and Learning.* New York: Information Age Publishing.

Learning Trajectories

Developmental Levels for Counting

The ability to count with confidence develops over the course of several years. Beginning in infancy, children show signs of understanding number. With instruction and number experience, most children can count fluently by age 8, with much progress in counting occurring in kindergarten and first grade. Most children follow a natural developmental progression in learning to count with recognizable stages or levels. This developmental path can be described as part of a learning trajectory.

Age Range	Level Name	Level	Description
1–2	Pre-Counter	1	A child at the earliest level of counting may name some numbers meaninglessly. The child may skip numbers and have no sequence.
1–2	Chanter	2	At this level a child may sing-song numbers, but without meaning.
2	Reciter	3	At this level the child verbally counts with separate words, but not necessarily in the correct order.
3	Reciter (10)	4	A child at this level can verbally count to 10 with some correspondence with objects. They may point to objects to count a few items but then lose track.
3	Corresponder	5	At this level a child can keep one-to-one correspondence between counting words and objects—at least for small groups of objects laid in a line. A corresponder may answer "how many" by recounting the objects starting over with one each time.
4	Counter (Small Numbers)	6	*At around 4 years children begin to count meaningfully. They accurately count objects to 5 and answer the "how many" question with the last number counted. When objects are visible, and especially with small numbers, begins to understand cardinality. These children can count verbally to 10 and may write or draw to represent 1–5.*
4	Producer—Counter To (Small Numbers)	7	The next level after counting small numbers is to count out objects up to 5 and produce a group of four objects. When asked to show four of something, for example, this child can give four objects.
4–5	Counter (10)	8	This child can count structured arrangements of objects to 10. He or she may be able to write or draw to represent 10 and can accurately count a line of nine blocks and says there are 9. A child at this level can also find the number just after or just before another number, but only by counting up from 1.
5–6	Counter and Producer—Counter to (10+)	9	Around 5 years of age children begin to count out objects accurately to 10 and then beyond to 30. They can keep track of objects that have and have not been counted, even in different arrangements. They can write or draw to represent 1 to 10 and then 20 and 30, and can give the next number to 20 or 30. These children can recognize errors in others' counting and are able to eliminate most errors in one's own counting.

Age Range	Level Name	Level	Description
5–6	Counter Backward from 10	10	Another milestone at about age 5 is being able to count backwards from 10.
6–7	Counter from N (N+1, N−1)	11	Around 6 years of age children begin to count on, counting verbally and with objects from numbers other than 1. Another noticeable accomplishment is that children can determine immediately the number just before or just after another number without having to start back at 1.
6–7	Skip-Counting by 10s to 100	12	A child at this level can count by tens to 100. They can count through decades knowing that 40 comes after 39, for example.
6–7	Counter to 100	13	A child at this level can count by ones through 100, including the decade transitions from 39 to 40, 49 to 50, and so on, starting at any number.
6–7	Counter On Using Patterns	14	At this level a child keeps track of counting acts by using numerical patterns such as tapping as he or she counts.
6–7	Skip Counter	15	The next level is when children can count by 5s and 2s with understanding.
6–7	Counter of Imagined Items	16	At this level a child can count mental images of hidden objects.
6–7	Counter On Keeping Track	17	A child at this level can keep track of counting acts numerically with the ability to count up one to four more from a given number.
6–7	Counter of Quantitative Units	18	At this level a child can count unusual units such as "wholes" when shown combinations of wholes and parts. For example when shown three whole plastic eggs and four halves, a child at this level will say there are five whole eggs.
6–7	Counter to 200	19	At this level a child counts accurately to 200 and beyond, recognizing the patterns of ones, tens, and hundreds.
7+	Number Conserver	20	A major milestone around age 7 is the ability to conserve number. A child who conserves number understands that a number is unchanged even if a group of objects is rearranged. For example, if there is a row of ten buttons, the child understands there are still ten without recounting, even if they are rearranged in a long row or a circle.
7+	Counter Forward and Back	21	A child at this level counts in either direction and recognizes that sequence of decades sequence mirrors single-digit sequence.

Developmental Levels for Comparing and Ordering Numbers

Comparing and ordering sets is a critical skill for children as they determine whether one set is larger than another to make sure sets are equal and "fair." Prekindergartners can learn to use matching to compare collections or to create equivalent collections. Finding out how many more or fewer in one collection is more demanding than simply comparing two collections. The ability to compare and order sets with fluency develops over the course of several years. With instruction and number experience, most children develop foundational understanding of number relationships and place value at ages 4 and 5. Most children follow a natural developmental progression in learning to compare and order numbers with recognizable stages or levels. This developmental path can be described as part of a learning trajectory.

Age Range	Level Name	Level	Description
2	Object Corresponder	1	At this early level a child puts objects into one-to-one correspondence, but with only intuitive understanding of resulting equivalence. For example, a child may know that each carton has a straw, but doesn't necessarily know there are the same numbers of straws and cartons.
2	Perceptual Comparer	2	At the next level a child can compare collections that are quite different in size (for example, one is at least twice the other) and know that one has more than the other. If the collections are similar, the child can compare very small collections.
2–3	First-Second Ordinal Counter	3	A child at this level can identify the first and often second objects in a sequence.
3	Nonverbal Comparer of Similar Items	4	At this level a child can identify that different organizations of the same number of small groups are equal and different from other sets. (1–4 items).
3	Nonverbal Comparer of Dissimilar Items	5	At the next level a child can match small, equal collections of dissimilar items, such as shells and dots, and show that they are the same number.
4	Matching Comparer	6	As children progress they begin to compare groups of 1–6 by matching. For example, a child gives one toy bone to every dog and says there are the same number of dogs and bones.
4	Knows-to-Count Comparer	7	A significant step occurs when the child begins to count collections to compare. At the early levels children are not always accurate when larger collection's objects are smaller in size than the objects in the smaller collection. For example, a child at this level may accurately count two equal collections, but when asked, says the collection of larger blocks has more.
4	Counting Comparer (Same Size)	8	At the next level children make accurate comparisons via counting, but only when objects are about the same size and groups are small (about 1–5).
5	Counting Comparer (5)	9	As children develop their ability to compare sets, they compare accurately by counting, even when larger collection's objects are smaller. A child at this level can figure out how many more or less.
5	Ordinal Counter	10	At the next level a child identifies and uses ordinal numbers from "first" to "tenth." For example, the child can identify who is "third in line."
5	Counting Comparer	11	At this level a child can compare by counting, even when the larger collection's objects are smaller. For example, a child can accurately count two collections and say they have the same number even if one has larger objects.
5	Mental Number Line to 10	12	At this level a child uses internal images and knowledge of number relationships to determine relative size and position. For example, the child can determine whether 4 or 9 is closer to 6.
5	Serial Orderer to 6+	13	Children demonstrate development in comparing when they begin to order lengths marked into units (1–6, then beyond). For example, given towers of cubes, this child can put them in order, 1 to 6. Later the child begins to order collections. For example, given cards with one to six dots on them, puts in order.
6	Counting Comparer (10)	14	The next level can be observed when the child compares sets by counting, even when larger collection's objects are smaller, up to 10. A child at this level can accurately count two collections of 9 each, and says they have the same number, even if one collection has larger blocks.
6	Mental Number Line to 10	15	As children move into the next level they begin to use mental rather than physical images and knowledge of number relationships to determine relative size and position. For example, a child at this level can answer which number is closer to 6, 4, or 9 without counting physical objects.
6	Serial Orderer to 6+	16	At this level a child can order lengths marked into units. For example, given towers of cubes the child can put them in order.
7	Place Value Comparer	17	Further development is made when a child begins to compare numbers with place value understandings. For example, a child at this level can explain that "63 is more than 59 because six tens is more than five tens even if there are more than three ones."

Learning Trajectories

Age Range	Level Name	Level	Description
7	Mental Number Line to 100	18	Children demonstrate the next level in comparing and ordering when they can use mental images and knowledge of number relationships, including ones embedded in tens, to determine relative size and position. For example, a child at this level when asked, "Which is closer to 45, 30 or 50?" says "45 is right next to 50, but 30 isn't."
8+	Mental Number Line to 1000s	19	About age 8 children begin to use mental images of numbers up to 1,000 and knowledge of number relationships, including place value, to determine relative size and position. For example, when asked, "Which is closer to 3,500—2,000 or 7,000?" a child at this level says "70 is double 35, but 20 is only fifteen from 35, so twenty hundreds, 2,000, is closer."

Developmental Levels for Recognizing Number and Subitizing (Instantly Recognizing)

The ability to recognize number values develops over the course of several years and is a foundational part of number sense. Beginning at about age 2, children begin to name groups of objects. The ability to instantly know how many are in a group, called *subitizing*, begins at about age 3. By age 8, with instruction and number experience, most children can identify groups of items and use place values and multiplication skills to count them. Most children follow a natural developmental progression in learning to count with recognizable stages or levels. This developmental path can be described as part of a learning trajectory.

Age Range	Level Name	Level	Description
2	Small Collection Namer	1	The first sign of a child's ability to subitize occurs when the child can name groups of one to two, sometimes three. For example, when shown a pair of shoes, this young child says, "Two shoes."
3	Nonverbal Subitizer	2	The next level occurs when shown a small collection (one to four) only briefly, the child can put out a matching group nonverbally, but cannot necessarily give the number name telling how many. For example, when four objects are shown for only two seconds, then hidden, child makes a set of four objects to "match."
3	Maker of Small Collections	3	At the next level a child can nonverbally make a small collection (no more than five, usually one to three) with the same number as another collection. For example, when shown a collection of three, makes another collection of three.
4	Perceptual Subitizer to 4	4	Progress is made when a child instantly recognizes collections up to four when briefly shown and verbally names the number of items. For example, when shown four objects briefly, says "four."
5	Perceptual Subitizer to 5	5	The next level is the ability to instantly recognize briefly shown collections up to five and verbally name the number of items. For example, when shown five objects briefly, says "five."
5	Conceptual Subitizer to 5+	6	At the next level the child can verbally label all arrangements to five shown only briefly. For example, a child at this level would say, "I saw 2 and 2 and so I saw 4."
5	Conceptual Subitizer to 10	7	The next step is when the child can verbally label most briefly shown arrangements to six, then up to ten, using groups. For example, a child at this level might say, "In my mind, I made two groups of 3 and one more, so 7."
6	Conceptual Subitizer to 20	8	Next, a child can verbally label structured arrangements up to twenty, shown only briefly, using groups. For example, the child may say, "I saw three 5s, so 5, 10, 15."
7	Conceptual Subitizer with Place Value and Skip Counting	9	At the next level a child is able to use skip counting and place value to verbally label structured arrangements shown only briefly. For example, the child may say, "I saw groups of tens and twos, so 10, 20, 30, 40, 42, 44, 46 . . . 46!"
8+	Conceptual Subitizer with Place Value and Multiplication	10	As children develop their ability to subitize, they use groups, multiplication, and place value to verbally label structured arrangements shown only briefly. At this level a child may say, "I saw groups of tens and threes, so I thought, five tens is 50 and four 3s is 12, so 62 in all."

Developmental Levels for Composing Number
(Knowing Combinations of Numbers)

Composing and decomposing are combining and separating operations that allow children to build concepts of "parts" and "wholes." Most prekindergartners can "see" that two items and one item make three items. Later, children learn to separate a group into parts in various ways and then to count to produce all of the number "partners" of a given number. Eventually children think of a number and know the different addition facts that make that number. Most children follow a natural developmental progression in learning to compose and decompose numbers with recognizable stages or levels. This developmental path can be described as part of a learning trajectory.

Age Range	Level Name	Level	Description
4	Pre-Part-Whole Recognizer	1	At the earliest levels of composing a child only nonverbally recognizes parts and wholes. For example, When shown four red blocks and two blue blocks, a young child may intuitively appreciate that "all the blocks" include the red and blue blocks, but when asked how many there are in all, may name a small number, such as 1.
5	Inexact Part-Whole Recognizer	2	A sign of development in composing is that the child knows that a whole is bigger than parts, but does not accurately quantify. For example, when shown four red blocks and two blue blocks and asked how many there are in all, names a "large number," such as 5 or 10.
5	Composer to 4, then 5	3	The next level is that a child begins to know number combinations. A child at this level quickly names parts of any whole, or the whole given the parts. For example, when shown four, then one is secretly hidden, and then is shown the three remaining, quickly says "1" is hidden.

Age Range	Level Name	Level	Description
6	Composer to 7	4	The next sign of development is when a child knows number combinations to totals of seven. A child at this level quickly names parts of any whole, or the whole given parts and can double numbers to 10. For example, when shown six, then four are secretly hidden, and shown the two remaining, quickly says "4" are hidden.
6	Composer to 10	5	The next level is when a child knows number combinations to totals of 10. A child at this level can quickly name parts of any whole, or the whole given parts and can double numbers to 20. For example, this child would be able to say "9 and 9 is 18."
7	Composer with Tens and Ones	6	At the next level the child understands two-digit numbers as tens and ones; can count with dimes and pennies; and can perform two-digit addition with regrouping. For example, a child at this level can explain, "17 and 36 is like 17 and 3, which is 20, and 33, which is 53."

Developmental Levels for Adding and Subtracting

Learning single-digit addition and subtraction is generally characterized as "learning math facts." It is assumed that children must memorize these facts, yet research has shown that addition and subtraction have their roots in counting, counting on, number sense, the ability to compose and decompose numbers, and place value. Research has shown that learning methods for adding and subtracting with understanding is much more effective than rote memorization of seemingly isolated facts. Most children follow an observable developmental progression in learning to add and subtract numbers with recognizable stages or levels. This developmental path can be described as part of a learning trajectory.

Age Range	Level Name	Level	Description
1	Pre +/−	1	At the earliest level a child shows no sign of being able to add or subtract.
3	Nonverbal +/−	2	The first inkling of development is when a child can add and subtract very small collections nonverbally. For example, when shown two objects, then one object going under a napkin, the child identifies or makes a set of three objects to "match."

Age Range	Level Name	Level	Description
4	Small Number +/−	3	The next level of development is when a child can find sums for joining problems up to 3 + 2 by counting all with objects. For example, when asked, "You have 2 balls and get 1 more. How many in all?" counts out 2, then counts out 1 more, then counts all 3: "1, 2, 3, 3!"

Learning Trajectories

Age Range	Level Name	Level	Description
5	Find Result +/−	4	**Addition** Evidence of the next level in addition is when a child can find sums for joining (you had 3 apples and get 3 more, how many do you have in all?) and part-part-whole (there are 6 girls and 5 boys on the playground, how many children were there in all?) problems by direct modeling, counting all, with objects. For example, when asked, "You have 2 red balls and 3 blue balls. How many in all?" the child counts out 2 red, then counts out 3 blue, then counts all 5. **Subtraction** In subtraction, a child at this level can also solve take-away problems by separating with objects. For example, when asked, "You have 5 balls and give 2 to Tom. How many do you have left?" the child counts out 5 balls, then takes away 2, and then counts the remaining 3.
5	Find Change +/−	5	**Addition** At the next level a child can find the missing addend ($5 + \_\_ = 7$) by adding on objects. For example, when asked, "You have 5 balls and then get some more. Now you have 7 in all. How many did you get?" the child counts out 5, then counts those 5 again starting at 1, then adds more, counting "6, 7," then counts the balls added to find the answer, 2. **Subtraction** Compares by matching in simple situations. For example, when asked, "Here are 6 dogs and 4 balls. If we give a ball to each dog, how many dogs won't get a ball?" a child at this level counts out 6 dogs, matches 4 balls to 4 of them, then counts the 2 dogs that have no ball.
5	Make It N +/−	6	A significant advancement in addition occurs when a child is able to count on. This child can add on objects to make one number into another, without counting from 1. For example, when asked, "This puppet has 4 balls but she should have 6. Make it 6," puts up 4 fingers on one hand, immediately counts up from 4 while putting up two fingers on the other hand, saying, "5, 6" and then counts or recognizes the two fingers.
6	Counting Strategies +/−	7	The next level occurs when a child can find sums for joining (you had 8 apples and get 3 more . . .) and part-part-whole (6 girls and 5 boys . . .) problems with finger patterns or by adding on objects or counting on. For example, when asked "How much is 4 and 3 more?" the child answers "4 . . . 5, 6, 7 [uses rhythmic or finger pattern]. 7!" Children at this level also can solve missing addend ($3 + \_\_ = 7$) or compare problems by counting on. When asked, for example, "You have 6 balls. How many more would you need to have 8?" the child says, "6, 7 [puts up first finger], 8 [puts up second finger]. 2!"
6	Part-Whole +/−	8	Further development has occurred when the child has part-whole understanding. This child can solve all problem types using flexible strategies and some derived facts (for example, "5 + 5 is 10, so 5 + 6 is 11"), sometimes can do start unknown ($\_\_ + 6 = 11$), but only by trial and error. This child when asked, "You had some balls. Then you get 6 more. Now you have 11 balls. How many did you start with?" lays out 6, then 3 more, counts and gets 9. Puts 1 more with the 3, says 10, then puts 1 more. Counts up from 6 to 11, then recounts the group added, and says, "5!"
6	Numbers-in-Numbers +/−	9	Evidence of the next level is when a child recognizes that a number is part of a whole and can solve problems when the start is unknown ($\_\_ + 4 = 9$) with counting strategies. For example, when asked, "You have some balls, then you get 4 more balls, now you have 9. How many did you have to start with?" this child counts, putting up fingers, "5, 6, 7, 8, 9." Looks at fingers, and says, "5!"
7	Deriver +/−	10	At the next level a child can use flexible strategies and derived combinations (for example, "7 + 7 is 14, so 7 + 8 is 15") to solve all types of problems. For example, when asked, "What's 7 plus 8?" this child thinks: $7 + 8 \square 7 + [7 + 1] \square [7 + 7] + 1 = 14 + 1 = 15$. A child at this level can also solve multidigit problems by incrementing or combining tens and ones. For example, when asked "What's 28 + 35?" this child thinks: 20 + 30 = 50; +8 = 58; 2 more is 60, 3 more is 63. Combining tens and ones: 20 + 30 = 50. 8 + 5 is like 8 plus 2 and 3 more, so, it's 13—50 and 13 is 63.
8+	Problem Solver +/−	11	As children develop their addition and subtraction abilities, they can solve all types of problems by using flexible strategies and many known combinations. For example, when asked, "If I have 13 and you have 9, how could we have the same number?" this child says, "9 and 1 is 10, then 3 more to make 13. 1 and 3 is 4. I need 4 more!"
8+	Multidigit +/−	12	Further development is evidenced when children can use composition of tens and all previous strategies to solve multidigit +/− problems. For example, when asked, "What's 37 − 18?" this child says, "I take 1 ten off the 3 tens; that's 2 tens. I take 7 off the 7. That's 2 tens and 0 . . . 20. I have one more to take off. That's 19." Another example would be when asked, "What's 28 + 35?" thinks, 30 + 35 would be 65. But it's 28, so it's 2 less . . . 63.

Developmental Levels for Multiplying and Dividing

Multiplication and division builds on addition and subtraction understandings and is dependent upon counting and place value concepts. As children begin to learn to multiply they make equal groups and count them all. They then learn skip counting and derive related products from products they know. Finding and using patterns aids in learning multiplication and division facts with understanding. Children typically follow an observable developmental progression in learning to multiply and divide numbers with recognizable stages or levels. This developmental path can be described as part of a learning trajectory.

Age Range	Level Name	Level	Description
2	Nonquantitive Sharer "Dumper"	1	Multiplication and division concepts begin very early with the problem of sharing. Early evidence of these concepts can be observed when a child dumps out blocks and gives some (not an equal number) to each person.
3	Beginning Grouper and Distributive Sharer	2	Progression to the next level can be observed when a child is able to make small groups (fewer than 5). This child can share by "dealing out," but often only between two people, although he or she may not appreciate the numerical result. For example, to share four blocks, this child gives each person a block, checks each person has one, and repeats this.
4	Grouper and Distributive Sharer	3	The next level occurs when a child makes small equal groups (fewer than 6). This child can deal out equally between two or more recipients, but may not understand that equal quantities are produced. For example, the child shares 6 blocks by dealing out blocks to herself and a friend 1 at a time.
5	Concrete Modeler ×/÷	4	As children develop, they are able to solve small-number multiplying problems by grouping—making each group and counting all. At this level a child can solve division/sharing problems with informal strategies, using concrete objects—up to twenty objects and two to five people—although the child may not understand equivalence of groups. For example, the child distributes twenty objects by dealing out two blocks to each of five people, then one to each, until blocks are gone.
6	Parts and Wholes ×/÷	5	A new level is evidenced when the child understands the inverse relation between divisor and quotient. For example, this child understands "If you share with more people, each person gets fewer."

Age Range	Level Name	Level	Description
7	Skip Counter ×/÷	6	As children develop understanding in multiplication and division they begin to use skip counting for multiplication and for measurement division (finding out how many groups). For example, given twenty blocks, four to each person, and asked how many people, the child skip counts by 4, holding up one finger for each count of 4. A child at this level also uses trial and error for partitive division (finding out how many in each group). For example, given twenty blocks, five people, and asked how many should each get, this child gives three to each, then one more, then one more.
8+	Deriver ×/÷	7	At the next level children use strategies and derived combinations and solve multidigit problems by operating on tens and ones separately. For example, a child at this level may explain "7 × 6, five 7s is 35, so 7 more is 42."
8+	Array Quantifier	8	Further development can be observed when a child begins to work with arrays. For example, given 7 × 4 with most of 5 × 4 covered, a child at this level may say, "There's eight in these two rows, and five rows of four is 20, so 28 in all."
8+	Partitive Divisor	9	The next level can be observed when a child is able to figure out how many are in each group. For example, given twenty blocks, five people, and asked how many should each get, a child at this level says "four, because 5 groups of 4 is 20."
8+	Multidigit ×/÷	10	As children progress they begin to use multiple strategies for multiplication and division, from compensating to paper-and-pencil procedures. For example, a child becoming fluent in multiplication might explain that "19 times 5 is 95, because twenty 5s is 100, and one less 5 is 95."

Learning Trajectories

Developmental Levels for Measuring

Measurement is one of the main real-world applications of mathematics. Counting is a type of measurement, determining how many items are in a collection. Measurement also involves assigning a number to attributes of length, area, and weight. Prekindergarten children know that mass, weight, and length exist, but they don't know how to reason about these or to accurately measure them. As children develop their understanding of measurement, they begin to use tools to measure and understand the need for standard units of measure. Children typically follow an observable developmental progression in learning to measure with recognizable stages or levels. This developmental path can be described as part of a learning trajectory.

Age Range	Level Name	Level	Description
3	Length Quantity Recognizer	1	At the earliest level children can identify length as an attribute. For example, they might say, "I'm tall, see?"
4	Length Direct Comparer	2	In the next level children can physically align two objects to determine which is longer or if they are the same length. For example, they can stand two sticks up next to each other on a table and say, "This one's bigger."
5	Indirect Length Comparer	3	A sign of further development is when a child can compare the length of two objects by representing them with a third object. For example, a child might compare length of two objects with a piece of string. Additional evidence of this level is that when asked to measure, the child may assign a length by guessing or moving along a length while counting (without equal length units). The child may also move a finger along a line segment, saying 10, 20, 30, 31, 32.
5	Serial Orderer to 6+	4	At the next level a child can order lengths, marked in one to six units. For example, given towers of cubes, a child at this level puts in order, 1 to 6.
6	End-to-End Length Measurer	5	At the next level the child can lay units end-to-end, although he or she may not see the need for equal-length units. For example, a child might lay 9-inch cubes in a line beside a book to measure how long it is.

Age Range	Level Name	Level	Description
7	Length Unit Iterater	6	A significant change occurs when a child can use a ruler and see the need for identical units.
7	Length Unit Relater	7	At the next level a child can relate size and number of units. For example, the child may explain, "If you measure with centimeters instead of inches, you'll need more of them, because each one is smaller."
8	Length Measurer	8	As children develop measurement ability they begin to measure, knowing the need for identical units, the relationships between different units, partitions of unit, and zero point on rulers. At this level the child also begins to estimate. The child may explain, "I used a meter stick three times, then there was a little left over. So, I lined it up from 0 and found 14 centimeters. So, it's 3 meters, 14 centimeters in all."
8	Conceptual Ruler Measurer	9	Further development in measurement is evidenced when a child possesses an "internal" measurement tool. At this level the child mentally moves along an object, segmenting it, and counting the segments. This child also uses arithmetic to measure and estimates with accuracy. For example, a child at this level may explain, "I imagine one meterstick after another along the edge of the room. That's how I estimated the room's length is 9 meters."

Developmental Levels for Recognizing Geometric Shapes

Geometric shapes can be used to represent and understand objects. Analyzing, comparing, and classifying shapes helps create new knowledge of shapes and their relationships. Shapes can be decomposed or composed into other shapes. Through their everyday activity, children build both intuitive and explicit knowledge of geometric figures. Most children can recognize and name basic two-dimensional shapes at 4 years of age. However, young children can learn richer concepts about shape if they have varied examples and nonexamples of shape, discussions about shapes and their characteristics, a wide variety of shape classes, and interesting tasks. Children typically follow an observable developmental progression in learning about shapes with recognizable stages or levels. This developmental path can be described as part of a learning trajectory.

Age Range	Level Name	Level	Description
2	Shape Matcher—	1	The earliest sign of understanding shape is when a child can match basic shapes (circle, square, typical triangle) with the same size and orientation. Example: Matches ☐ to ☐. A sign of development is when a child can match basic shapes with different sizes. Example: Matches ☐ to ☐. The next sign of development is when a child can match basic shapes with different orientations. Example: Matches ☐ to ◇.
3	Shape Prototype Recognizer and Identifier	2	A sign of development is when a child can recognize and name prototypical circle, square, and, less often, a typical triangle. For example, the child names this a square ☐. Some children may name different sizes, shapes, and orientations of rectangles, but also accept some shapes that look rectangular but are not rectangles. Children name these shapes "rectangles" (including the non-rectangular parallelogram).
3	Shape Matcher— More Shapes	3	As children develop understanding of shape, they can match a wider variety of shapes with the same size and orientation. —4 Matches wider variety of shapes with different sizes and orientations. Matches these shapes ╲ ╱ . —5 Matches combinations of shapes to each other. Matches these shapes ⊛ ⊛ .
4	Shape Recognizer— Circles, Squares, and Triangles	4	The next sign of development is when a child can recognize some nonproto-typical squares and triangles and may recognize some rectangles, but usually not rhombi (diamonds). Often, the child doesn't differentiate sides/corners. The child at this level may name these as triangles .
4	Constructor of Shapes from Parts – Looks Like	5	A significant sign of development is when a child represents a shape by making a shape "look like" a goal shape. For example, when asked to make a triangle with sticks, the child creates the following ☐.
5	Shape Recognizer— All Rectangles	6	As children develop understanding of shape, they recognize more rectangle sizes, shapes, and orientations of rectangles. For example, a child at this level correctly names these shapes "rectangles" .
5	Side Recognizer	7	A sign of development is when a child recognizes parts of shapes and identifies sides as distinct geometric objects. For example, when asked what this shape is ⋀, the child says it is a quadrilateral (or has four sides) after counting and running a finger along the length of each side.
5	Angle Recognizer	8	At the next level a child can recognize angles as separate geometric objects. For example, when asked, "Why is this a triangle," says, "It has three angles" and counts them, pointing clearly to each vertex (point at the corner).
5	Shape Recognizer	9	As children develop they are able to recognize most basic shapes and prototypical examples of other shapes, such as hexagon, rhombus (diamond), and trapezoid. For example, a child can correctly identify and name all the following shapes.
6	Shape Identifier	10	At the next level the child can name most common shapes, including rhombi, "ellipses-is-not-circle." A child at this level implicitly recognizes right angles, so distinguishes between a rectangle and a parallelogram without right angles. Correctly names all the following shapes:
6	Angle Matcher	11	A sign of development is when the child can match angles concretely. For example, given several triangles, finds two with the same angles by laying the angles on top of one another.

Learning Trajectories

Age Range	Level Name	Level	Description
7	Parts of Shapes Identifier	12	At the next level the child can identify shapes in terms of their components. For example, the child may say, "No matter how skinny it looks, that's a triangle because it has three sides and three angles."
7	Constructor of Shapes from Parts Exact	13	A significant step is when the child can represent a shape with completely correct construction, based on knowledge of components and relationships. For example, asked to make a triangle with sticks, creates the following:
8	Shape Class Identifier	14	As children develop, they begin to use class membership (for example, to sort), not explicitly based on properties. For example, a child at this level may say, "I put the triangles over here, and the quadrilaterals, including squares, rectangles, rhombi, and trapezoids, over there."
8	Shape Property Identifier	15	At the next level a child can use properties explicitly. For example, a child may say, "I put the shapes with opposite sides parallel over here, and those with four sides but not both pairs of sides parallel over there."

Age Range	Level Name	Level	Description
8	Angle Size Comparer	16	The next sign of development is when a child can separate and compare angle sizes. For example, the child may say, "I put all the shapes that have right angles here, and all the ones that have bigger or smaller angles over there."
8	Angle Measurer	17	A significant step in development is when a child can use a protractor to measure angles.
8	Property Class Identifier	18	The next sign of development is when a child can use class membership for shapes (for example, to sort or consider shapes "similar") explicitly based on properties, including angle measure. For example, the child may say, "I put the equilateral triangles over here, and the right triangles over here."
8	Angle Synthesizer	19	As children develop understanding of shape, they can combine various meanings of angle (turn, corner, slant). For example, a child at this level could explain, "This ramp is at a 45° angle to the ground."

Developmental Levels for Composing Geometric Shapes

Children move through levels in the composition and decomposition of two-dimensional figures. Very young children cannot compose shapes but then gain ability to combine shapes into pictures, synthesize combinations of shapes into new shapes, and eventually substitute and build different kinds of shapes. Children typically follow an observable developmental progression in learning to compose shapes with recognizable stages or levels. This developmental path can be described as part of a learning trajectory.

Age Range	Level Name	Level	Description
2	Pre-Composer	1	The earliest sign of development is when a child can manipulate shapes as individuals, but is unable to combine them to compose a larger shape. Make a Picture　　Outline Puzzle
3	Pre-DeComposer	2	At the next level a child can decompose shapes, but only by trial and error. For example, given only a hexagon, the child can break it apart to make this simple picture by trial and error:

Age Range	Level Name	Level	Description
4	Piece Assembler	3	Around age 4 a child can begin to make pictures in which each shape represents a unique role (for example, one shape for each body part) and shapes touch. A child at this level can fill simple outline puzzles using trial and error. Make a Picture　　Outline Puzzle

Age Range	Level Name	Level	Description
5	Picture Maker	4	As children develop they are able to put several shapes together to make one part of a picture (for example, two shapes for one arm). A child at this level uses trial and error and does not anticipate creation of the new geometric shape. The child can choose shapes using "general shape" or side length and fill "easy" outline puzzles that suggest the placement of each shape (but note below that the child is trying to put a square in the puzzle where its right angles will not fit). Make a Picture Outline Puzzle
5	Simple Decomposer	5	A significant step occurs when the child is able to decompose ("take apart" into smaller shapes) simple shapes that have obvious clues as to their decomposition.
5	Shape Composer	6	A sign of development is when a child composes shapes with anticipation ("I know what will fit!"). A child at this level chooses shapes using angles as well as side lengths. Rotation and flipping are used intentionally to select and place shapes. For example, in the outline puzzle below, all angles are correct, and patterning is evident. Make a Picture Outline Puzzle
6	Substitution Composer	7	A sign of development is when a child is able to make new shapes out of smaller shapes and uses trial and error to substitute groups of shapes for other shapes to create new shapes in different ways. For example, the child can substitute shapes to fill outline puzzles in different ways.

Age Range	Level Name	Level	Description
6	Shape Decomposer (with Help)	8	As children develop they can decompose shapes by using imagery that is suggested and supported by the task or environment. For example, given hexagons, the child at this level can break it apart to make this shape:
7	Shape Composite Repeater	9	The next level is demonstrated when the child can construct and duplicate units of units (shapes made from other shapes) intentionally, and understands each as being both multiple small shapes and one larger shape. For example, the child may continue a pattern of shapes that leads to tiling.
7	Shape Decomposer with Imagery	10	A significant sign of development is when a child is able to decompose shapes flexibly by using independently generated imagery. For example, given hexagons, the child can break it apart to make shapes such as these:
8	Shape Composer— Units of Units	11	Children demonstrate further understanding when they are able to build and apply units of units (shapes made from other shapes). For example, in constructing spatial patterns the child can extend patterning activity to create a tiling with a new unit shape—a unit of unit shapes that he or she recognizes and consciously constructs. For example, the child builds Ts out of four squares, uses four Ts to build squares, and uses squares to tile a rectangle.
8	Shape DeComposer with Units of Units	12	As children develop understanding of shape they can decompose shapes flexibly by using independently generated imagery and planned decompositions of shapes that themselves are decompositions. For example, given only squares, a child at this level can break them apart—and then break the resulting shapes apart again— to make shapes such as these:

Developmental Levels for Comparing Geometric Shapes

As early as 4 years of age children can create and use strategies, such as moving shapes to compare their parts or to place one on top of the other for judging whether two figures are the same shape. From Pre-K to Grade 2 they can develop sophisticated and accurate mathematical procedures for comparing geometric shapes. Children typically follow an observable developmental progression in learning about how shapes are the same and different with recognizable stages or levels. This developmental path can be described as part of a learning trajectory.

Age Range	Level Name	Level	Description
3	"Same Thing" Comparer	1	The first sign of understanding is when the child can compare real-world objects. For example, the child says two pictures of houses are the same or different.
4	"Similar" Comparer	2	The next sign of development occurs when the child judges two shapes the same if they are more visually similar than different. For example, the child may say, "These are the same. They are pointy at the top."
4	Part Comparer	3	At the next level a child can say that two shapes are the same after matching one side on each. For example, "These are the same" (matching the two sides).
4	Some Attributes Comparer	4	As children develop they look for differences in attributes, but may examine only part of a shape. For example, a child at this level may say, "These are the same" (indicating the top halves of the shapes are similar by laying them on top of each other).

Age Range	Level Name	Level	Description
5	Most Attributes Comparer	5	At the next level the child looks for differences in attributes, examining full shapes, but may ignore some spatial relationships. For example, a child may say, "These are the same."
7	Congruence Determiner	6	A sign of development is when a child determines congruence by comparing all attributes and all spatial relationships. For example, a child at this level says that two shapes are the same shape and the same size after comparing every one of their sides and angles.
7	Congruence Superposer	7	As children develop understanding they can move and place objects on top of each other to determine congruence. For example, a child at this level says that two shapes are the same shape and the same size after laying them on top of each other.
8	Congruence Representer	8	Continued development is evidenced as children refer to geometric properties and explain transformations. For example, a child at this level may say, "These must be congruent, because they have equal sides, all square corners, and I can move them on top of each other exactly."

Developmental Levels for Spatial Sense and Motions

Infants and toddlers spend a great deal of time exploring space and learning about the properties and relations of objects in space. Very young children know and use the shape of their environment in navigation activities. With guidance they can learn to "mathematize" this knowledge. They can learn about direction, perspective, distance, symbolization, location, and coordinates. Children typically follow an observable developmental progression in developing spatial sense with recognizable stages or levels. This developmental path can be described as part of a learning trajectory.

Age Range	Level Name	Level	Description
4	Simple Turner	1	An early sign of spatial sense is when a child mentally turns an object to perform easy tasks. For example, given a shape with the top marked with color, correctly identifies which of three shapes it would look like if it were turned "like this" (90 degree turn demonstrated) before physically moving the shape.

Age Range	Level Name	Level	Description
5	Beginning Slider, Flipper, Turner	2	The next sign of development is when a child can use the correct motions, but is not always accurate in direction and amount. For example, a child at this level may know a shape has to be flipped to match another shape, but flips it in the wrong direction.

Age Range	Level Name	Level	Description
6	Slider, Flipper, Turner	3	As children develop spatial sense they can perform slides and flips, often only horizontal and vertical, by using manipulatives. For example, a child at this level can perform turns of 45, 90, and 180 degrees and knows a shape must be turned 90 degrees to the right to fit into a puzzle.

Age Range	Level Name	Level	Description
7	Diagonal Mover	4	A sign of development is when a child can perform diagonal slides and flips. For example, a child at this level knows a shape must be turned or flipped over an oblique line (45 degree orientation) to fit into a puzzle.
8	Mental Mover	5	Further signs of development occur when a child can predict results of moving shapes using mental images. A child at this level may say, "If you turned this 120 degrees, it would be just like this one."

Developmental Levels for Patterning and Early Algebra

Algebra begins with a search for patterns. Identifying patterns helps bring order, cohesion, and predictability to seemingly unorganized situations and allows one to make generalizations beyond the information directly available. The recognition and analysis of patterns are important components of the young child's intellectual development because they provide a foundation for the development of algebraic thinking. Although prekindergarten children engage in pattern-related activities and recognize patterns in their everyday environment, research has revealed that an abstract understanding of patterns develops gradually during the early childhood years. Children typically follow an observable developmental progression in learning about patterns with recognizable stages or levels. This developmental path can be described as part of a learning trajectory.

Age Range	Level Name	Level	Description
2	Pre-Patterner	1	A child at the earliest level does not recognize patterns. For example, a child may name a striped shirt with no repeating unit a "pattern."
3	Pattern Recognizer	2	At the next level the child can recognize a simple pattern. For example, a child at this level may say, "I'm wearing a pattern" about a shirt with black, white, black, white stripes.
3–4	Pattern Fixer	3	A sign of development is when the child fills in a missing element of a pattern. For example, given objects in a row with one missing, the child can identify and fill in the missing element.
4	Pattern Duplicator AB	3	A sign of development is when the child can duplicate an ABABAB pattern, although the child may have to work close to the model pattern. For example, given objects in a row, ABABAB, makes their own ABBABBABB row in a different location.

Age Range	Level Name	Level	Description
4	Pattern Extender AB	4	At the next level the child is able to extend AB repeating patterns.
4	Pattern Duplicator	4	At this level the child can duplicate simple patterns (not just alongside the model pattern). For example, given objects in a row, ABBABBABB, makes their own ABBABBABB row in a different location.
5	Pattern Extender	5	A sign of development is when the child can extend simple patterns. For example, given objects in a row, ABBABBABB, adds ABBABB to the end of the row.
7	Pattern Unit Recognizer	7	At this level a child can identify the smallest unit of a pattern. For example, given objects in a ABBAB_BABB patterns, identifies the core unit of the pattern as ABB.

Learning Trajectories

Developmental Levels for Classifying and Analyzing Data

Data analysis contains one big idea: classifying, organizing, representing, and using information to ask and answer questions. The developmental continuum for data analysis includes growth in classifying and counting to sort objects and quantify their groups. . . . Children eventually become capable of simultaneously classifying and counting, for example, counting the number of colors in a group of objects.

Children typically follow an observable developmental progression in learning about patterns with recognizable stages or levels. This developmental path can be described as part of a learning trajectory.

Age Range	Level Name	Level	Description
2	Similarity Recognizer	1	The first sign that a child can classify is when he or she recognizes, intuitively, two or more objects as "similar" in some way. For example, "that's another doggie."
2	Informal Sorter	2	A sign of development is when a child places objects that are alike on some attribute together, but switches criteria and may use functional relationships are the basis for sorting. A child at this level might stack blocks of the same shape or put a cup with its saucer.
3	Attribute Identifier	3	The next level is when the child names attributes of objects and places objects together with a given attribute, but cannot then move to sorting by a new rule. For example, the child may say, "These are both red."
4	Attribute Sorter	4	At the next level the child sorts objects according to a given attributes, forming categories, but may switch attributes during the sorting. A child at this stage can switch rules for sorting if guided. For example, the child might start putting red beads on a string, but switches to the spheres of different colors.
5	Consistent Sorter	5	A sign of development is when the child can sort consistently by a given attribute. For example, the child might put several identical blocks together.
6	Exhaustive Sorter	6	At the next level, the child can sort consistently and exhaustively by an attribute, given or created. This child can use terms "some" and "all" meaningfully. For example, a child at this stage would be able to find all the attribute blocks of a certain size and color.
6	Multiple Attribute Sorter	7	A sign of development is when the child can sort consistently and exhaustively by more than one attribute, sequentially. For example, a child at this level, can put all the attribute blocks together by color, then by shape.
7	Classifier and Counter	8	At the next level, the child is capable of simultaneously classifying and counting. For example, the child counts the number of colors in a group of objects.

Age Range	Level Name	Level	Description
7	List Grapher	9	In the early stage of graphing, the child graphs by simply listing all cases. For example, the child may list each child in the class and each child's response to a question.
8+	Multiple Attribute Classifier	10	A sign of development is when the child can intentionally sort according to multiple attributes, naming and relating the attributes. This child understands that objects could belong to more than one group. For example, the child can complete a two-dimensional classification matrix or forming subgroups within groups.
8+	Classifying Grapher	11	At the next level the child can graph by classifying data (e.g., responses) and represent it according to categories. For example, the child can take a survey, classify the responses, and graph the result.
8+	Classifier	12	At sign of development is when the child creates complete, conscious classifications logically connected to a specific property. For example, a child at this level gives definition of a class in terms of a more general class and one or more specific differences and begins to understand the inclusion relationship.
8+	Hierarchical Classifier	13	At the next level, the child can perform hierarchical classifications. For example, the child recognizes that all squares are rectangles, but not all rectangles are squares.
8+	Data Representer	14	Signs of development are when the child organizes and displays data through both simple numerical summaries such as counts, tables, and tallies, and graphical displays, including picture graphs, line plots, and bar graphs. At this level the child creates graphs and tables, compares parts of the data, makes statements about the data as a whole, and determines whether the graphs answer the questions posed initially.

Trajectory Progress Chart

Student's Name _____

Number

Age Range	Counting	Comparing and Ordering Number	Recognizing Number and Subitizing (instantly recognizing)	Composing Number (knowing combinations of numbers)	Adding and Subtracting	Multiplying and Dividing (sharing)
1 year	____ Pre-Counter ____ Chanter				____ Pre +/−	____ Nonquantitative Sharer
2	____ Reciter	____ Object Corresponder ____ Perceptual Comparer	____ Small Collection Namer			____ Beginning Grouper and Distributive Sharer
3	____ Reciter (10) ____ Corresponder	____ First-Second Ordinal Counter ____ Nonverbal Comparer of Similar Items (1–4 items)	____ Nonverbal Subitizer ____ Maker of Small Collections		____ Nonverbal +/−	
4	____ Counter (small numbers) ____ Producer (small numbers) ____ Counter (10)	____ Nonverbal Comparer of Dissimilar Items ____ Matching Comparer ____ Knows-to-Count Comparer ____ Counting Comparer (same size)	____ Perceptual Subitizer to 4	____ Pre-Part-Whole Recognizer	____ Small Number +/−	____ Grouper and Distributive Sharer
5	____ Counter and Producer (10+) ____ Counter Backward from 10	____ Counting Comparer (5) ____ Ordinal Counter	____ Perceptual Subitizer to 5 ____ Conceptual Subitizer to 5+ ____ Conceptual Subitizer to 10	____ Inexact Part-Whole Recognizer ____ Composer to 4, then 5	____ Find Result +/− ____ Find Change +/− ____ Make It N +/−	____ Concrete Modeler ×/÷
6	____ Counter from N (N+1, N−1) ____ Skip Counter by tens to 100 ____ Counter to 100 ____ Counter On Using Patterns ____ Skip Counter ____ Counter of Imagined Items ____ Counter On Keeping Track ____ Counter of Quantitative Units ____ Counter to 200	____ Counting Comparer (10) ____ Mental Number Line to 10 ____ Serial Orderer to 6+	____ Conceptual Subitizer to 20	____ Composer to 7 ____ Composer to 10	____ Counting Strategies +/− ____ Part-Whole +/−	____ Parts and Wholes ×/÷
7	____ Number Conserver ____ Counter Forward and Back	____ Place Value Comparer ____ Mental Number Line to 100	____ Conceptual Subitizer with Place Value and Skip Counting	____ Composer with Tens and Ones	____ Numbers-in-Numbers +/− ____ Deriver +/−	____ Skip Counter ×/÷
8+	____ Mental Number Line to 1,000s		____ Conceptual Subitizer with Place Value and Multiplication		____ Problem Solver +/− ____ Multidigit +/−	____ Deriver ×/÷ ____ Array Quantifier ____ Partitive Divisor ____ Multidigit ×/÷

Student's Name _____

Geometry

Age Range	Shapes	Composing Shapes	Comparing Shapes	Motions and Spatial Sense	Measuring	Patterning	Classifying and Analyzing Data
2 years	— Shape Matcher—Identical —Sizes —Orientations					— Pre-Patterner	— Similarity Recognizer — Informal Sorter
3	— Shape Recognizer—Typical — Shape Matcher—More Shapes —Sizes and Orientations —Combinations	— Pre-Composer — Pre-Decomposer	— "Same Thing" Comparer		— Length Quantity Recognizer	— Pattern Recognizer	— Attribute Identifier
4	— Shape Recognizer—Circles, Squares, and Triangles+ — Constructor of Shapes from Parts—Looks Like Representing	— Piece Assembler	— "Similar" Comparer — Part Comparer — Some Attributes Comparer	— Simple Turner	— Length Direct Comparer	— Pattern Fixer — Pattern Duplicator AB — Pattern Extender AB — Pattern Duplicator	— Attribute Sorter
5	— Shape Recognizer—All Rectangles — Side Recognizer — Angle Recognizer — Shape Recognizer—More Shapes	— Picture Maker — Simple Decomposer — Shape Composer	— Most Attributes Comparer	— Beginning Slider, Flipper, Turner	— Indirect Length Comparer	— Pattern Extender	— Consistent Sorter
6	— Shape Identifier — Angle Matcher Parts	— Substitution Composer — Shape Decomposer (with help)		— Slider, Flipper, Turner	— Serial Orderer to 6+ — End-to-End Length Measurer		— Exhaustive Sorter — Multiple Attribute Sorter
7	— Parts of Shapes Identifier — Constructor of Shapes from Parts—Exact Representing	— Shape Composite Repeater — Shape Decomposer with Imagery	— Congruence Determiner — Congruence Superposer	— Diagonal Mover	— Length Unit Iterater — Length Unit Relater	— Pattern Unit Recognizer	— Classifier and Counter — List Grapher
8+	— Shape Class Identifier — Shape Property Identifier — Angle Size Comparer — Angle Measurer — Property Class Identifier — Angle Synthesizer	— Shape Composer—Units of Units — Shape Decomposer with Units of Units	— Congruence Representer	— Mental Mover	— Length Measurer — Conceptual Ruler Measurer		— Multiple Attribute Classifier — Classifying Grapher — Classifier — Hierarchical Classifier — Data Representer

Glossary

A

acute angle An angle with a measure greater than 0 degrees and less than 90 degrees.

addend One of the numbers being added in an addition sentence. In the sentence 41 + 27 = 68, the numbers 41 and 27 are addends.

addition A mathematical operation based on "putting things together." Numbers being added are called *addends*. The result of addition is called a *sum*. In the number sentence 15 + 63 = 78, the numbers 15 and 63 are addends.

additive inverses Two numbers whose sum is 0. For example, 9 + −9 = 0. The additive inverse of 9 is −9, and the additive inverse of −9 is 9.

adjacent angles Two angles with a common side that do not otherwise overlap. In the diagram, angles 1 and 2 are adjacent angles; so are angles 2 and 3, angles 3 and 4, and angles 4 and 1.

algorithm A step-by-step procedure for carrying out a computation or solving a problem.

angle Two rays with a common endpoint. The common endpoint is called the vertex of the angle.

area A measure of the surface inside a closed boundary. The formula for the area of a rectangle or parallelogram is $A = b \times h$, where A represents the area, b represents the length of the base, and h is the height of the figure.

array A rectangular arrangement of objects in rows and columns in which each row has the same number of elements, and each column has the same number of elements.

attribute A feature such as size, shape, or color.

average See **mean**. The **median** and **mode** are also sometimes called the *average*.

axis (plural **axes**) A number line used in a coordinate grid.

B

bar graph A graph in which the lengths of horizontal or vertical bars represent the magnitude of the data represented.

base ten The commonly used numeration system, in which the ten digits 0, 1, 2,..., 9 have values that depend on the place in which they appear in a numeral (ones, tens, hundreds, and so on, to the left of the decimal point; tenths, hundredths, and so on, to the right of the decimal point).

bisect To divide a segment, angle, or figure into two parts of equal measure.

C

capacity A measure of how much liquid or substance a container can hold. See also **volume**.

centi- A prefix for units in the metric system meaning one hundredth.

centimeter (cm) In the metric system, a unit of length defined as 1/100 of a meter; equal to 10 millimeters or 1/10 of a decimeter.

circle The set of all points in a plane that are a given distance (the radius) from a given point (the center of the circle).

circle graph A graph in which a circular region is divided into sectors to represent the categories in a set of data. The circle represents the whole set of data.

circumference The distance around a circle or sphere.

closed figure A figure that divides the plane into two regions, inside and outside the figure. A closed space figure divides space into two regions in the same way.

common denominator Any nonzero number that is a multiple of the denominators of two or more fractions.

common factor Any number that is a factor of two or more numbers.

complementary angles Two angles whose measures total 90 degrees.

composite function A function with two or more operations. For example, this function multiplies the input number by 5 then adds 3.

composite number A whole number that has 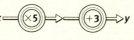 more than two whole number factors. For example, 14 is a composite number because it has more than two whole number factors.

cone A space figure having a circular base, curved surface, and one vertex.

congruent Having identical sizes and shapes. Congruent figures are said to be congruent to each other.

coordinate One of two numbers used to locate a point on a coordinate grid. See also **ordered pair**.

coordinate grid A device for locating points in a plane by means of ordered pairs or coordinates. A coordinate grid is formed by two number lines that intersect at their 0-points.

corresponding angles Two angles in the same relative position in two figures, or in similar locations in relation to a transversal intersecting two lines. In the diagram above, angles 1 and 5, 3 and 7, 2 and 6, and 4 and 8 are corresponding angles. If the lines are parallel, then the corresponding angles are congruent.

corresponding sides Two sides in the same relative position in two figures. In the diagram AB and A'B', BC and B'C', and AC and A'C' are corresponding sides.

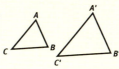

cube A space figure whose six faces are congruent squares that meet at right angles.

cubic centimeter (cm³) A metric unit of volume; the volume of a cube 1 centimeter on an edge. 1 cubic centimeter is equal to 1 milliliter.

cubic unit A unit used in a volume and capacity measurement.

customary system of measurement The measuring system used most often in the United States. Units for linear measure (length, distance) include inch, foot, yard, and mile; units for weight include ounce and pound; units for capacity (amount of liquid or other substance a container can hold) include fluid ounce, cup, pint, quart, and gallon.

cylinder A space figure having a curved surface and parallel circular or elliptical bases that are congruent.

D

decimal A number written in standard notation, usually one containing a decimal point, as in 3.78.

decimal approximation A decimal that is close to the value of a rational number. By extending the decimal approximation to additional digits, it is possible to come as close as desired to the value of the rational number. For example, decimal approximations of $\frac{1}{12}$ are 0.083, 0.0833, 0.08333, and so on.

decimal equivalent A decimal that names the same number as a fraction. For example, the decimal equivalent of $\frac{3}{4}$ is 0.75. The only rational numbers with decimal equivalents are those that can be written as fractions whose denominators have prime factors only of 2 and 5. For example, $\frac{1}{2}, \frac{1}{4},$ and $\frac{1}{20}$ have decimal equivalents, but $\frac{1}{6}, \frac{1}{7},$ and $\frac{1}{9}$ have only decimal approximations.

degree (°) A unit of measure for angles; based on dividing a circle into 360 equal parts. Also, a unit of measure for temperature.

degree Celsius (°C) In the metric system, the unit for measuring temperature. Water freezes at 0°C and boils at 100°C.

degree Fahrenheit (°F) In the U.S. customary system, the unit for measuring temperature. Water freezes at 32°F and boils at 212°F.

denominator The number of equal parts into which a whole is divided. In the fraction $\frac{a}{b}$, b is the denominator. See also **numerator**.

diameter A line segment, going through the center of a circle, that starts at one point on the circle and ends at the opposite point on the circle; also, the length of such a line segment. The diameter of a circle is twice its radius. AB is a diameter of this circle. See also **circle**.

difference The result of subtraction. In the subtraction sentence 40 − 10 = 30, the difference is 30.

digit In the base-ten numeration system, one of the symbols 0, 1, 2, 3, 4, 5, 6, 7, 8, 9. Digits can be used to write a numeral for any whole number in the base-ten numbering system. For example, the numeral 145 is made of the digits 1, 4, and 5.

distributive law A law that relates two operations on numbers, usually multiplication and addition, or multiplication and subtraction. Distributive law of multiplication over addition: $a \times (b + c) = (a \times b) + (a \times c)$

dividend See **division**.

division A mathematical operation based on "equal sharing" or "separating into equal parts." The *dividend* is the total before sharing. The divisor is the number of equal parts or the number in each equal part. The *quotient* is the result of division. For example, in 35 ÷ 5 = 7, 35 is the dividend, 5 is the divisor, and 7 is the quotient. If 35 objects are separated into 5 equal parts, there are 7 objects in each part. If 35 objects are separated into parts with 5 in each part, there are 7 equal parts. The number left over when a set of objects is shared equally or separated into equal groups is called the *remainder*. For 35 ÷ 5, the quotient is 7 and the remainder is 0. For 36 ÷ 5, the quotient is 7 and the remainder is 1.

Glossary

divisor See **division.**

E

edge The line segment where two faces of a polyhedron meet.

endpoint The point at either end of a line segment; also, the point at the end of a ray. Line segments are named after their endpoints; a line segment between and including points A and B is called segment AB or segment BA.

equation A mathematical sentence that states the equality of two expressions. For example, $3 + 7 = 10$, $y = x + 7$, and $4 + 7 = 8 + 3$ are equations.

equilateral polygon A polygon in which all sides are the same length.

equivalent Equal in value, but in a different form. For example, $\frac{1}{2}$, $\frac{2}{4}$, 0.5, and 50% are equivalent forms of the same number.

equivalent fractions Fractions that have different numerators and denominators but name the same number. For example, $\frac{2}{3}$ and $\frac{6}{9}$ are equivalent fractions.

estimate A judgment of time, measurement, number, or other quantity that may not be exactly right.

evaluate an algebraic expression To replace each variable in an algebraic expression with a particular number and then to calculate the value of the expression.

evaluate a numerical expression To carry out the operations in a numerical expression to find the value of the expression.

even number A whole number such as 0, 2, 4, 6, and so on, that can be divided by 2 with no remainder. See also **odd number.**

event A happening or occurrence. The tossing of a coin is an event.

expression A group of mathematical symbols (numbers, operation signs, variables, grouping symbols) that represents a number (or can represent a number if values are assigned to any variables it contains).

F

face A flat surface on a space figure.

fact family A group of addition or multiplication facts grouped together with the related subtraction or division facts. For example, $4 + 8 = 12$, $8 + 4 = 12$, $12 - 4 = 8$, and $12 - 8 = 4$ form an addition fact family. The facts $4 \times 3 = 12$, $3 \times 4 = 12$, $12 \div 3 = 4$, and $12 \div 4 = 3$ form a multiplication fact family.

factor (noun) One of the numbers that is multiplied in a multiplication expression. For example, in $4 \times 1.5 = 6$, the factors are 4 and 1.5. See also **multiplication.**

factor (verb) To represent a quantity as a product of factors. For example, 20 factors to 4×5, 2×10, or $2 \times 2 \times 5$.

factor of a whole number n A whole number, which, when multiplied by another whole number, results in the number n. The whole number n is divisible by its factors. For example, 3 and 5 are factors of 15 because $3 \times 5 = 15$, and 15 is divisible by 3 and 5.

factor tree A method used to obtain the prime factorization of a number. The original number is represented as a product of factors, and each of those factors is represented as a product of factors, and so on, until the factor string consists of prime numbers.

formula A general rule for finding the value of something. A formula is usually written as an equation with variables representing unknown quantities. For example, a formula for distance traveled at a constant rate of speed is $d = r \times t$, where d stands for distance, r is for rate, and t is for time.

fraction A number in the form $\frac{a}{b}$, where a and b are integers and b is not 0. Fractions are used to name part of a whole object or part of a whole collection of objects, or to compare two quantities. A fraction can represent division; for example, $\frac{2}{5}$ can be thought of as 2 divided by 5.

frequency The number of times an event or value occurs in a set of data.

function machine An imaginary machine that processes numbers according to a certain rule. A number (input) is put into the machine and is transformed into a second number (output) by application of the rule.

G

greatest common factor The largest factor that two or more numbers have in common. For example, the common factors of 24 and 30 are 1, 2, 3, and 6. The greatest common factor of 24 and 30 is 6.

H

height (of a parallelogram) The length of the line segment between the base of the parallelogram and the opposite side (or an extension of the opposite side), running perpendicular to the base.

height (of a polyhedron) The perpendicular distance between the bases of the polyhedron or between a base and the opposite vertex.

height (of a rectangle) The length of the side perpendicular to the side considered the base of the rectangle. (Base and height of a rectangle are interchangeable.)

height (of a triangle) The length of the line segment perpendicular to the base of the triangle (or an extension of the base) from the opposite vertex.

hexagon A polygon with six sides.

histogram A bar graph in which the labels for the bars are numerical intervals.

hypotenuse In a right triangle, the side opposite the right angle.

I

improper fraction A fraction that names a number greater than or equal to 1; a fraction whose numerator is equal to or greater than its denominator. Examples of improper fractions are $\frac{4}{3}$, $\frac{10}{8}$, and $\frac{4}{4}$.

inch (in.) In the U. S. customary system, a unit of length equal to $\frac{1}{12}$ of a foot.

indirect measurement Methods for determining heights, distances, and other quantities that cannot be measured or are not measured directly.

inequality A number sentence stating that two quantities are not equal. Relation symbols for inequalities include < (is less than), > (is greater than), and ≠ (is not equal to).

integers The set of integers is $\{..., -4, -3, -2, -1, 0, 1, 2, 3, 4, ...\}$. The set of integers consists of whole numbers and their opposites.

intersect To meet (at a point, a line, and so on), sharing a common point or points.

interior The set of all points in a plane "inside" a closed plane figure, such as a polygon or circle. Also, the set of all points in space "inside" a closed space figure, such as a polyhedron or sphere.

isosceles Having two sides of the same length; commonly used to refer to triangles and trapezoids.

K

kilo- A prefix for units in the metric system meaning one thousand.

L

least common denominator The least common multiple of the denominators of every fraction in a given set of fractions. For example, 12 is the least common denominator of $\frac{2}{3}$, $\frac{1}{4}$, and $\frac{5}{6}$. See also **least common multiple.**

least common multiple The smallest number that is a multiple of two or more numbers. For example, some common multiples of 6 and 8 are 24, 48, and 72. 24 is the least common multiple of 6 and 8.

leg of a right triangle A side of a right triangle that is not the hypotenuse.

line A straight path that extends infinitely in opposite directions.

line graph (broken-line graph) A graph in which points are connected by line segments to represent data.

line of symmetry A line that separates a figure into halves. The figure can be folded along this line into two parts which exactly fit on top of each other.

line segment A straight path joining two points, called endpoints of the line segment. A straight path can be described as the shortest distance between two points.

line symmetry A figure has line symmetry (also called bilateral symmetry) if a line of symmetry can be drawn through the figure.

liter (L) A metric unit of capacity, equal to the volume of a cube 10 centimeters on an edge. $1 \text{ L} = 1,000 \text{ mL} = 1,000 \text{ cm}^3$. A liter is slightly larger than a quart. See also **milliliter (mL).**

M

map scale A ratio that compares the distance between two locations shown on a map with the actual distance between them.

mean A typical or central value that may be used to describe a set of numbers. It can be found by adding the numbers in the set and dividing the sum by the number of numbers. The mean is often referred to as the average.

median The middle value in a set of data when the data are listed in order from least to greatest (or greatest to least). If the number of values in the set is even (so that there is no "middle" value), the median is the mean of the two middle values.

meter (m) The basic unit of length in the metric system, equal to 10 decimeters, 100 centimeters, and 1,000 millimeters.

metric system of measurement A measurement system based on the base-ten numeration system and used in most countries in the world. Units for linear measure (length, distance) include millimeter, centimeter, meter, kilometer; units for mass (weight) include gram and kilogram; units for capacity (amount of liquid or other substance a container can hold) include milliliter and liter.

Glossary

midpoint A point halfway between two points.

milli- A prefix for units in the metric system meaning one thousandth.

milliliter (mL) A metric unit of capacity, equal to 1/1,000 of a liter and 1 cubic centimeter.

millimeter (mm) In the metric system, a unit of length equal to 1/10 of a centimeter and 1/1,000 of a meter.

minuend See **subtraction.**

mixed number A number greater than 1, written as a whole number and a fraction less than 1. For example, $5\frac{1}{2}$ is equal to $5 + \frac{1}{2}$.

mode The value or values that occur most often in a set of data.

multiple of a number _n_ The product of a whole number and the number _n_. For example, the numbers 0, 4, 8, 12, and 16 are all multiples of 4 because $4 \times 0 = 0$, $4 \times 1 = 4$, $4 \times 2 = 8$, $4 \times 3 = 12$, and $4 \times 4 = 16$.

multiplication A mathematical operation used to find the total number of things in several equal groups, or to find a quantity that is a certain number of times as much or as many as another number. Numbers being multiplied are called _factors_. The result of multiplication is called the _product_. In $8 \times 12 = 96$, 8 and 12 are the factors and 96 is the product.

multiplicative inverses Two numbers whose product is 1. For example, the multiplicative inverse of $\frac{2}{5}$ is $\frac{5}{2}$, and the multiplicative inverse of 8 is $\frac{1}{8}$. Multiplicative inverses are also called _reciprocals_ of each other.

N

negative number A number less than 0; a number to the left of 0 on a horizontal number line.

number line A line on which equidistant points correspond to integers in order.

number sentence A sentence that is made up of numerals and a relation symbol ($<$, $>$, or $=$). Most number sentences also contain at least one operation symbol. Number sentences may also have grouping symbols, such as parentheses.

numeral The written name of a number.

numerator In a whole divided into a number of equal parts, the number of equal parts being considered. In the fraction $\frac{a}{b}$, _a_ is the numerator.

O

obtuse angle An angle with a measure greater than 90 degrees and less than 180 degrees.

octagon An eight-sided polygon.

odd number A whole number that is not divisible by 2, such as 1, 3, 5, and so on. When an odd number is divided by 2, the remainder is 1. A whole number is either an odd number or an even number.

opposite of a number A number that is the same distance from 0 on the number line as the given number, but on the opposite side of 0. If _a_ is a negative number, the opposite of _a_ will be a positive number. For example, if $a = -5$, then $-a$ is 5. See also **additive inverses.**

ordered pair Two numbers or objects for which order is important. Often, two numbers in a specific order used to locate a point on a coordinate grid. They are usually written inside parentheses; for example, (2, 3). See also **coordinate.**

ordinal number A number used to express position or order in a series, such as first, third, tenth. People generally use ordinal numbers to name dates; for example, "May fifth" rather than "May five."

origin The point where the _x_- and _y_-axes intersect on a coordinate grid. The coordinates of the origin are (0, 0).

outcome The result of an event. Heads and tails are the two outcomes of the event of tossing a coin.

P

parallel lines (segments, rays) Lines (segments, rays) going in the same direction that are the same distance apart and never meet.

parallelogram A quadrilateral that has two pairs of parallel sides. Pairs of opposite sides and opposite angles of a parallelogram are congruent.

parentheses A pair of symbols, (and), used to show in which order operations should be done. For example, the expression $(3 \times 5) + 7$ says to multiply 5 by 3 then add 7. The expression $3 \times (5 + 7)$ says to add 5 and 7 and then multiply by 3.

pattern A model, plan, or rule that uses words or variables to describe a set of shapes or numbers that repeat in a predictable way.

pentagon A polygon with five sides.

percent A rational number that can be written as a fraction with a denominator of 100. The symbol % is used to represent percent. 1% means 1/100 or 0.01. For example, "53% of the students in the school are girls" means that of every 100 students in the school, 53 are girls.

perimeter The distance along a path around a plane figure. A formula for the perimeter of a rectangle is $P = 2 \times (B + H)$, where _B_ represents the base and _H_ is the height of the rectangle. Perimeter may also refer to the path itself.

perpendicular Two rays, lines, line segments, or other figures that form right angles are said to be perpendicular to each other.

pi The ratio of the circumference of a circle to its diameter. Pi is the same for every circle, approximately 3.14 or $\frac{22}{7}$. Also written as the Greek letter π.

pictograph A graph constructed with pictures or icons, in which each picture stands for a certain number. Pictographs make it easier to visually compare quantities.

place value A way of determining the value of a digit in a numeral, written in standard notation, according to its position, or place, in the numeral. In base-ten numbers, each place has a value ten times that of the place to its right and one-tenth the value of the place to its left.

plane A flat surface that extends forever.

plane figure A figure that can be contained in a plane (that is, having length and width but no height).

point A basic concept of geometry; usually thought of as a location in space, without size.

polygon A closed plane figure consisting of line segments (sides) connected endpoint to endpoint. The interior of a polygon consists of all the points of the plane "inside" the polygon. An _n_-gon is a polygon with _n_ sides; for example, an 8-gon has 8 sides.

polyhedron A closed space figure, all of whose surfaces (faces) are flat. Each face consists of a polygon and the interior of the polygon.

power A product of factors that are all the same. For example, $6 \times 6 \times 6$ (or 216) is called 6 to the third power, or the third power of 6, because 6 is a factor three times. The expression $6 \times 6 \times 6$ can also be written as 6^3.

power of 10 A whole number that can be written as a product using only 10 as a factor. For example, 100 is equal to 10×10 or 10^2, so 100 is called 10 squared, the second power of 10, or 10 to the second power. Other powers of 10 include 10^1, or 10, and 10^3, or 1,000.

prime factorization A whole number expressed as a product of prime factors. For example, the prime factorization of 18 is $2 \times 3 \times 3$. A number has only one prime factorization (except for the order in which the factors are written).

prime number A whole number greater than 1 that has exactly two whole number factors, 1 and itself. For example, 13 is a prime number because its only factors are 1 and 13. A prime number is divisible by 1 and itself. The first five prime numbers are 2, 3, 5, 7, and 11. See also **composite number.**

prism A polyhedron with two parallel faces (bases) that are the same size and shape. Prisms are classified according to the shape of the two parallel bases. The bases of a prism are connected by parallelograms that are often rectangular.

probability A number between 0 and 1 that indicates the likelihood that something (an event) will happen. The closer a probability is to 1, the more likely it is that an event will happen.

product See **multiplication.**

protractor A tool for measuring or drawing angles. When measuring an angle, the vertex of the angle should be at the center of the protractor and one side should be aligned with the 0 mark.

pyramid A polyhedron in which one face (the base) is a polygon and the other faces are formed by triangles with a common vertex (the apex). A pyramid is classified according to the shape of its base, as a triangular pyramid, square pyramid, pentagonal pyramid, and so on.

Pythagorean Theorem A mathematical theorem, proven by the Greek mathematician Pythagoras and known to many others before and since, that states that if the legs of a right triangle have lengths _a_ and _b_, and the hypotenuse has length _c_, then $a^2 + b^2 = c^2$.

Q

quadrilateral A polygon with four sides.

quotient See **division.**

Glossary

R

radius A line segment that goes from the center of a circle to any point on the circle; also, the length of such a line segment.

random sample A sample taken from a population in a way that gives all members of the population the same chance of being selected.

range The difference between the maximum and minimum values in a set of data.

rate A ratio comparing two quantities with unlike units. For example, a measure such as 23 miles per gallon of gas compares mileage with gas usage.

ratio A comparison of two quantities using division. Ratios can be expressed with fractions, decimals, percents, or words. For example, if a team wins 4 games out of 5 games played, the ratio of wins to total games is $\frac{4}{5}$, 0.8, or 80%.

rational number Any number that can be represented in the form $a \div b$ or $\frac{a}{b}$, where a and b are integers and b is positive. Some, but not all, rational numbers have exact decimal equivalents.

ray A straight path that extends infinitely in one direction from a point, which is called its *endpoint*.

reciprocal See **multiplicative inverses.**

rectangle A parallelogram with four right angles.

reduced form A fraction in which the numerator and denominator have no common factors except 1.

reflection A transformation in which a figure "flips" so that its image is the reverse of the original.

regular polygon A convex polygon in which all the sides are the same length and all the angles have the same measure.

relation symbol A symbol used to express the relationship between two numbers or expressions. Among the symbols used in number sentences are = for "is equal to," < for "is less than," > for "is greater than," and ≠ for "is not equal to."

remainder See **division.**

rhombus A parallelogram whose sides are all the same length.

right angle An angle with a measure of 90 degrees, representing a quarter of a full turn.

right triangle A triangle that has a right angle.

rotation A transformation in which a figure "turns" around a center point or axis.

rotational symmetry Property of a figure that can be rotated around a point (less than a full, 360-degree turn) in such a way that the resulting figure exactly matches the original figure. If a figure has rotational symmetry, its order of rotational symmetry is the number of different ways it can be rotated to match itself exactly. "No rotation" is counted as one of the ways.

rounding Changing a number to another number that is easier to work with and is close enough for the purpose. For example, 12,924 rounded to the nearest thousand is 13,000 and rounded to the nearest hundred is 12,900.

S

sample A subset of a group used to represent the whole group.

scale The ratio of the distance on a map or drawing to the actual distance.

scalene triangle A triangle in which all three sides have different lengths.

scale drawing An accurate picture of an object in which all parts are drawn to the same scale. If an actual object measures 32 by 48 meters, a scale drawing of it might measure 32 by 48 millimeters.

scale model A model that represents an object or display in proportions based on a determined scale.

similar figures Figures that are exactly the same shape but not necessarily the same size.

space figure A figure which cannot be contained in a plane. Common space figures include the rectangular prism, square pyramid, cylinder, cone, and sphere.

sphere The set of all points in space that are a given distance (the radius) from a given point (the center). A ball is shaped like a sphere.

square number A number that is the product of a whole number and itself. The number 36 is a square number, because 36 = 6 × 6.

square of a number The product of a number multiplied by itself. For example, 2.5 squared is $(2.5)^2$.

square root The square root of a number n is a number which, when multiplied by itself, results in the number n. For example, 8 is a square root of 64, because 8 × 8 = 64.

square unit A unit used to measure area—usually a square that is 1 inch, 1 centimeter, 1 yard, or other standard unit of length on each side.

standard notation The most familiar way of representing whole numbers, integers, and decimals by writing digits in specified places; the way numbers are usually written in everyday situations.

statistics The science of collecting, classifying, and interpreting numerical data as it is related to a particular subject.

stem-and-leaf plot A display of data in which digits with larger place values are named as stems, and digits with smaller place values are named as leaves.

straight angle An angle of 180 degrees; a line with one point identified as the vertex of the angle.

subtraction A mathematical operation based on "taking away" or comparing ("How much more?"). The number being subtracted is called the *subtrahend*; the number it is subtracted from is called the *minuend*; the result of subtraction is called the *difference*. In the number sentence 63 − 45 = 18, 63 is the minuend, 45 is the subtrahend, and 18 is the difference.

subtrahend See **subtraction.**

supplementary angles Two angles whose measures total 180 degrees.

surface area The sum of the areas of the faces of a space figure.

symmetrical Having the same size and shape across a dividing line or around a point.

T

tessellation An arrangement of closed shapes that covers a surface completely without overlaps or gaps.

tetrahedron A space figure with four faces, each formed by an equilateral triangle.

transformation An operation that moves or changes a geometric figure in a specified way. Rotations, reflections, and translations are types of transformations.

translation A transformation in which a figure "slides" along a line.

transversal A line which intersects two or more other lines.

trapezoid A quadrilateral with exactly one pair of parallel sides.

tree diagram A tool used to solve probability problems in which there is a series of events. This tree diagram represents a situation where the first event has three possible outcomes and the second event has two possible outcomes.

triangle A polygon with three sides. An *equilateral* triangle has three sides of the same length. An *isosceles* triangle has two sides of the same length. A *scalene* triangle has no sides of the same length.

U

unit (of measure) An agreed-upon standard with which measurements are compared.

unit fraction A fraction whose numerator is 1. For example, $\frac{1}{2}$, $\frac{1}{3}$, and $\frac{1}{10}$ are unit fractions.

unit cost The cost of one item or one specified amount of an item. If 20 pencils cost 60¢, then the unit cost is 3¢ per pencil.

unlike denominators Unequal denominators, as in $\frac{3}{4}$ and $\frac{5}{6}$.

V

variable A letter or other symbol that represents a number, one specific number, or many different values.

Venn diagram A picture that uses circles to show relationships between sets. Elements that belong to more than one set are placed in the overlap between the circles.

vertex The point at which the rays of an angle, two sides of a polygon, or the edges of a polyhedron meet.

vertical angles Two intersecting lines form four adjacent angles. In the diagram, angles 2 and 4 are vertical angles. They have no sides in common. Their measures are equal. Similarly, angles 1 and 3 are vertical angles.

volume A measure of the amount of space occupied.

W

whole number Any of the numbers 0, 1, 2, 3, 4, and so on. Whole numbers are the numbers used for counting and zero.

Scope and Sequence

The topics addressed at each grade level were determined after extensive analysis of national and state mathematics teaching expectations, standardized assessments, and topics covered in mathematics basal programs. Across all levels, the program follows the optimal sequence outlined by the learning trajectories of primary mathematics.

	A	B	C	D	E	F	G	H
Addition (whole numbers)								
Basic facts	•	•	•	•	•	•	•	•
Three or more addends			•	•	•	•	•	•
Two-digit numbers				•	•	•	•	•
Three-digit numbers					•	•	•	•
Greater numbers							•	•
Estimating sums						•		•
Algebra								
Properties of whole numbers	•	•	•	•	•	•	•	•
Integers (negative numbers)							•	•
Operations with integers							•	•
Make and solve number sentences and equations		•	•	•	•	•		
Variables							•	•
Order of operations							•	•
Writing variable expressions							•	•
Evaluating expressions							•	•
Solving one-step equations	•	•	•	•	•	•		
Solving two-step equations					•	•	•	•
Combining like terms								•
Solving inequalities							•	•
Function machines/tables					•	•	•	
Coordinate graphing					•	•	•	
Graphing linear functions					•	•	•	
Using formulas						•	•	•
Decimals and Money								
Place value						•	•	•
Adding						•	•	•
Subtracting						•	•	•
Multiplying by a whole number							•	•
Multiplying by a decimal								•
Multiplying by powers of 10						•	•	•

Decimals and Money

	A	B	C	D	E	F	G	H
Dividing by a whole number							•	•
Dividing by a decimal								•
Identifying and counting currency				•	•	•		•
Exchanging money				•	•	•		•
Computing with money				•	•	•		•

Division

	A	B	C	D	E	F	G	H
Basic facts							•	•
Remainders							•	•
One-digit divisors							•	•
Two-digit divisors							•	•
Greater divisors							•	•
Estimating quotients							•	•

Fractions

	A	B	C	D	E	F	G	H
Fractions of a whole				•	•	•	•	•
Comparing/ordering				•	•	•	•	•
Equivalent fractions						•	•	•
Reduced form						•	•	•
Mixed numbers/improper fractions								•
Adding–like denominators							•	•
Adding–unlike denominators								•

Geometry

	A	B	C	D	E	F	G	H
Plane figures	•	•	•	•	•	•	•	•
Classifying figures	•	•	•	•	•	•	•	•
Solid figures	•	•	•	•	•	•	•	•
Congruence				•	•	•	•	•
Symmetry				•	•	•	•	•
Symmetry (rotational)					•	•	•	•
Slides/Flips/Turns					•	•	•	•
Angles					•	•	•	•
Classifying triangles				•	•	•	•	•
Classifying quadrilaterals						•	•	•
Parallel and perpendicular lines						•	•	•
Perimeter				•	•	•	•	•
Radius and diameter								•
Circumference								•
Surface Area				•	•	•	•	•
Volume				•	•	•	•	•

Scope and Sequence

Measurement

	A	B	C	D	E	F	G	H
Length								
Use customary units				•	•	•	•	•
Use metric units				•	•	•	•	•
Mass/Weight								
Use customary units					•	•	•	•
Use metric units					•	•	•	•
Capacity								
Use customary units					•	•	•	•
Use metric units				•	•	•	•	•
Temperature								
Use degrees Fahrenheit					•	•	•	•
Use degrees Celsius				•	•	•	•	•
Converting within customary system						•	•	
Converting within metric system						•		•
Telling Time								
to the hour	•	•			•			
to the half hour					•			
to the quarter hour					•			
to the minute					•			
Converting units of time						•		
Reading a calendar					•			

Multiplication

	A	B	C	D	E	F	G	H
Basic facts					•	•	•	•
One-digit multipliers						•	•	•
Two-digit multipliers						•	•	•
Greater multipliers							•	•
Estimating products							•	•

Number and Numeration

	A	B	C	D	E	F	G	H
Reading and writing numbers	•	•	•	•	•	•	•	•
Counting	•	•	•	•	•	•	•	•
Skip counting			•	•	•	•		
Ordinal numbers			•	•	•	•		
Place value			•	•	•	•		
Even/odd numbers				•	•			
Comparing and ordering numbers	•	•	•	•	•	•		•
Rounding						•	•	•
Estimation/Approximation				•	•	•	•	•

Scope and Sequence

	A	B	C	D	E	F	G	H
Comparing and ordering integers							•	•
Integers (negative numbers)							•	•
Prime and composite numbers							•	•
Factors and prime factorization							•	•
Common factors							•	•
Common multiples						•	•	•

Subtraction (whole numbers)

	A	B	C	D	E	F	G	H
Basic facts	•	•	•	•	•	•	•	•
Two-digit numbers		•		•	•	•	•	•
Three-digit numbers					•	•	•	•
Greater numbers						•	•	•
Estimating differences	•	•	•	•	•	•	•	•

Patterns, Relations, and Functions

	A	B	C	D	E	F	G	H
Number patterns	•	•	•	•	•	•	•	•
Geometric patterns	•	•	•	•	•	•	•	•
Inequalities	•							

Ratio and Proportion

	A	B	C	D	E	F	G	H
Meaning/Use							•	•
Similar Figures							•	•
Meaning of Percent							•	•
Percent of a Number							•	•

Statistics and Graphing

	A	B	C	D	E	F	G	H
Real and picture graphs	•	•	•	•	•	•	•	•
Bar graphs	•	•	•	•	•	•	•	•
Line graphs			•	•	•	•		
Circle graphs					•		•	•
Analyzing graphs	•	•	•	•	•	•	•	•
Finding the mean							•	•
Finding the median							•	•
Finding the mode							•	•
Finding the range							•	•